Urban Sustainability

Editor-in-Chief

The Urban Sustainability Book Series is a valuable resource for sustainability and urban-related education and research. It offers an inter-disciplinary platform covering all four areas of practice, policy, education, research, and their nexus. The publications in this series are related to critical areas of sustainability, urban studies, planning, and urban geography.

This book series aims to put together cutting-edge research findings linked to the overarching field of urban sustainability. The scope and nature of the topic are broad and interdisciplinary and bring together various associated disciplines from sustainable development, environmental sciences, urbanism, etc. With many advanced research findings in the field, there is a need to put together various discussions and contributions on specific sustainability fields, covering a good range of topics on sustainable development, sustainable urbanism, and urban sustainability. Despite the broad range of issues, we note the importance of practical and policy-oriented directions, extending the literature and directions and pathways towards achieving urban sustainability.

The series will appeal to urbanists, geographers, planners, engineers, architects, governmental authorities, policymakers, researchers of all levels, and to all of those interested in a wide-ranging overview of urban sustainability and its associated fields. The series includes monographs and edited volumes, covering a range of topics under the urban sustainability topic, which can also be used for teaching materials.

Junyi Zhang

Urbanimmunology

Rethinking Cities and Society through Immunological Principles

Springer

Junyi Zhang
School of Transportation
Southeast University
Nanjing, China

ISSN 2731-6483 ISSN 2731-6491 (electronic)
Urban Sustainability
ISBN 978-981-95-4877-4 ISBN 978-981-95-4878-1 (eBook)
https://doi.org/10.1007/978-981-95-4878-1

This Springer imprint is published by the registered company Springer Nature Singapore Pte Ltd.
The registered company address is: 152 Beach Road, #21-01/04 Gateway East, Singapore 189721, Singapore

Preface

My journey toward *Urbanimmunology* began in the 1990s with studies on travel behavior, where I first noticed parallels to the human circulatory system. This initial insight expanded into research on time use, tourism, sustainability, health, and life-course analysis, gradually revealing the interconnections of human activities. By the 2010s, these explorations culminated in the life-oriented approach, emphasizing interdependence across multiple life domains, much like the systems that sustain the human body.

The pandemic marked a turning point. In 2020, as co-chair of the WCTRS[1] COVID-19 Task Force, I developed the PASS and LASTING approaches to address systemic risks. In parallel, looking beyond the 2030 horizon of the Sustainable Development Goals (SDGs), I proposed Harmonized Development Goals (HDGs) as a future-oriented framework to align human progress with planetary well-being. I also introduced the DIRECT approach to highlight the importance of process management in policymaking. Despite these broader developments, transportation remained central to my work, leading to the Transport-in-All-Policies (TiAP) approach—a framework shaped by the pandemic's revelation of systemic interdependence.

I have developed the concept of Urbanimmunology, which did not arise from a single idea but evolved through my decades of work responding to complex challenges with new frameworks. It reflects principles of resilience, adaptation, and integration, drawing inspiration from the human immune system to offer a lens for understanding and navigating contemporary urban and social dynamics.

In a world shaped by climate change, pandemics, resource scarcity, and widening global divides, it is essential to rethink the foundations of urban and social life. *Urbanimmunology: Rethinking Cities and Society Through Immunological Principles* presents a new paradigm for developing transformative theories by merging immunological insights with urban and social studies. It views urban and social

[1] WCTRS: World Conference on Transport Research Society.

systems not as static entities but as evolving ecosystems—capable of learning, adapting, and flourishing under uncertainty.

At the core of this work is a powerful analogy: urban and social systems, like the human body, are intricate and interdependent. The well-being of each part is directly connected with the whole. When one element weakens, its effects spread across the system. Inspired by the immune system's ability to adapt and defend, *Urbanimmunology* advocates for urban and social environments designed to evolve and respond to both external pressures and internal change. This book moves beyond theory to chart a new course for inquiry, calling for frameworks capable of capturing the complexity and fluidity of modern systems. It is intended as a foundational text, aiming to redefine the boundaries of the field and open transformative avenues for future research.

Urbanimmunology ultimately belongs to everyone invested in our collective future—the planners and policymakers shaping our spaces, the scholars and students seeking new tools and ideas, and all citizens who care for their communities. It is an invitation to see the world differently: to envision our cities and society not as fragile constructs, but as dynamic, adaptive systems capable of learning, healing, and thriving. It is our hope that this book provides not just a new set of ideas, but a foundational lens through which to build a more resilient and equitable world for generations to come.

A day celebrating the wisdom of Confucius

Nanjing, China Junyi Zhang
September 28, 2025

Competing Interests The author has no competing interests to declare that are relevant to the content of this manuscript.

Contents

Chapter 1
Reimagining Cities and Society Through Human Immune Intelligence

1.1 Cities and Society as Adaptive, Living Systems

Cities, shaped by rapid urbanization across global society, stand as enduring symbols of human achievement, embodying the ingenuity, creativity, and collective efforts of generations (Mumford, 1961; Calvino, 1978; Lefebvre, 2003). They function as epicenters of innovation, culture, commerce, and progress (Glaeser, 2012; Brenner, 2018; Jacobs, 2022). However, beneath their impressive façades and seemingly robust infrastructures lies an often-overlooked fragility—an intrinsic vulnerability to the very forces they aim to manage and control (Muggah, 2014; de Boer, 2015; Patel et al., 2020). Much like living organisms, cities and society exist in a state of continuous evolution, constantly adapting and responding to external and internal dynamics (Burdett & Sudjic, 2010; McFarlane, 2011; Hall, 2014). Despite these advancements, they persistently grapple with recurring challenges that threaten their harmony and stability (Hall, 2014; Campbell et al., 2018).

The challenges confronting cities and society—natural disasters, pandemics, socio-economic inequities, environmental degradation, and political instability—are not random occurrences. They stem from deep-seated systemic imbalances and interconnected vulnerabilities embedded within the fabric of urban and social systems (McFarlane, 2011). One can observe the stark disparity between the growing wealth of urban centers and the widening gap between affluent and marginalized populations, or the advanced technological achievements in urban and social infrastructure contrasted with the hidden societal and environmental costs they impose. These challenges, while varying in form or intensity, rarely find resolution; instead, they evolve and reemerge, often in more harmful ways, continually testing the limits of the resilience of cities and society. The cyclical recurrence of these urban and social crises underscores an unsettling reality: despite the relentless drive toward modernity and innovation, cities and society have yet to fully internalize the lessons of past failures or develop the mechanisms necessary to break the cycles of

J. Zhang, *Urbanimmunology*, Urban Sustainability,
https://doi.org/10.1007/978-981-95-4878-1_1

recurring challenges they strive to overcome (Temenos & Lauermann, 2020; Zhang & Hayashi, 2022).

The recurring cycle of crisis, response, recovery, and resurgence underscores the limitations of prevailing approaches to urban and social governance. With each disruption, cities and society hastily implement policies and interventions, only to find themselves unprepared for the next wave of uncertainty (Temenos & Lauermann, 2020; Zhang & Hayashi, 2022). This cycle exposes a fundamental flaw in the prevailing conception of the resilience of cities and society: cities and society are often perceived as static entities, defined by their physical structures, and supporting infrastructures. However, cities and society, much like living organisms, should be understood as dynamic, ever-evolving systems that require continuous adaptation to sustain themselves.

Addressing these recurring challenges necessitates rethinking how cities and society respond to external pressures. Like the human immune system, which is designed to identify and neutralize threats, cities and society should cultivate systems of recognition, adaptation, and recovery to secure their long-term viability. Yet the resilience of cities and society cannot remain a reactionary process; it should be proactive, adaptive, and grounded in continual learning. Just as the immune system not only combats infections but also learns from each encounter to strengthen future defenses, urban and social systems should evolve—drawing insights from past crises, addressing inherent vulnerabilities, and preparing to navigate future challenges. This calls for a holistic understanding of cities and society—one that incorporates the lessons of past disruptions, embraces diverse voices and stakeholders, and promotes ongoing improvement.

Urbanimmunology emerges as a transformative framework in this context. Drawing inspiration from the immune system's intelligent mechanisms (Janeway et al., 2001; Delves et al., 2017; Murphy et al., 2022; Abbas et al., 2023), *Urbanimmunology* offers a novel perspective on how cities and society can build the capacity to resist, adapt to, and recover from the complex challenges they face. Much like the immune system, which relies on a network of cells, organs, and signaling pathways to detect, respond to, and retain memory of threats, cities and society should cultivate interconnected systems—spanning environmental, political, economic, and technological domains—that function cohesively to enhance resilience. This framework underscores the critical importance of early detection, rapid response, continuous learning, and collaborative efforts in addressing both immediate risks and long-term uncertainties.

Urbanimmunology is not solely about defending against immediate threats; it is about shaping cities and society that evolve in response to adversity. It prompts a reevaluation of the city and society, moving beyond the notion of a static entity to be shielded from external forces and instead envisioning cities and society as dynamic systems that continually learn from experience, refine their mechanisms, and adapt to shifting conditions. Just as the immune system builds memory, enabling quicker responses to recurring threats, cities and society should also internalize lessons from past disruptions. They should cultivate memory through systems, policies, and institutions that not only respond to crises but also anticipate them, preparing for future challenges before they fully emerge.

Similarly, just as the immune system relies on an intricate network of specialized cells, each fulfilling a distinct function, urban and social systems comprise a web of interdependent components: infrastructure, governance, community engagement, technology, and the economy. The vitality of a city or society is not defined by the strength of any single element but by the coordinated interplay of these diverse components. Just as the immune system depends on communication among its cells, the resilience of cities and society hinges on collaboration among the public, private, and civil sectors, as well as the integration of expertise across disciplines.

In conclusion, drawing inspiration from *human immune intelligence* (HII), cities and society should be understood as adaptive, evolving systems—constantly responding to internal and external forces. As the world confronts increasingly complex challenges such as climate change, social inequality, and public health crises, it is essential to move beyond reactive policies toward a proactive, adaptive approach. *Urbanimmunology* provides a novel framework for understanding and enhancing the resilience and vitality of cities and society. It emphasizes the interconnectedness in urban and social systems, the importance of learning from past disruptions, and the imperative of continuous adaptation. In doing so, it offers a pathway for cities and society not only to withstand the challenges of today but to thrive amidst the uncertainties of tomorrow.

1.2 Leveraging Human Immune Intelligence to Inform the Transformation of Cities and Society

The immune system, with its intricate mechanisms and remarkable adaptability, stands as a profound metaphor for the resilience and vitality that cities and society should cultivate in the face of ever-evolving challenges. It is a system that not only reacts to immediate threats but, through learning and adaptation, strengthens its defenses against future incursions. In much the same way, cities and society, as dynamic organisms, should develop the capacity to recognize, respond to, and recover from a multitude of disruptions—whether they arise from natural disasters, pandemics, socio-economic inequalities, or environmental degradation. The lessons embedded within the immune system's operations offer a compelling blueprint for urban and social systems seeking to navigate the complexities and uncertainties of both today and the future.

Just as the immune system's defense mechanisms evolve in response to pathogens, cities and society should also evolve in response to the challenges that continuously emerge in the urban landscape. The immune system's extraordinary ability to learn from past infections, "remember" them, and mount a swifter, more effective response upon re-exposure offers an important lesson for cities and society: resilience is not simply about survival; it is about learning from adversity, strengthening the system in the process, and emerging more robust. The resilience of cities and society, therefore, should not be understood merely as a capacity to withstand crises. Rather, it should be viewed as an ongoing process of growth, adaptation, and continual learning from both past successes and failures.

One of the fundamental principles of the immune system is its capacity for recognition—recognizing the presence of a pathogen and distinguishing between harmful invaders and benign entities. This capacity for detection forms the first line of defense and is crucial in allowing the immune system to respond appropriately. In urban and social systems, this principle of detection is mirrored in the need for cities and society to have effective mechanisms in place for identifying emerging risks. The challenges in cities and society, much like pathogens, evolve over time. Climate change, for example, manifests in ever-more frequent and severe weather events, while pandemics—like the one we have just witnessed—emerge rapidly and often unpredictably. Likewise, socio-economic disparities, environmental degradation, and the vulnerability of infrastructure to failure are persistent threats that require vigilant monitoring. Cities and society should develop systems of early warning, data collection, and surveillance—tools that can help detect threats before they reach a critical point. Just as the immune system depends on specialized cells to identify foreign invaders, urban and social systems need mechanisms to identify risks as soon as they arise, allowing for timely intervention.

Response, the next phase in the immune system's defense process, is equally significant for the resilience of cities and society. Once a threat is detected, the immune system activates a series of responses to neutralize or remove the pathogen. These responses can range from the immediate action of innate immune cells to the more targeted efforts of adaptive immune responses. In urban and social systems, this is akin to how cities and society mobilize their resources and networks to mitigate or resolve a crisis. For instance, in the face of a natural disaster, cities and society rely on emergency response teams, healthcare systems, transportation networks, and communication technologies to minimize damage and ensure public safety. However, much like the immune system, this response cannot be static. Cities and society should not only react to the immediate consequences of a crisis but also refine their response strategies, learning from each event to improve future preparedness. The more adaptive a city or society is in its response, the better equipped it will be to handle future challenges. Policymakers, planners, and community leaders should embrace the idea that responses to crises are not merely about managing the present—they should also involve preparing for the unknowns of the future, fostering resilience through anticipation, flexibility, and innovation.

The recovery phase of the immune system is perhaps one of its most remarkable features. After successfully neutralizing a pathogen, the immune system does not simply return to its previous state but instead strengthens its defenses and improves its memory, ensuring a more effective response should the same pathogen appear again. This process of recovery is not about returning to the status quo but about evolving into a stronger, more capable version of itself. In urban and social terms, recovery after a crisis—whether it be a natural disaster, economic collapse, or public health emergency—should not simply be about rebuilding what was lost, but about rebuilding in ways that improve the ability of a city or society to withstand future challenges. Recovery provides a unique opportunity for growth and transformation. Cities and society should, therefore, view recovery not as a passive return to normalcy but as an active process of reimagining urban and social systems, improving resilience, and adapting policies to prevent future vulnerabilities.

For instance, after a city experiences a major flood, the recovery process should involve not just repairing damaged infrastructure but rethinking urban planning to mitigate future flood risks—perhaps by enhancing green spaces that absorb water, improving drainage systems, or revising zoning laws to discourage building in flood-prone areas. This approach turns recovery into an opportunity to build a more resilient urban and social environment, one that anticipates future risks and incorporates adaptive strategies.

Ongoing collaboration is also integral to this model of the resilience of cities and society. In the immune system, communication among cells—through chemical signals, cytokines, and other forms of molecular interaction—is essential for coordinating a unified and effective response. Similarly, in urban and social systems, a truly resilient city or society is one in which multiple stakeholders—government agencies, businesses, nongovernmental organizations, and citizens—work together, share information, and collaborate in the pursuit of common goals. Just as immune cells do not act in isolation but coordinate efforts to address threats, the resilience of cities and society requires a collaborative network in which different sectors and communities come together to pool resources, share knowledge, and devise coordinated responses to emerging challenges.

The immune system also thrives on a delicate balance—an equilibrium between responding to external threats and maintaining internal harmony. Similarly, the resilience of cities and society requires a balance between immediate, short-term responses to crises and the long-term strategies needed to build a sustainable, adaptive city or society. Urban and social policies should be flexible enough to respond to urgent issues but also forward-thinking enough to anticipate the needs of future generations. This requires a shift in how cities and society approach urban and social governance—moving from reactive crisis management to proactive, adaptive policymaking that is continuously informed by data, research, and real-time feedback.

In summary, while the examples above are just a starting point, the immune system offers far more than a mere metaphor for the resilience and vitality of cities and society. It provides a robust framework for how cities and society can recognize, respond to, and recover from challenges, all while continuously evolving and growing stronger in the process. By adopting principles such as detection, response, recovery, and collaboration, policymakers of cities and society can design systems that not only react to immediate threats but anticipate and prepare for future disruptions. Through this process of adaptive learning and proactive intervention, cities and society can enhance their resilience and vitality, ensuring that they not only survive but thrive in the face of uncertainty and change. As we move forward, it is imperative that we view cities and society not as static entities but as evolving organisms—capable of learning from their experiences, growing stronger through adversity, and ultimately emerging more resilient in the face of whatever challenges the future may hold.

1.3 *Urbanimmunology*: Catalyzing a Paradigm Shift for Future Cities and Society

The contemporary landscape of cities and society is a complex, living organism, shaped by an intricate web of interactions between its physical, social, and ecological components. As cities and society grow in size and complexity, they increasingly face a range of unprecedented challenges—challenges that transcend traditional disciplinary boundaries. From climate change to socio-economic inequality, from pandemics to infrastructural collapse, urban and social systems are confronted by multifaceted risks that demand innovative, holistic approaches to governance and policymaking. It is within this context that we recognize the urgent necessity of a transformative new discipline: *Urbanimmunology*.

Drawing inspiration from *human immune intelligence* (HII), *Urbanimmunology* seeks to redefine how cities and society understand and respond to threats, encouraging a shift from reactive crisis management to proactive, adaptive, and continuous resilience-building. The immune system, as a marvel of nature's design, offers a model of intelligent, adaptive response to threats—one that is not merely about surviving the present but about learning, evolving, and becoming stronger through each encounter with disruption. Cities and society, like organisms, should not only endure external stresses but grow through them. They should be capable of both protecting themselves and evolving into more robust, self-sustaining entities capable of addressing the challenges of an unpredictable future.

1.3.1 A Response to the Fragmentation of Urban Knowledge

At the heart of the necessity for *Urbanimmunology* lies a pressing concern: the fragmentation of urban and social knowledge. The challenges in cities and society—whether environmental, economic, or political—are not isolated problems; they are interconnected, often emerging as a result of deeply ingrained systemic failures. Yet, urban and social policymaking has traditionally relied on siloed approaches, treating each issue in isolation. This piecemeal approach, though well-intentioned, fails to address the complexity of urban and social life and the ways in which various challenges interlock.

Urbanimmunology proposes an integrated, systemic view of cities and society—one that draws upon the interconnectedness of their various components, much like the immune system's network of cells, organs, and signaling pathways. Just as the immune system does not treat a pathogen in isolation but recognizes it as part of a broader ecological context, *Urbanimmunology* encourages urban and social policy-makers to view cities and society not as static entities but as dynamic, evolving systems. By understanding the interdependencies between urban infrastructure, governance, social networks, and the natural environment, we can more effectively recognize and respond to the complex threats facing cities and society both today and in the future.

This interconnected perspective is essential, for modern challenges in cities and society rarely occur in isolation. Consider the confluence of climate change and socio-economic inequality, where the most vulnerable populations often bear the brunt of environmental disasters. Similarly, pandemics like COVID-19 expose the fragility of interconnected systems—healthcare, transportation, economics—and the cascading effects of disruptions in one area upon others. *Urbanimmunology* advocates for an urban and social policy approach that recognizes and addresses these interdependencies, proposing a unified model that enables cities and society to respond to complex, multifaceted challenges in an integrated and adaptive manner.

1.3.2 A Shift from Reaction to Proactivity: Preventing Crises Before They Arise

One of the most significant limitations of traditional urban and social governance is its reactive nature. Cities and society often wait for crises to unfold before responding, and even when they do act, their efforts are often short-term and patchwork in nature. The immune system, by contrast, is built upon the capacity to anticipate and react swiftly to threats—this anticipatory aspect is a critical function that *Urbanimmunology* seeks to instill in urban and social governance. The immune system does not simply react to pathogens when they are detected; it has learned to identify patterns, respond more efficiently upon re-exposure, and even prevent future threats by adapting its defense mechanisms over time.

Urbanimmunology offers a paradigm shift from reactive crisis management to proactive resilience-building. By integrating early warning systems, real-time data collection, and predictive modeling, cities and society can detect emerging threats before they escalate into full-blown crises. Through the application of *Urbanimmunology* principles, cities and society can move beyond the traditional cycle of crisis-response-recovery, instead focusing on adaptive learning, continuous improvement, and the anticipation of future risks.

For instance, a city faced with the threat of climate change may invest in adaptive infrastructure—such as flood defenses, green spaces, and sustainable energy systems—that not only addresses immediate risks but also anticipates future shifts in climate patterns. Similarly, in the context of public health, cities and society can leverage data to detect early signs of a pandemic, thereby activating targeted interventions that reduce the spread of disease before it reaches a critical threshold. The principles of *Urbanimmunology* encourage such forward-thinking strategies, enabling cities and society to not only withstand shocks but proactively create conditions that minimize the likelihood of those shocks in the first place.

1.3.3 Building Resilience Through Adaptation and Learning

Another essential aspect of *Urbanimmunology* is its emphasis on adaptive learning. The immune system's ability to "remember" previous threats and mount a more effective response upon subsequent exposures offers a powerful lesson for the resilience of cities and society. Rather than seeing each crisis as a unique and isolated event, *Urbanimmunology* encourages cities and society to learn from each experience, adapting their systems and strategies to become more resilient over time.

Urban and social systems, like the immune system, should be designed to learn from the past. Each urban and social crisis—whether a natural disaster, an economic collapse, or a public health emergency—offers an opportunity to assess weaknesses, refine strategies, and enhance the overall resilience of the city or society. This process of learning and adaptation requires continuous feedback loops, data-driven decision-making, and a willingness to evolve in response to changing circumstances.

In the realm of urban and social policy, this adaptive approach would not only improve responses to future crises but would also foster a culture of innovation and experimentation. For example, planners of cities and society might test new solutions on a small scale—whether in infrastructure, transportation, or social services—before scaling them up, much like how the immune system adapts its response based on repeated exposure. In this way, cities and society can develop a more fluid, dynamic capacity for resilience, one that is capable of evolving in tandem with an ever-changing world.

1.3.4 A Unified Response to Global and Local Challenges

Urbanimmunology stands out for its ability to transcend the limitations of individual disciplines, offering a unified approach to both global and local challenges. It acknowledges that the issues confronting cities and society today—such as climate change, urban sprawl, social inequality, and pandemics—are interconnected and cannot be addressed in isolation. These challenges are not merely urban or social in nature but are global issues that demand a global perspective. Consequently, *Urbanimmunology* draws from a wide range of disciplines, including immunology, public health, economics, and governance, to develop a holistic framework that addresses the full spectrum of urban and social challenges.

Just as the immune system responds to threats that may have global origins but manifest locally—such as diseases that spread from one part of the world to another—*Urbanimmunology* seeks to address challenges that span both local and global contexts. Climate change, for instance, is a global phenomenon with local consequences, and urban and social systems should adapt to both the local manifestations of these global issues as well as their larger, more systemic causes.

1.3.5 Summary: A Call to Action

In sum, *Urbanimmunology* offers a transformative framework for addressing the multifaceted challenges that cities and society face today. It calls for a shift in perspective—one that views cities and society not as static constructs but as living, evolving organisms that should adapt and learn in response to ever-changing conditions. By drawing from the immune system's principles of detection, response, recovery, and adaptation, *Urbanimmunology* offers a new way of thinking about the resilience of cities and society—one that emphasizes proactive, continuous learning and collaborative action across all sectors.

As cities and society continue to grow and evolve, it is essential to equip them with the tools and strategies needed to address future challenges. *Urbanimmunology* offers a path forward—a discipline that unites diverse stakeholders, fosters adaptive learning, and empowers cities and society not only to survive today's crises but to thrive amid tomorrow's uncertainties. Through the lens of *Urbanimmunology*, we can reimagine cities and society as dynamic, resilient organisms—capable of learning, adapting, and growing stronger with each challenge they face.

1.4 Navigating the Journey: A Chapter-by-Chapter Guide

1.4.1 **Urbanimmunology***: Laying the Foundations for a New Discipline in Urban and Social Transformation*

Chapter 2 embarks on a journey through the "Great Acceleration" and the concept of planetary health, highlighting their profound and undeniable connections to urban and social environments. These concepts lay the solid foundation for *Urbanimmunology*, a trailblazing discipline inspired by the intricacies of the human immune system. Following this, the chapter offers a comprehensive exploration of the immune system, illuminating its core functions and mechanisms, which serve as the bedrock for *Urbanimmunology*'s development. From this foundation, we venture into how the principles of immunity offer fresh perspectives on solving the pressing challenges of urban and social development. Moreover, the chapter links *Urbanimmunology* with the visionary HEARTY City planning concept and the broader mission of planetary health, all while engaging with the theoretical challenges that this emerging field faces. In its conclusion, we contemplate how

Urbanimmunology could illuminate the path forward for cities and society, providing guidance for navigating the complexities of an ever-evolving world.

1.4.2 *Energy-Disaster-Underground Integration Policymaking Informed by* **Urbanimmunology**

Chapter 3 introduces a groundbreaking integration framework essential for the resilience of cities and society: the integration of energy systems, disaster prevention and management, and underground infrastructure (EDU integration). These three systems, vital to the urban and social infrastructure, are harmonized through the principles of the immune system. Beginning with an individual examination of each system, the chapter reorients policymaking in each domain, drawing profound insights from the key functions and mechanisms of the immune system. The unique characteristics of each system are brought to the forefront, culminating in a discussion on how to effectively guide policymaking for EDU integration, followed by key findings and recommendations for the future of urban and social development.

1.4.3 *Driving* **Urbanimmunology** *Forward with Integrated Perspectives*

The preceding chapters have delved into the application of human immune intelligence to diverse urban and social challenges, underscoring *Urbanimmunology*'s potential to drive effective solutions. Integrating these insights, Chap. 4 derives nine immune-inspired principles foundational to the evolution of *Urbanimmunology*: Critical Dependency, Avoidance, Tolerance, Adaptation, Regulation, Coordination, Diversity, Robustilient Regeneration, and Coherence. These principles are then rigorously evaluated against three key criteria for a robust scientific system: validity, significance, and comprehensiveness. While addressing the limitations of existing urban theories, Chap. 10 also harnesses their strengths to enhance the scientific rigor and practical applicability of *Urbanimmunology*. Lastly, the chapter examines the synergistic and complementary interplay between these nine principles and the strengths of existing urban theories. This analysis demonstrates that *Urbanimmunology* is not only conceptually sound but also firmly positioned as a scientifically robust framework capable of addressing complex urban and social challenges.

1.4.4 **Spatial Interference Hazards** ***and Solutions Informed by*** **Urbanimmunology**

Chapter 5 serves as the sole and comprehensive application of the nine immune-inspired principles derived in *Urbanimmunology*, introducing the concept of spatial interference hazards, which arise from the complex interplay of physical, environmental, and infrastructural characteristics impacting both urban spaces and societal systems. While the chapter focuses on hazards such as those related to underground spaces, critical infrastructure, natural disasters, human-wildlife interactions, mixed traffic, skyfall, low-altitude air mobility, and overcrowding, these examples span diverse urban and social contexts, illustrating how the principles address immediate challenges while offering transferable insights. By analyzing the root causes and systemic factors behind these hazards, the chapter demonstrates the principles' versatility in fostering resilience, adaptability, coordination, and coherence, while emphasizing diversity and regeneration. Through this comprehensive exploration, Chap. 5 not only showcases the practical utility of *Urbanimmunology* but also provides a foundation for addressing interconnected urban and societal challenges explored throughout the book and beyond.

1.4.5 The HEARTY City: Drawing Parallels with the Human Immune System for Urban Futures

A city, much like a complex organism, thrives when its components work in unison. When isolated, its individual parts begin to falter. Guided by this philosophy, the HEARTY City concept—comprising six essential elements: Healthy, Eco, AI-driven, Resilient, Tourism, and Yearning—emerges as an innovative model for urban development. This chapter explores the foundations of the HEARTY City, drawing inspiration from *Urbanimmunology* to create a scientifically grounded framework. Each of the six components is analogized to aspects of the human immune system, where every element is interconnected and interdependent. This analogy envisions the HEARTY City as a dynamic, evolving entity—one that adapts and thrives amid the ever-changing challenges of the modern world. Through its interconnected functions, the city fosters environments that are resilient, sustainable, and conducive to the well-being of its residents, while also addressing the critical global challenges cities and society face today.

1.4.6 *The DIRECT Approach: Navigating Policymaking Process Management*

Chapter 7 begins by underscoring the critical importance of effective policymaking process management, before introducing the DIRECT approach, which is derived from the principles of the human immune system. The approach unfolds in six key components: "D: Detect," "I: Inform/Intervene," "R: React," "E: Enlighten/Enforce/Evaluate," "C: Collaborate," and "T: Transfer," each representing a crucial step in managing complex processes. The DIRECT approach is then applied to key areas such as pandemics, decarbonization in the transport sector, and public health promotion. Notably, the innovative application of this approach to public health offers a new framework for health promotion, explored further through tangible examples. The chapter concludes by considering potential future expansions of this approach to address the broader array of urban and social challenges.

1.4.7 **Urbanimmunology***: Catalyst for the Evolution of Cities and Society*

This concluding chapter positions *Urbanimmunology* as a transformative lens through which urban and social existence can be reimagined. By drawing parallels between immune intelligence and urban and social governance, it proposes strategies that embrace the organic complexity of cities and society, transcending rigid methodologies. The chapter envisions cities and society as interconnected, self-renewing organisms that evolve in response to change, redefining resilience not as mere endurance, but as the art of renewal and adaptation. Instead of critiquing existing paradigms, it integrates diverse perspectives into a cohesive vision, inviting a fresh understanding of urban and social development. As humanity moves beyond traditional frameworks like the Sustainable Development Goals (SDGs), this chapter advocates for an approach centered on self-renewal, adaptability, and continuous regeneration, ensuring cities and society thrive in harmony with both human and ecological systems.

1.5 Beyond the Horizon: Reimagining the Future of Cities and Society

As we face a rapidly changing world, cities are where human dreams and challenges come together. Climate change, social inequality, pandemics, and resource shortages are not just problems—they're powerful forces shaping the future of urban and social life. But within these challenges lies an opportunity for transformation. *Urbanimmunology*, inspired by how the human immune system adapts and survives, offers a new way of thinking about cities and society. By focusing on detection, adaptation, collaboration, and response, *Urbanimmunology* provides more than just a way to survive—it's a guide for cities and society to grow stronger, adapt, and thrive through challenges.

This work invites exploration into the boundless potential of *Urbanimmunology* and its profound implications for the future of cities and society. Like living organisms, cities and society have the capacity to learn, adapt, and transform, and through the chapters ahead, readers will delve into how this analogy unfolds across critical domains. The strategies illuminated in this book aim not to patch up the fractures of urban and social systems but to reimagine cities and society as thriving ecosystems—agile, foresighted, and alive with potential.

Urban and social governance, through the lens of *Urbanimmunology*, transcends isolated problem-solving. It becomes a living, interconnected process—an ecosystem that mirrors the intricate adaptability of life itself. Cities and society are no longer static constructs but dynamic organisms, evolving in response to external disruptions and internal rhythms. This perspective challenges us to rethink urban and social development, offering a narrative of cities and society as learning entities, where crises become catalysts for innovation and growth.

Within these pages lies not just a conceptual framework but a call to action. Whether you are a policymaker, urban planner, technologist, or engaged citizen, this book provides practical insights into applying the principles of *Urbanimmunology* to create cities and society capable of thriving in the face of uncertainty. This holistic approach integrates ecological stewardship, social cohesion, technological advancement, and economic vitality, crafting adaptive cities and society that do not merely endure but prosper.

The future of cities and society isn't about holding onto outdated systems or fixing what's broken. It's about embracing change, drawing from collective wisdom, and learning from nature's designs. *Urbanimmunology* challenges us to see cities and society as places where challenges become opportunities for innovation, where crises lead to valuable lessons, and where each success brings us closer to a more sustainable, fair, and vibrant world.

As you read this book, let its ideas encourage you to think boldly, act with purpose, and imagine cities and society that shine with resilience and vitality. Cities and society will adapt, grow, and respond to changing needs. *Urbanimmunology* gives us the tools to create cities and society where people and nature live in harmony—a future where cities and society not only survive but flourish, turning challenges into opportunities and ensuring prosperity for generations to come.

References

Abbas, A. K., Lichtman, A. H., & Pillai, S. (2023). *Basic immunology: Functions and disorders of the immune system* (7th ed.). Elsevier.

Brenner, N. (2018). Debating planetary urbanization: For an engaged pluralism. *Environment and Planning D: Society and Space, 36*(3), 570–590.

Burdett, R., & Sudjic, D. (2010). *The endless city*. Phaidon Press.

Calvino, I. (1978). *Invisible cities*. Mariner Books Classics.

Campbell, H., Forester, J., & Sanyal, B. (2018). Can we learn from our mistakes? *Planning Theory & Practice, 19*(3), 425–446.

de Boer, J. (2015). Resilience and the fragile city. *Stability: International Journal of Security & Development, 4*(1), 17–17.

Delves, P. J., Martin, J. E., Burton, D. R., & Roitt, I. M. (2017). *Janeway's immunobiology* (9th ed.). Garland Science.

Glaeser, E. (2012). *Triumph of the city: How our greatest invention makes us richer, smarter, greener, healthier, and happier*. Penguin Books.

Hall, P. (2014). *Cities of tomorrow: An intellectual history of urban planning and design since 1880*. Wiley-Blackwell.

Jacobs, J. (2022). *Cities and the wealth of nations: Principles of economic life*. Blackstone Publishing.

Janeway, C. A., Travers, P., Walport, M., & Shlomchik, M. J. (2001). *Immunobiology: The immune system in health and disease* (5th ed.). Garland Science.

Lefebvre, H. (2003). *The urban revolution*. University of Minnesota Press.

McFarlane, C. (2011). The city as assemblage: Dwelling and urban space. *Environment and Planning D, 29*(4), 649–671.

Muggah, R. (2014). Deconstructing the fragile city: Exploring insecurity, violence and resilience. *Environment and Urbanization, 26*(2), 345–358.

Mumford, L. (1961). *The city in history: Its origins, its transformations, and its prospects*. Harcourt, Brace & World.

Murphy, K. M., Weaver, C., & Berg, L. J. (2022). *Janeway's immunobiology* (10th ed.). W. W. Norton & Company.

Patel, R., Sanderson, D., Sitko, P., & De Boer, J. (2020). Investigating urban vulnerability and resilience: A call for applied integrated research to reshape the political economy of decision-making. *Environment and Urbanization, 32*(2), 589–598.

Temenos, C., & Lauermann, J. (2020). The urban politics of policy failure. *Urban Geography, 41*(9), 1109–1118.

Zhang, J., & Hayashi, Y. (2022). *Transportation amid pandemics: Lessons learned from COVID-19*. Elsevier.

Chapter 2
Urbanimmunology: Laying the Foundations for a New Discipline in Urban and Social Transformation

2.1 Great Acceleration and Planetary Health

2.1.1 Understanding the Great Acceleration: Challenges and Opportunities

The term "Great Acceleration" (Steffen et al., 2007, 2015) evokes an era of profound transformation, a defining chapter in human history marked by an unprecedented surge in activity and influence over the Earth. Since the mid-twentieth century, this epoch has seen humanity's aspirations soar to dizzying heights, yet the very momentum of this ascent has cast a long and perilous shadow over the planet's fragile systems.

This era, born in the aftermath of World War II, is fueled by the forces of industrialization and globalization, whose convergence has reshaped the contours of human existence and the Earth's ecology. Population growth has surged exponentially, driven by the relentless pull of urbanization. Cities, as hubs of innovation and opportunity, have swelled to accommodate millions, placing immense demands on infrastructure, resources, and ecosystems. This rapid urban expansion symbolizes

---

The author's intellectual journey into a new discipline was sparked by the unprecedented crisis of the COVID-19 pandemic. When invited to co-chair the WCTRS COVID-19 Task Force in early 2020, he was thrust into the urgent challenge of responding to a rapidly evolving global crisis. In this context, he developed PASS, a methodology designed to address the pandemic's cascading impacts (Zhang, 2020). As the virus spread relentlessly, it became clear that the policymaking process itself was of existential importance. This insight led to the creation of the DIRECT approach—a model inspired by the adaptive nature of the human immune system, emphasizing its profound interconnectedness with human resilience. From these intellectual breakthroughs emerged the discipline of "***Urbanimmunology***," a field shaped by careful reflection and enriched by global academic dialogues. Through this synthesis, the author reimagined urban resilience, grounded in the principles of the immune system, as a framework for navigating both present and future crises.

J. Zhang, *Urbanimmunology*, Urban Sustainability,
https://doi.org/10.1007/978-981-95-4878-1_2

humanity's ingenuity, yet it also underscores the strain of maintaining equilibrium within the boundaries of what the Earth can sustain.

Economic growth, the beacon of progress, has become a driving force of the Great Acceleration. Yet its pursuit often prioritizes immediacy over foresight, charting a course of expansion that undermines the very resources upon which it depends. The relentless extraction of fossil fuels, the draining of aquifers, and the depletion of raw materials have pushed ecosystems to their limits, leaving a legacy of scarcity in their wake.

Technological advancements, hailed as the crown jewels of modernity, have offered extraordinary possibilities to elevate the human condition. Yet, paradoxically, they have deepened our dependence on finite resources, accelerating environmental degradation. Industrial processes churn out pollution and waste with devastating efficiency, while the unintended consequences of innovation ripple across ecosystems, disrupting their delicate balances.

The unintended toll of this era is stark. Biodiversity dwindles as species vanish at rates unseen in millennia. Climate change tightens its grip, altering landscapes and weather patterns with a force that defies comprehension. Pollution permeates the air, water, and soil, posing threats not just to ecosystems but to human health and survival itself.

The Great Acceleration is, therefore, a paradox: a time of human ingenuity and triumph that simultaneously undermines the very foundations of life on Earth. It stands as both a testament to human ambition and a warning of its consequences—a stark reminder that the path of unchecked progress comes at a cost, one that demands urgent reckoning and a reimagining of what it means to thrive on this shared planet.

2.1.2 *The Symphony of Planetary Health: Envisioning Interconnected Well-being*

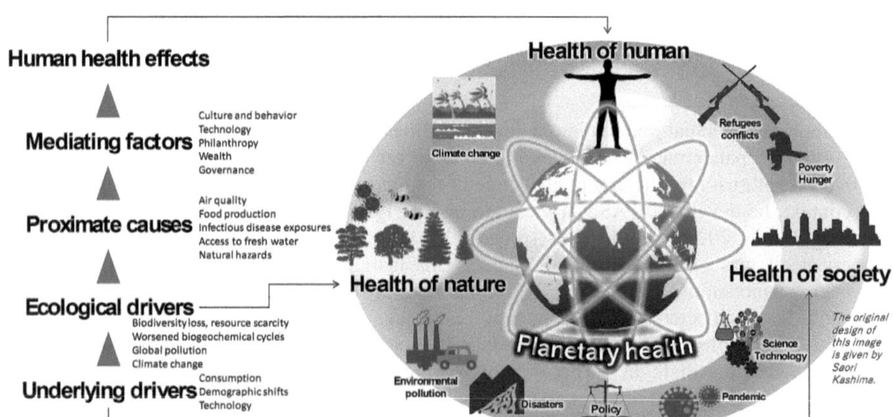

Modified by Junyi Zhang, based on https://www.thelancet.com/action/showPdf?pii=S0140-6736%2817%2932846-5

Planetary health (Whitmee et al., 2015) unfolds as a profound and urgent narrative, binding the fates of humanity and Earth into a singular, inseparable thread. It proclaims a truth both elemental and transformative: that the health of our planet is not a distant abstraction but a cornerstone of human survival and flourishing. In this vision, the disruptions wrought upon the environment—climate change, biodiversity loss, and pollution—are not mere ecological crises; they are crises of human well-being, each ripple of planetary imbalance reverberating through the fabric of our existence.

At its heart, planetary health champions an integrated philosophy of sustainability, one that illuminates the profound interdependence of human and ecological systems. It invites us to see humanity not as an entity apart from nature but as an intimate participant in its rhythms. Two guiding principles anchor this framework: interdependence, which underscores the intricate connections between human health and the vitality of ecosystems, and sustainability, a call for stewardship that safeguards the enduring health of both planet and people.

Expanding this vision, the concept of co-health, introduced by Zhang (2024),[1] enriches the dialogue with a deeper focus on the social dimensions of well-being. Co-health beckons us toward a holistic paradigm in which human, societal, and ecological health are not parallel aspirations but interwoven realities, evolving in harmony. This approach demands the fusion of diverse disciplines—ecology, medicine, social sciences—acknowledging the complexity and nuance of planetary health as a shared endeavor.

Central to this enriched perspective is the call for equity and justice, recognizing the unequal toll environmental and health crises impose on marginalized communities. It emphasizes the moral imperative to dismantle systemic inequities, ensuring fair access to resources, opportunities, and health services for all. Co-health thus envisions a world where every life, regardless of geography or circumstance, shares in the promise of a flourishing planet.

In embracing planetary health and its co-health counterpart, we are invited to reimagine humanity's role not as conquerors of the Earth but as stewards of its enduring vitality. Together, these frameworks offer a symphony of interconnected well-being—a harmonious vision where the health of the Earth and its inhabitants crescendos into a shared future of resilience, justice, and sustainability.

[1] The concept of "co-health," as a re-interpretation of planetary health, was first introduced by the author within the context of the "Trans-disciplinary Planetary Health Doctoral Education Program" (2020–2023). This program, which received support from the Ministry of Education, Culture, Sports, Science, and Technology of Japan, emerged during his tenure as Director of the Transdisciplinary Science and Engineering Program at Hiroshima University. In this intellectual endeavor, Zhang sought to deepen our understanding of the interconnectedness between human well-being and the health of the planet, inviting a rethinking of our collective responsibility in an age where the fates of both are inextricably linked.

2.1.3 Cities at the Crossroads: Navigating Challenges and Possibilities

The convergence of the Great Acceleration and planetary health finds its most striking manifestation in the urban and social landscapes of our world. Here, amid the hum of activity and the pulse of growth, cities stand as both the architects and the inheritors of a rapidly changing Earth. They are the arenas where the complex interplay of environmental, societal, and health challenges unfolds, demanding innovative visions and transformative action (Myers & Frumkin, 2021).

2.1.3.1 Urbanization and Environmental Reverberations

Cities, the beating hearts of the Great Acceleration, are home to over half the world's population—a density of life that magnifies their impact on the planet. They consume resources voraciously and generate waste prodigiously, their influence rippling far beyond their boundaries. Pollution, deforestation, and greenhouse gas emissions emanate from urban centers, reshaping landscapes and ecosystems (Alberti, 2023). Yet within this maelstrom lies the promise of recalibration: by embracing sustainable growth, cities can temper their environmental footprint and harmonize with Earth's delicate rhythms.

2.1.3.2 Public Health in the Urban Crucible

Within their crowded confines, cities reveal the stark vulnerabilities of human health. The noxious air of pollution, the oppressive heat of urban islands, and the silent weight of mental health struggles, all echo the need for a planetary health perspective (Ellena et al., 2020). Expanding green spaces and purifying urban air become acts of restoration, transforming cities into sanctuaries that nurture both body and soul.

2.1.3.3 Infrastructure as a Canvas for Sustainability

Urban infrastructure holds the brush that paints the future. Through green building practices, renewable energy systems, and efficient waste management, cities can craft a masterpiece of sustainability (Cheshmehzangi et al., 2021). Nature-based solutions—verdant forests, tranquil wetlands, and flourishing green corridors—infuse urban spaces with resilience and biodiversity, cultivating ecosystems that thrive alongside humanity (Bush & Doyon, 2019).

2.1.3.4 Biodiversity's Urban Haven

Contrary to their image as concrete jungles, cities can emerge as havens for biodiversity. By integrating wildlife corridors and safeguarding natural habitats, urban areas become sanctuaries where the wild and the human coexist. These flourishing ecosystems not only enrich urban life but also sustain essential services that humanity cannot do without (Castelli et al., 2021).

2.1.3.5 Justice as the Compass of Urban Progress

Amid the challenges of planetary health, the call for social equity rings clear. Marginalized communities often bear the brunt of environmental harm, and their voices should guide urban decision-making. Justice and inclusivity are not mere ideals but imperatives, forging cities that uplift all inhabitants and reflect the shared values of humanity (Bogard et al., 2017).

2.1.3.6 Adapting to a Warming World

The specter of climate change looms large over urban landscapes, necessitating resilience and foresight. Cities should fortify their systems against the onslaught of storms, floods, and heatwaves, crafting plans that protect their people and safeguard their futures. In adaptation lies survival; in resilience lies hope (Datola, 2023).

2.1.3.7 Governance for a Planetary Future

The governance of cities should evolve to match the magnitude of the challenges they face. Policies that incentivize sustainability and empower citizens are the scaffolding upon which resilient cities are built. Governance should be dynamic, flexible, and infused with the principles of planetary health, aligning ecological preservation with economic vitality and social well-being (Clark, 2015).

2.1.4 Call for a New Discipline: Toward Urban and Social Transformation

The entwined tales of the Great Acceleration and planetary health call for a new discipline—one that reimagines cities and society not as islands of human endeavor but as ecosystems in symbiosis with nature. This vision draws inspiration from the human immune system, a marvel of resilience and adaptability. Like this intricate network, cities and society should learn to detect, respond, and adapt, fostering balance and harmony amid challenges.

Through this lens, urban and social spaces transcend their traditional confines, becoming dynamic entities that protect, nourish, and flourish. They stand as testaments to the ingenuity of humanity and the wisdom of nature, merging growth with preservation, and progress with equity. In this reimagined urban future, cities and society rise as resilient, harmonious organisms, thriving in tandem with the natural world.

2.2 The Human Immune System

The human immune system (Janeway et al., 2001; Delves et al., 2017; Murphy et al., 2022; Abbas et al., 2023)[2] is a complex network of cells, tissues, and organs that work together to protect the body from pathogens, including bacteria, viruses, fungi, and parasites. Its primary role is to differentiate between self and non-self entities, enabling the body to neutralize harmful invaders while preserving its own cells. This highly coordinated defense mechanism reflects what can be described as human immune intelligence—an adaptive and dynamic capability that allows the system to learn from past encounters and mount increasingly effective responses. This system is crucial for maintaining homeostasis and overall health.

2.2.1 The Human Immune System: Key Components and Functions

2.2.1.1 White Blood Cells (Leukocytes)

White blood cells (leukocytes) are key components of the immune system, circulating throughout the body to defend against pathogens. Produced in the bone marrow, they are primarily classified into two main groups: phagocytes, which engulf and digest pathogens, and lymphocytes, which play a critical role in the adaptive immune response.

[2] The author's exploration of the human immune system is based on extensive learning, including numerous non-English sources. With intellectual humility, the information presented here is a synthesis of widely accepted knowledge on the immune system. To guide the reader, only several carefully selected references are provided throughout the book, recognizing that the depth of this subject far exceeds what can be covered in this brief overview.

2.2.1.2 Innate Immunity

It is the body's first line of defense, consisting of physical barriers like the skin and mucous membranes, as well as immune cells that respond quickly to a broad range of pathogens. This type of immunity is non-specific, meaning it does not adapt or enhance its response with repeated exposure to the same pathogen.

Physical barriers, such as the skin and mucous membranes, play a crucial role in preventing pathogen entry. The skin, as the body's largest organ, provides a strong outer layer of protection. Meanwhile, the mucous membranes in the respiratory and gastrointestinal tracts secrete mucus, which traps pathogens and helps eliminate them from the body.

Chemical barriers, such as lysozyme in saliva, gastric acid in the stomach, and antimicrobial peptides, offer chemical defenses by either destroying pathogens or inhibiting their growth. For instance, lysozyme breaks down the cell walls of certain bacteria, effectively neutralizing them. These chemical barriers are essential for protecting the body from infections.

Cellular Components Phagocytes, including macrophages and neutrophils, play a vital role in the immune response through phagocytosis, in which they engulf and digest pathogens, effectively eliminating infections and supporting the body's defense mechanisms. Natural killer (NK) cells, a type of lymphocyte, are responsible for recognizing and inducing apoptosis, or programmed cell death, in damaged, infected, or cancerous cells. This function is essential for preventing disease spread and preserving the integrity of the immune system.

Inflammatory Response When tissue is damaged or infected, the immune system triggers an inflammatory response, marked by redness, heat, swelling, and pain. This response is driven by cytokines, signaling molecules released by immune cells that recruit additional immune components to the site of infection, helping to facilitate the healing process.

2.2.1.3 Adaptive Immunity

Unlike innate immunity, adaptive immunity is a specific response that develops over time, especially after exposure to particular pathogens. This type of immunity provides long-lasting protection and involves the activation of lymphocytes, particularly T cells and B cells.

B cells are responsible for humoral immunity, which involves the production of antibodies—proteins that specifically target and neutralize pathogens. Upon encountering a specific antigen (a substance that triggers an immune response), B cells differentiate into plasma cells that produce these antibodies. Antibodies bind specifically to antigens, neutralizing them or marking them for destruction by other immune cells.

T cells are central to cell-mediated immunity, orchestrating targeted immune responses to eliminate infected or malignant cells. These lymphocytes are broadly categorized into two functional subsets: Helper T cells (CD4+) and Cytotoxic T cells (CD8+). Helper T cells amplify immune activity by releasing cytokines that activate macrophages, B cells, and cytotoxic T cells to enhance pathogen clearance. In contrast, Cytotoxic T cells directly identify and destroy compromised host cells by inducing apoptosis through molecular mechanisms such as perforin and granzyme release. Together, these subsets ensure precise coordination of immune defenses, balancing effective pathogen elimination with regulated tissue protection.

2.2.1.4 Antibodies

Antibodies, also known as immunoglobulins, are specialized proteins produced by B cells in response to pathogens or foreign substances. They bind to specific antigens—molecules that trigger an immune response—on the surface of pathogens, marking them for destruction and neutralization. Once bound, antibodies can prevent the pathogen from entering host cells or assist other immune cells in eliminating the threat. Additionally, antibodies can "remember" previous invaders, providing long-term immunity and faster responses upon re-exposure to the same pathogen.

2.2.1.5 Antigens

Antigens are molecules or components of pathogens that the immune system recognizes as foreign and potentially harmful. These can include proteins, polysaccharides, or other substances found on the surface of bacteria, viruses, or other pathogens. When detected, antigens trigger an immune response, prompting the production of antibodies and the activation of immune cells to neutralize or eliminate the threat. Antigens are also key in the development of vaccines, as they help the immune system "learn" to recognize and respond to specific pathogens without causing disease.

2.2.1.6 Cytokines

Cytokines are signaling proteins secreted by immune cells that facilitate communication and coordination among cells during immune responses. They play crucial roles in cell activation, proliferation, and the regulation of immune responses, ensuring that the immune system acts efficiently and appropriately. Cytokines help to orchestrate the body's defense mechanisms by promoting inflammation, recruiting immune cells to sites of infection, and modulating the intensity of immune responses. However, an imbalance in cytokine production can lead to inflammatory disorders or an insufficient immune response, highlighting their importance in maintaining immune system homeostasis.

2.2.1.7 Major Histocompatibility Complex (MHC)

Major histocompatibility complex (MHC) molecules are proteins located on cell surfaces that play a critical role in presenting antigens to T cells, enabling the immune system to recognize and respond to foreign substances. There are two classes of MHC molecules: Class I MHC molecules present endogenous antigens, which originate from within the cell, such as viral proteins, to cytotoxic T cells. Class II MHC molecules present exogenous antigens, derived from outside the cell, to helper T cells, which assist in activating other immune responses. This antigen presentation is crucial for initiating adaptive immune responses and distinguishing self from non-self, preventing the immune system from attacking the body's own tissues.

2.2.2 Understanding Immune Response: The Phases of Protection

The immune response can be divided into several distinct phases, each crucial for protecting the body from pathogens. These phases include recognition, activation, response, and memory. Each phase works together to detect, eliminate, and remember invaders, ensuring a swift and efficient defense. Understanding these phases helps explain how the immune system adapts to and overcomes infections.

2.2.2.1 Recognition

The immune system detects the presence of a pathogen by identifying specific antigens—molecules that are foreign or abnormal—through specialized receptors on immune cells, such as B cells and T cells. This initial recognition is crucial because it helps the immune system distinguish between harmful invaders and the body's own cells. The process of antigen recognition often triggers a cascade of signaling events that alert the immune system to activate its defense mechanisms, ensuring an appropriate and timely response to the threat.

2.2.2.2 Activation

Once an antigen is recognized, the immune system's T cells and B cells are activated. Helper T cells (CD4+ T cells) play a central role in coordinating the immune response by releasing cytokines that stimulate B cells to produce antibodies and activate cytotoxic T cells (CD8+ T cells) to target and destroy infected cells. This phase of activation amplifies the immune response, ensuring that both humoral immunity (antibody-mediated) and cell-mediated immunity (T cell-mediated) work in concert to eliminate the pathogen effectively.

2.2.2.3 Response

In the response phase, activated B cells proliferate and differentiate into plasma cells, which are specialized in producing large quantities of antibodies that specifically target the pathogen. These antibodies bind to the pathogen, neutralizing it or marking it for destruction by other immune cells. Meanwhile, cytotoxic T cells directly recognize and kill infected host cells, preventing the pathogen from replicating and spreading. This coordinated action eliminates the infection, restoring the body to health.

2.2.2.4 Memory

After the pathogen is cleared from the body, a subset of T cells and B cells persist as memory cells. These memory cells "remember" the specific antigens of the pathogen, allowing the immune system to respond more rapidly and effectively if the same pathogen is encountered in the future. This memory is the basis of long-term immunity, which is the principle behind vaccines, as they help train the immune system to recognize and fight off future infections without causing illness.

2.2.3 *When Immunity Fails: Disorders of the Immune System*

Immunological disorders can impair the immune system in various ways, leading to three key types of conditions: autoimmune diseases, immunodeficiency disorders, and allergies. These disorders can result in chronic health issues, increased vulnerability to infections, or harmful responses to non-threatening agents.

2.2.3.1 Autoimmune Diseases

In autoimmune diseases, the immune system mistakenly identifies the body's own tissues as foreign and attacks them, leading to inflammation and tissue damage. The cause of these conditions is not entirely understood but may involve genetic and environmental factors. For example, in rheumatoid arthritis, the immune system targets the joints, causing pain, swelling, and potential joint damage over time. In lupus, the immune system can affect multiple organs and systems, including the skin, kidneys, and heart, leading to a wide range of symptoms that can be challenging to manage.

2.2.3.2 Immunodeficiency Disorders

Immunodeficiency disorders occur when the immune system is unable to respond effectively to pathogens, leaving the body vulnerable to infections. These disorders can be inherited, such as in primary immunodeficiency diseases, where individuals are born with genetic mutations that impair immune function. Secondary immunodeficiencies, such as those caused by the HIV virus, result from external factors that compromise immune cells, particularly CD4+ T cells. This leaves individuals with a weakened immune system that struggles to fight off even common infections, putting them at increased risk of severe illness.

2.2.3.3 Allergies

Allergies occur when the immune system mounts an exaggerated response to harmless substances known as allergens. This overreaction can trigger a wide range of symptoms, from mild ones like sneezing and itching to more severe reactions, such as anaphylaxis, a life-threatening condition. In allergic reactions, the immune system releases chemicals like histamines, which cause inflammation and tissue damage in an attempt to eliminate the perceived threat. Common allergens include pollen, pet dander, and certain foods, and while these substances are usually harmless, the immune system treats them as dangerous invaders.

To prevent these disorders, the immune system is meticulously regulated to avoid overreactions that could lead to autoimmune diseases, where the body mistakenly attacks its own cells. This process, known as "Immune System Regulation," involves Regulatory T cells (Tregs), which are essential for maintaining immune tolerance and preventing excessive immune responses. Tregs help the body distinguish between self and non-self entities. Additionally, various feedback mechanisms are in place to ensure immune responses are appropriate and proportional.

In summary, the human immune system is a complex defense mechanism that integrates various components, including both innate and adaptive responses, to protect the body from infections and diseases. Similarly, urban and social systems share characteristics with the immune system, such as the ability to recognize, respond to, and adapt to various challenges and disruptions. By studying the principles and functions of the human immune system, urban and social development theories and practices can gain valuable insights. This interdisciplinary approach can strengthen urban and social resilience, enabling cities and society to better protect themselves against potential threats and adapt to changing conditions.

2.3 Proposing a New Discipline "*Urbanimmunology*"

In the complex and uncertain world with ever-increasing urbanization, cities and society—despite their progress—still face ongoing challenges. These issues, much like recurring shadows, change shape but often have the same underlying causes that remain unaddressed. Even when one problem appears to be solved, another arises, or the original issue resurfaces in a new form, despite the implementation of policies and interventions. This cycle suggests that cities and society, despite their continuous growth, have not fully learned from past mistakes.

In this persistent pattern, we can draw a helpful comparison to the human immune system. Just as the immune system learns and adapts to external threats, cities and society, in their pursuit of resilience and vitality, could benefit from a similar process of adaptive learning. The immune system's ability to recognize, respond to, and recover from threats offers a powerful metaphor for urban and social systems, emphasizing the need for cities and society not only to survive disruptions but to learn from them and grow stronger as a consequence.

2.3.1 Definition

Why is "immunity" a fitting term to describe the state of cities and society? Due to the diminishing effects of both soft measures (such as policies, strategies, and awareness initiatives) and hard measures (like buildings, facilities, and infrastructure), once these measures are implemented, it is essential to maintain their intended performance. This is similar to the human body—once a surgery is performed, recovery is the first phase, but the body's condition should be maintained post-surgery, with immunity playing a protective role. Even in the case of non-surgical conditions, such as lifestyle diseases, after treatment or medication, recovery is followed by the need to sustain that improved state through the body's immune system.

In this sense, strengthening the immunity of cities and society goes beyond a mere analogy to human immune intelligence; it is a fundamental characteristic of urban systems that we have not yet fully recognized.

Thus, the author introduces the concept of *Urbanimmunology*—an emerging field that focuses on how cities and society can develop the capacity to resist and adapt to disruptions. By designing strategies that enhance resilience and vitality, cities and society can transform into self-sustaining systems. Just as the immune system adapts to threats, cities and society should find ways to identify, respond to, and recover from various stressors, including natural disasters, pandemics, economic changes, and climate impacts as well as impacts of geopolitical tensions. Like living organisms, cities and society should be understood as dynamic entities that evolve in response to the challenges they encounter.

2.3.2 *Drawing Inspirations from Human Immune Intelligence*

The specialized cells of the human immune system—such as T cells and B cells—each serve distinct roles, much like the various components of urban and social systems: for example, residential areas, commercial districts, public spaces, infrastructure, and services. Each component, whether physical or social, contributes to the overall health of the city and society, and all are interconnected in a delicate web of influence. The resilience and vitality of a city or society rely not only on its physical infrastructure but also on the relationships and interactions between its people, organizations, and governing bodies.

As we continue exploring *Urbanimmunology*, we will examine how the principles of the immune system—recognition, response, and recovery—can inspire the development of cities and society that are not only able to withstand disruptions but can thrive and evolve through them. In this way, urban and social systems, much like the immune system, can learn from challenges, fostering a more sustainable, adaptable, and resilient future.

In this section, we provide a brief analogy to human immune intelligence, with more detailed comparisons discussed in the chapters that follow. These are merely examples; the full scope of *Urbanimmunology* can only be fully understood by exploring the entire book, where various facets of cities and society will be examined in depth.

2.3.2.1 Detection of Urban and Social Threats by Mimicking Immune Recognition

In the human body, immune cells use specialized pattern recognition receptors (PRRs) like toll-like receptors (TLRs) to detect different types of threats. For example, TLRs in epithelial cells recognize bacterial components, while phagocytes identify and engulf pathogens or damaged cells. These immune cells each serve a specific function in detecting distinct threats, triggering appropriate immune responses such as inflammation or pathogen elimination.

Similarly, urban and social systems, with their inherent complexity and dynamism, should develop mechanisms to detect and recognize various types of disruptive forces—whether environmental pollutants, economic declines, disease outbreaks, or social unrest. Mimicking immune cells’ specific roles, different detection tools can monitor distinct threats in cities and society. For instance, environmental monitoring systems function like “sensor cells” in the body, detecting pollutants or hazardous substances in the air or water. Public health surveillance platforms, akin to immune cells detecting pathogens, track disease outbreaks and monitor population health, allowing for early responses to epidemics. Crime and social unrest tracking mechanisms act like the body’s detection of harmful social behaviors or instability, helping authorities identify rising tensions or criminal activities before they escalate. Real-time data analytics networks, much like the body's ability to integrate information from various immune receptors, process multiple sources of data—from traffic patterns to social media activity—helping urban systems detect a broad range of emerging issues.

In this way, urban and social systems, much like the immune system, rely on specialized detection mechanisms tailored to identify specific threats, ensuring that they can respond swiftly and effectively to maintain resilience and stability.

2.3.2.2 Empowering Actors in Cities and Society by Learning from Cellular Components

Cities and society require governance by a wide range of actors, both formal and informal. Key actors include governmental bodies, the private sector, civil society (such as community groups and citizens), academic and research institutions, the technology and innovation sector, international organizations, media and communication platforms, financial institutions, and cultural institutions. The roles and responsibilities of these actors should be clearly defined, taking into account both their formal structures and their informal dynamics, to effectively leverage their capacities and empower the growth and resilience of cities and society.

The following three paragraphs provide examples of how the roles of cellular components can be mimicked, offering a clearer understanding of how human immune intelligence can be applied to define the actors in cities and society in a scientifically grounded manner.

Phagocytes and Emergency Services Just as phagocytes engulf and neutralize pathogens, emergency services—such as firefighters, police, and medical responders—act as the frontline during crises. Their role is not only reactive but can be strengthened through ongoing training, strategic resource allocation, and active community engagement, ensuring they are well-prepared to tackle the unpredictable challenges of the urban and social environment.

T Lymphocytes and Community Networks T lymphocytes are key players in the immune system's adaptive response, recognizing and counteracting specific threats. In the urban and social realm, community networks, local organizations, and grassroots movements mirror this function. These entities are attuned to the unique needs of the city and society, mobilizing resources and support in response to particular disturbances, thus fostering resilience and adaptability in the face of change.

B Lymphocytes and Local Governance B lymphocytes generate antibodies that neutralize pathogens, providing a safeguard against future threats. Similarly, local governance structures—such as city councils and municipal authorities—serve as the urban equivalent of antibodies. They craft policies and implement programs that prevent harm, mitigate risks, and ultimately strengthen the resilience and vitality of the city, ensuring that it can withstand both immediate and long-term disruptions.

2.3.2.3 Enhancing Adaptive Capacities of Cities and Society Through the Lens of Immune Memory

Cities and society are inherently dynamic and full of uncertainties, constantly evolving in response to shifting needs, challenges, and opportunities. As a result, no single, fixed policy or plan can effectively address the wide array of urban and social issues in a one-size-fits-all manner. Solutions should be flexible and adaptable, tailored to the unique context and circumstances of each situation, allowing cities and society to respond proactively to emerging trends and disruptions. The following two examples demonstrate how human immune intelligence can be mirrored to enhance the adaptive capacities of cities and society.

Innate and Acquired Immunity Innate immunity is the body’s first line of defense, providing a rapid, non-specific response to a wide range of threats. Similarly, cities and society possess innate resilience—basic systems, structures, and behaviors that provide an immediate, foundational response to disruptions. These might include established emergency response frameworks, public health systems, or social networks that quickly mobilize in times of crisis. This innate resilience forms the backbone of a city’s ability to absorb and cope with challenges, offering an initial layer of protection. However, the strength of innate immunity alone is not enough for long-term resilience. Just as the immune system can adapt and strengthen through acquired immunity, cities and society should learn from their past experiences, using them to enhance their adaptive capacities. By reflecting on past disruptions—whether it’s through adapting infrastructure after a natural disaster or adjusting policies following an economic downturn—cities and society can build a more robust and adaptive framework. The combination of innate resilience, strengthened by continuous learning and adaptation, enables cities and society to not only withstand shocks but to evolve and thrive in the face of future challenges.

Thymus and Training Programs The thymus is the site where T lymphocytes (T cells) mature and are trained to recognize and respond to threats. It plays a critical role in shaping the immune system’s ability to differentiate between the body’s own cells and foreign invaders. Through a process called positive and negative selection, the thymus ensures that T cells are equipped to effectively combat pathogens while preventing them from attacking the body’s own tissues. The thymus also provides a platform for immune cells to adapt to new and emerging threats, ensuring that the body’s defenses remain flexible and responsive to a constantly changing environment. In the context of cities and society, the thymus analogy can be extended to the broader network of actors—governmental bodies, the private sector, civil society, educational institutions, and other stakeholders—that contribute to the resilience and adaptability of urban and social systems. Cities and society need interconnected systems that enable all actors to recognize and respond to emerging challenges. These systems, including education, public health infrastructures, policy frameworks, and community engagement, play a role in continuously preparing and refining the collective capacity of cities and society to confront future uncertainties. By fostering collaborative learning, resource-sharing, and flexibility across all sectors, cities and society can evolve in response to disruptions, ensuring sustained adaptability and resilience in the face of complex challenges.

2.3.2.4 Antibody Production and Policy Development for Urban Immunity

Antibodies are uniquely designed to target and neutralize specific pathogens, ensuring that the body can swiftly respond to recurrent threats and providing long-term protection against future infections. In the urban and social context, this concept mirrors the creation of targeted policies and strategies tailored to address specific vulnerabilities within the city and society. Urban policies—such as zoning laws, environmental regulations, and public health initiatives—should be carefully designed to address identified risks and enhance the overall resilience and vitality of the city and society. These policy "antibodies" should be formulated to neutralize pressing challenges like air pollution, traffic congestion, or public health crises, offering protective measures to ensure the continued health of the city and society. By identifying and targeting specific issues, these strategies not only mitigate existing risks but also help prevent future disruptions. For example, policies that promote green spaces or regulate waste management work to neutralize environmental pollutants, while urban mobility regulations or emergency preparedness plans can reduce the impact of natural disasters or social unrest. Policies in cities should also evolve, continuously learning from past experiences to strengthen the immune system of the city and society. Through this dynamic, responsive approach, cities and society can enhance their capacity to thrive in a world marked by constant change and uncertainty.

2.3.3 Illustrative Strategies Enhancing the Immunity Level of Cities and Society

To transform both cities and society into resilient, adaptive systems through the principles of *Urbanimmunology*, a range of strategies can be implemented, each rooted in the foundational principles of the immune system. Below are a few key examples, with more in-depth analogies and insights provided in the chapters that follow.

2.3.3.1 Integrated Monitoring Systems for Evidence-Based Policymaking

Just as the immune system continuously monitors for pathogens, urban and social systems can similarly benefit from the development of robust monitoring systems. These systems would incorporate “sensors” to collect diverse data across multiple time scales, including real-time information, and “assessors” for threat evaluation. Such systems would provide valuable evidence for policymaking, tracking the health of the city and society—including their environment, social dynamics, and overall vitality. This would enable timely, informed responses to emerging threats and disruptions.

2.3.3.2 Infrastructure and Functions for Resilient Cities and Society

The immune system relies on physical barriers like the skin and mucous membranes to defend against invaders, acting as frontiers that protect the body. Similarly, urban and social environments should be designed with resilient infrastructure and functions that serve as frontiers, absorbing shocks and safeguarding against disruptions. This may include flood-resistant riverbanks, earthquake-proof buildings, and facilities essential for transport, energy, and other vital services. Green spaces can serve dual purposes, offering both recreational areas and emergency evacuation zones. Additionally, robust public health response mechanisms are necessary to address both epidemics and general health threats, ensuring the resilience and vitality of cities and society.

2.3.3.3 Community Engagement: The Pulse of Thriving Cities and Society

The immune system thrives through the cooperation and coordination of various cells, each playing a specialized role in defending the body. Similarly, urban and social resilience and vitality are greatly enhanced by fostering strong community networks where individuals are engaged and work together in solidarity. When citizens are connected and actively involved, they can respond more effectively to crises, much like immune cells collaborating to protect the body. Community cohesion not only strengthens a city’s immediate response but also amplifies its long-term capacity to adapt to and recover from challenges, fostering a dynamic and resilient society. In a world of uncertainty, empowered and unified communities can create a robust foundation for social, economic, and environmental recovery.

2.3.3.4 The Feedback Revolution: Empowering Adaptive Policymaking

The immune system continuously assesses and adjusts its response to new threats through adaptive feedback loops. Similarly, cities and society should integrate feedback mechanisms throughout every step of the policymaking process. This includes identifying issues, collecting data, analyzing underlying causes, generating alternatives, forecasting, planning based on rigorous evaluation and public involvement, implementing policies, and monitoring outcomes. Feedback should flow between all these steps, ensuring that each phase informs and adjusts the others. By creating a system of continuous feedback, urban systems can evolve and adapt in real time, promoting ongoing learning and improvement.

2.3.3.5 The Art of Preparedness: Safeguarding Against the Unexpected

The human body defends itself through both innate immunity, present from birth, and adaptive immunity, which develops and strengthens over a lifetime, allowing the body to recognize and respond effectively to a wide range of invaders. Additionally, the thymus trains specialized cells, such as T cells, to distinguish between the body's own healthy cells and harmful foreign invaders, maintaining precision and efficiency in the immune system. Similarly, urban and social systems should prioritize preparedness by developing contingency plans and ensuring that key actors—whether public officials, emergency responders, or community leaders—are well-trained. By fostering a culture of readiness and continuous learning, these systems can react with greater agility and effectiveness when unexpected challenges arise, supporting a resilient and adaptive society.

2.4 Revisiting Responses to Urban Development Issues

The human immune system, with its intricate balance of responsiveness and adaptation, offers a profound metaphor for rethinking urban development. It invites us to see cities not as static entities, but as living systems—complex, interdependent, and in constant flux. Just as the immune system maintains the body's equilibrium through resilience and adaptability, so too should our urban environments evolve to sustain themselves in the face of growing challenges. The emergence of *Urbanimmunology* as a discipline calls for us to critically reimagine the frameworks of traditional urban policymaking, urging a shift from reactive approaches to those that cultivate long-term resilience and equity. It compels us to learn from past failures, to build systems that are not only robust but also adaptive, and to ensure that no community is left vulnerable in the face of urbanization's complexities.

Just as an effective immune system protects the body in all its parts, urban development should prioritize the well-being of all its inhabitants, particularly those marginalized by economic, social, or environmental factors. The analogy with the immune system provides a holistic approach to urban resilience—one that addresses both immediate threats and lays the foundation for sustainable, equitable futures. Below, we explore several strategies rooted in immune system principles that can inspire and guide the evolution of more resilient, responsive urban policymaking.

2.4.1 Recognizing Threats in Cities: Pathogen Detection

The immune system's ability to recognize and neutralize pathogens hinges on its capacity to remember past infections. This "memory" allows the body to respond preemptively to familiar threats. Likewise, cities should cultivate the foresight to recognize potential disruptions—be they natural disasters, economic crises, or public health emergencies. To achieve this, cities should invest in advanced monitoring systems, incorporating sensors, data analytics, and community feedback mechanisms. These tools can help identify emerging risks early, empowering cities to intervene before threats fully materialize, thus mitigating their impact.

2.4.2 Collaborative Responses to Urban Challenges: Immune Mobilization

Upon identifying a threat, the immune system mobilizes a coordinated response, deploying various cells to address the invader. Urban environments, too, require dynamic and adaptable response mechanisms. Cities should be capable of quickly marshaling resources—both human and material—across sectors. Effective responses depend on collaboration: between government agencies, private enterprises, and local communities. This collective approach strengthens the overall urban immune response, enabling swifter and coordinated action during times of crisis.

2.4.3 Building Urban Immunity Over Time: Immunological Memory

The immune system's ability to "remember" previous infections enhances its efficiency in future encounters. This principle can be translated to the urban context, where cities, too, should develop institutional memory. By learning from past disruptions—such as past natural disasters, financial crises, or public health emergencies—cities can refine their strategies and improve their resilience over time. This process of continual learning and adaptation ensures that cities are not doomed to repeat their mistakes, but rather can build upon their experiences to enhance future response capabilities.

2.4.4 Adaptive Urban Systems: Learning from Immune Cells' Flexibility

Just as the immune system evolves through mutation and selection to confront new threats, urban systems should demonstrate similar adaptability. Cities are dynamic entities, subject to shifts in population, climate, and economic conditions. Urban environments should adopt flexible planning strategies that allow them to evolve in response to these changes. This may include embracing adaptive infrastructure, transforming land-use policies, and prioritizing green technologies. Furthermore, fostering robust community networks is essential, as these too act as critical nodes in the system's capacity to adapt.

2.4.5 Building Resilient Cities: Embracing Diversity and Redundancy

A healthy immune system thrives on diversity, with multiple types of immune cells working together to ensure a robust defense. Similarly, urban resilience flourishes in diversity—whether in the form of varied populations, mixed economies, or diverse cultures. Diversity in urban environments fosters creativity, innovation, and the capacity to problem-solve in novel ways. Just as immune systems rely on redundancy to ensure defense against a range of pathogens, cities benefit from multiple systems of resilience, such as alternative transportation options or diverse energy sources. By fostering inclusive communities and maintaining redundancy in critical urban systems, cities can build resilience that is adaptable and robust in the face of disruption.

2.4.6 Maintaining Balance in Urban Systems: Homeostasis

Homeostasis—the ability of the immune system to maintain balance—is key to its function. Similarly, urban systems should strike a balance between development and sustainability. Cities should integrate practices that promote self-sufficiency and environmental stewardship, such as developing local food systems, harnessing renewable energy, and implementing water conservation initiatives. The balance extends beyond ecological concerns; cities should also find equilibrium between economic growth, social equity, and environmental health. Striving for such balance ensures that urban environments are capable of responding to disruptions while fostering long-term sustainability.

2.4.7 Long-Term Urban Vitality: Immunity Over Time

A well-functioning immune system not only defends the body but also supports long-term health. In the same way, urban systems should be designed to sustain long-term resilience, nurturing their capacity to evolve and thrive over time. This requires a forward-thinking approach to urban development that prioritizes sustainability and future-proofing. Investments in renewable energy, sustainable transportation systems, and efficient waste management are key elements of this vision. By ensuring that cities are designed with resilience in mind, we can create urban environments that are not only capable of surviving but flourishing in an uncertain future.

Thus, by reimagining urban development through the lens of the immune system, we open up new avenues for understanding how cities can build resilience. In doing so, we may construct not only more robust urban systems but ones that are more adaptive, equitable, and sustainable—systems that, like the human body, can thrive amidst uncertainty and change.

2.5 Deriving a Policymaking Process Management Methodology

The human immune system continuously responds to invasive pathogens without interruption. Similarly, policymaking in urban settings, as well as general policies, should be subject to ongoing improvement and continuous monitoring. To effectively address the impacts of COVID-19 and similar threats in the future, while facilitating post-pandemic recovery, the author developed the DIRECT approach (Zhang, 2021, 2022). In this framework, "D" stands for Detect, "I" for Inform/Intervene, "R" for React, "E" for Enlighten or Enforce, "C" for Collaborate, and "T" for Transfer. The following will demonstrate how the human immune system can provide theoretical support for this novel approach. More details refer to Zhang (2025).

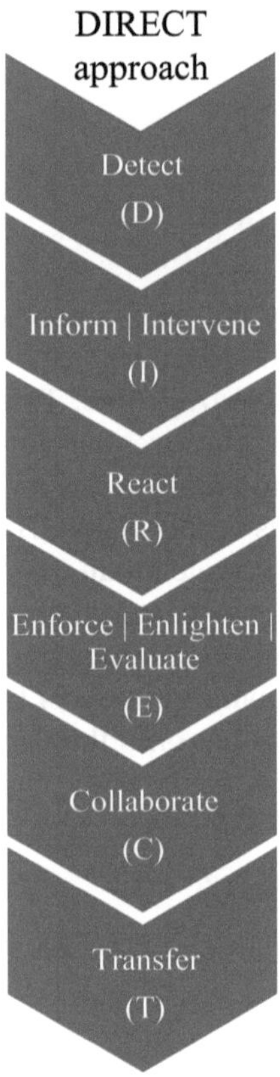

2.5.1 D: Detect (Recognition Phase of the Immune System)

The immune system detects pathogens using pattern recognition receptors (PRRs) on immune cells, such as macrophages and dendritic cells. These receptors recognize harmful patterns (PAMPs/DAMPs), triggering a response. Similarly, in urban policymaking, smart technologies, including sensors and data analytics, detect emerging issues like pollution and inefficient systems. These technologies act as "immune receptors," identifying and informing timely interventions while balancing ethical concerns like privacy.

2.5.2 I: Inform/Intervene (Activation Phase of the Immune System)

Once a pathogen is detected, antigen-presenting cells (APCs) activate immune responses by presenting antigens to T cells. In urban policy, the Inform/Intervene phase involves sharing risk information and initiating behavior-changing interventions. Clear communication helps engage the public, guiding them toward sustainable actions. Just like the immune system activates coordinated defense, this phase ensures effective mobilization of policy efforts through informed decision-making and public involvement.

2.5.3 R: React (Response Phase of the Immune System)

In the immune system, after recognition, a rapid defense is activated with innate cells responding immediately, and adaptive cells targeting specific threats. In urban policymaking, the React phase focuses on how individuals and communities respond to interventions. Awareness, trust, and social factors influence the effectiveness of policies. As with immune responses adapting to new pathogens, urban policies should evolve based on real-time feedback to remain effective and ensure long-term resilience.

2.5.4 E: Enlighten/Enforce/Evaluate (Memory, Activation, and Adaptive Functions of the Immune System)

The immune system retains memory of past infections, enabling faster, more efficient responses upon re-exposure. In urban policy, the Enlighten function uses past experiences to guide current decisions, promoting sustainable practices. The Enforce function involves legal measures to ensure compliance when voluntary actions fall short, while Evaluate ensures continuous monitoring and adaptation of policies. This dynamic, adaptive approach improves responses to emerging urban challenges, enhancing long-term sustainability.

2.5.5 C: Collaborate (Cooperation of Immune Cells and Organs)

The immune system relies on coordinated collaboration among various cells, such as helper T cells, B cells, and macrophages, to mount an effective defense. In urban policymaking, the Collaborate phase emphasizes the importance of coordination among stakeholders like governments, businesses, and communities. Just as immune cells communicate through cytokines, urban stakeholders should share data and resources to address complex issues. This cooperative approach ensures comprehensive policies, enhancing resilience in urban ecosystems.

2.5.6 T: Transfer (Acquired Immunity and Adaptation)

The immune system's Transfer function involves memory cells that enhance responses to previously encountered pathogens, making future defenses more efficient. In urban policymaking, the Transfer phase draws on past experiences and data to adapt and refine strategies. By learning from previous successes and failures, cities can respond more effectively to emerging challenges, improving resilience. This adaptive approach ensures that urban systems are better equipped to handle future issues, fostering long-term sustainability.

2.6 The Significance of *Urbanimmunology*

2.6.1 Practical Significance of Urbanimmunology

The emergence of *Urbanimmunology* arrives at a moment when cities and society face significant challenges, ranging from climate crises to social and economic upheaval. *Urbanimmunology* provides a practical and transformative framework for addressing these complex urban and societal issues. By emphasizing resilience, collaboration, data-driven decision-making, community involvement, and sustainable practices, it offers a necessary approach for cities and society to not only survive but thrive in an uncertain future.

2.6.1.1 Growing Vulnerabilities

Modern cities and society are increasingly vulnerable to climate change, natural disasters, economic instability, and public health crises. Traditional urban planning models often fall short in addressing these dynamic and interrelated risks. *Urbanimmunology* responds by advocating for adaptable strategies that recognize the interdependence of urban systems and society, emphasizing the need for resilient, responsive mechanisms integrated into both the physical environment and social structures.

2.6.1.2 Resilience Building

A core aspect of *Urbanimmunology* is fostering resilience, not just within urban areas but across society as a whole. This means evolving and adapting to challenges, rather than merely returning to a previous state after a crisis. By enabling both cities and society to endure and thrive in the face of ongoing change, this approach ensures long-term sustainability and livability.

2.6.1.3 Interdisciplinary Collaboration

Urbanimmunology encourages collaboration across multiple disciplines, including urban planning, public health, sociology, and environmental science. Building resilience in both cities and society requires a collective effort, where diverse perspectives and expertise come together to create holistic solutions to urban and social challenges. This collaborative spirit is crucial for making cities and society adaptable and responsive to their unique needs.

2.6.1.4 Data-Driven Decision-Making

Urbanimmunology emphasizes the use of data to monitor the health and resilience of both urban areas and society. Real-time data helps predict and address emerging issues before they become critical, enabling cities and society to take proactive, effective actions. This approach allows for better resource allocation and long-term resilience.

2.6.1.5 Sustainable Development Goals

Urbanimmunology aligns with the global Sustainable Development Goals (SDGs), addressing the interconnected needs of environmental sustainability, social equity, and economic vitality. It offers a comprehensive approach to urban and societal challenges, integrating these elements to create spaces and communities that are not only sustainable but also just, equitable, and capable of fostering thriving, inclusive society.

2.6.1.6 Community Empowerment and Social Cohesion

A key tenet of *Urbanimmunology* is recognizing that resilience is shaped by strong social bonds. By empowering communities to engage in resilience-building efforts, both urban spaces and society as a whole benefit. This participatory approach strengthens social cohesion and ensures that policies reflect the diverse needs of people, ultimately enhancing societal well-being.

2.6.1.7 Adapting to Climate Change

Both urban areas and society should adapt to climate change to minimize risks and promote sustainability. *Urbanimmunology* provides strategies that help cities and society learn from environmental disruptions, building adaptive capacities and regenerating in response to challenges. This resilience-oriented approach ensures that cities and communities remain viable despite shifting environmental conditions.

2.6.1.8 Policy Innovation and Governance

Urbanimmunology supports the development of flexible governance structures that can respond to complex challenges in both urban and societal contexts. It encourages policies that prioritize resilience, long-term sustainability, and collaboration, ensuring that both cities and society can navigate expected and unforeseen challenges.

2.6.2 *Theoretical Significance of* Urbanimmunology

Urbanimmunology is expected to become a groundbreaking theoretical framework that fundamentally transforms how we understand cities and society. Drawing inspiration from human immune intelligence, it proposes a radical shift from static models of urban and social systems to dynamic, adaptive, and interconnected entities. By conceptualizing cities and society as living, evolving organisms, *Urbanimmunology* challenges traditional theoretical assumptions and offers a new paradigm for urban and social theory.

2.6.2.1 Reframing Cities and Society as Dynamic, Adaptive Systems

Traditional urban and social theories have often treated cities as fixed entities, governed by linear growth and predefined structures. *Urbanimmunology* disrupts this view by emphasizing that cities and society are inherently dynamic, interconnected systems. Like the immune system, which continuously adapts to emerging threats, cities should be understood as evolving systems that respond and adapt to both external shocks and internal changes. This shift encourages a view of cities and society not as static, top-down systems but as flexible, self-organizing entities that continuously learn and regenerate in response to new challenges.

2.6.2.2 A Holistic Approach to Urban and Social Theory

The traditional theoretical approach often isolates various urban and social components—such as infrastructure, governance, and social systems—treating them as separate entities. *Urbanimmunology*, however, presents a holistic view, drawing parallels to the immune system's integrated functions. It suggests that urban and social systems cannot be fully understood by focusing on individual components in isolation. Instead, the theory emphasizes the interconnectedness of these systems, encouraging theorists to explore how different components interact and shape each other over time. This theoretical shift advocates for systems thinking in urban and social theory, highlighting the importance of understanding how all aspects of urban life—social, environmental, economic, and infrastructural—function in dynamic, reciprocal relationships.

2.6.2.3 Evolving Concept of Urban and Social Resilience

Urbanimmunology introduces a dynamic and evolving concept of resilience, in contrast to traditional theories that often treat resilience as a static or fixed state. Rather than simply returning to equilibrium after a disruption, resilience, according to *Urbanimmunology*, is about continuous adaptation and the capacity of cities and society to evolve and regenerate in the face of ongoing change. This theoretical reframing invites scholars to reconsider resilience not as an endpoint but as a continuous process of learning, adaptation, and transformation. Cities, therefore, should be theorized not merely as survivors of crises but as systems capable of adapting, transforming, and thriving in the face of uncertainty and flux.

2.6.2.4 Shifting the Focus of Urban and Social Development Goals

Urbanimmunology suggests a radical rethinking of the goals of urban and social development. Conventional urban theories have often been driven by goals of growth, expansion, and optimization, focusing on metrics like population size, economic output, and resource consumption. *Urbanimmunology*, however, refocuses these goals on long-term sustainability, social equity, and ecological health, recognizing that cities and society are not merely defined by their growth but by their capacity to adapt, regenerate, and foster equitable, inclusive development. This theoretical shift calls for the development of cities that are self-regulating, sustainable, and capable of evolving alongside the challenges of a rapidly changing world.

2.6.2.5 A New Framework for Governance and Policy Development

Urbanimmunology also offers a theoretical shift in governance and policymaking, emphasizing the need for adaptive, decentralized, and collaborative governance structures. Traditional urban theories often assume top-down, control-based models of governance, but *Urbanimmunology* suggests that urban and social systems should be governed by flexible, responsive, and distributed decision-making structures that mirror the adaptive immunity of the human body. The theory proposes that cities need governance models that can anticipate future challenges, foster collective action, and respond dynamically to changing conditions, enabling cities to evolve in a way that remains both equitable and sustainable.

2.6.2.6 Reevaluating Metrics of Urban and Social Success

Urbanimmunology challenges traditional metrics used to assess the health and success of cities and society. Conventional measures—such as GDP, population growth, or infrastructural expansion—fail to capture the full complexity of urban vitality. *Urbanimmunology* introduces a new conceptualization of urban health, one that includes metrics of resilience, adaptability, social cohesion, and environmental sustainability. This theoretical shift encourages the development of more holistic, multidimensional measures of success, focusing on cities that can thrive in the face of change, rather than simply grow in size or wealth. It advocates for the integration of dynamic indicators that allow cities to be continually assessed based on their capacity for self-correction and long-term well-being.

In summary, *Urbanimmunology* represents a theoretical revolution in how we understand cities and society. By shifting from static, growth-oriented models to dynamic, adaptive, and integrated systems, it offers a profound rethinking of urban and social theory. This new theoretical approach calls for cities to be seen as living organisms, capable of continuous evolution and self-regulation. It challenges traditional goals of urban development and governance, emphasizing resilience, sustainability, and equity. Through this transformative lens, *Urbanimmunology* redefines the foundational concepts of urban and social theory, creating a new path forward for theorizing cities that can adapt, thrive, and regenerate in an ever-changing world.

2.7 Connecting HEARTY City and Planetary Health

2.7.1 HEARTY City: A New Model for Urban Futures

In addressing the pressing issue of planetary health within urban and social settings, the author introduces the concept of a *HEARTY City* (Zhang et al., 2025), a synthesis of existing ideas that captures the multidimensional complexities of urban and social development. In this framework, each letter embodies a distinct yet interconnected aspect of urban life: "H" stands for a Healthy city, "E" integrates both Eco-city and Energy-efficient city, "A" signifies an AI-driven city, "R" embodies a Resilient city, "T" represents a Tourism city, and "Y" speaks to a Yearning city. Together, these components illustrate a dynamic interplay where both dilemmas and co-benefits emerge in the pursuit of sustainable urban and social futures.

For example, in an AI-driven city (A), the digital infrastructure designed to enhance efficiency and connectivity can, paradoxically, undermine resilience (R) in the face of technological disruptions. Furthermore, efforts to fortify resilience might inadvertently erode the natural beauty that contributes to the city's *Yearning* (Y)—a longing to preserve its aesthetic and cultural essence. Similarly, the economic objectives of an AI-driven model may conflict with the ecological aspirations of an Eco-city (E), whose goal is to minimize the environmental footprint of urban life. In tourism cities (T), the burgeoning influx of visitors often creates tension between the needs of tourists and locals, potentially diminishing the sense of *Yearning* (Y) for tranquility and harmony with nature.

However, these tensions are not insurmountable. Promoting the interdependent concepts of the HEARTY city offers significant co-benefits. An AI-driven city (A), for instance, can amplify resilience (R) by deploying intelligent systems that anticipate and mitigate risks, while also preserving the city's natural landscape, thus fostering a deeper *Yearning* (Y) for the inherent beauty of the environment. Similarly, a *Yearning* city (Y), where people seek connection to the natural world, can become an attractive destination for tourists (T), bolstering its cultural vibrancy. Moreover, AI-driven solutions can help manage the complexities of over-tourism, ensuring a balanced relationship between visitors and residents.

By nurturing a Healthy (H) and Yearning (Y) city, urban spaces can more effectively support eco-friendly practices (E), fostering sustainability and enhancing the collective well-being of all residents. In the broader context of planetary health, the principles of *Urbanimmunology* can guide cities in addressing these interconnected challenges. Through careful integration of ecological, technological, and social considerations, *Urbanimmunology* offers a path toward a more sustainable and equitable urban future—one that contributes positively to the health of the planet itself.

2.7.2 The Intrinsic Link Between People and Nature in HEARTY City

The relationship between individuals and nature is a profound, reciprocal bond, marked by a continuous and dynamic exchange of influence. On one hand, human activities—especially those related to housing, transportation, tourism, waste management, and resource consumption—often give rise to unsustainable practices that degrade the very environment that sustains us. Urban sprawl, driven by the insatiable demand for housing, fragments ecosystems and destroys habitats, while the overreliance on fossil fuels to power transportation systems exacerbates air pollution and contributes to the accelerating crisis of climate change. Tourism, when unchecked, puts undue pressure on fragile local ecosystems, while poor waste management practices poison both wildlife and human health alike.

On the other hand, nature exerts its own powerful influence over humanity, often with negative consequences. Climate change, for example, intensifies weather extremes and amplifies resource scarcity, while the emergence of zoonotic diseases forces us to confront the undeniable interconnectedness between human beings and the natural world. The COVID-19 pandemic stands as a stark reminder of how disruptions in our relationship with nature can trigger profound public health crises, reverberating across the globe and highlighting our vulnerability when harmony is disrupted.

This growing disconnection from nature has fueled unsustainable ways of life, particularly in urban environments that strive to embody the ideals of Healthy, Eco-city, Tourism, and Yearning (H-E-T-Y) cities. The H-E-T-Y framework acknowledges that urban areas should not only accommodate human activity but do so in a way that respects and integrates the integrity of natural ecosystems. By centering health, sustainability, tourism, and a deep yearning for connection to nature, the framework emphasizes the urgent necessity for individuals and communities to reevaluate their relationship with the environment.

To address these challenges, the H-E-T-Y model provides a roadmap for creating urban spaces that foster sustainable practices, such as green building initiatives, efficient public transportation systems, eco-friendly tourism strategies, and comprehensive waste management solutions. Equally important is the cultivation of a culture that encourages individuals to reconnect with nature through urban green spaces, community gardens, and nature-based recreation, thus nurturing a collective sense of stewardship and responsibility for the environment.

In aligning urban development with principles of sustainability, we can craft a future where human existence flourishes in harmony with nature, ensuring the well-being of both people and the planet. Through the H-E-T-Y framework, cities are not merely spaces of habitation but catalysts for transformation, guiding individuals toward lifestyles that are not only sustainable but also enriching—where every step taken in the urban landscape brings us closer to the rhythms of the natural world.

2.7.3 The Intrinsic Link Between Nature and Society in HEARTY City

The relationship between nature and human society unfolds as a dynamic interplay—one where the actions of society exert profound influence over the environment, and, in turn, nature responds with challenges that threaten human flourishing. In urban settings, particularly within Resilient, Tourism, and Yearning (R-T-Y) cities, unsustainable development practices have become increasingly prevalent, resulting in detrimental land-use changes and land-cover transformations. These alterations not only degrade local ecosystems but also exacerbate climate change, amplify pollution, and contribute to the rise of pandemics.

In R-T-Y cities, rapid urbanization frequently prioritizes immediate economic benefits over long-term environmental sustainability. Such shortsightedness leads to deforestation, biodiversity loss, and heightened vulnerability to natural disasters—all of which compromise the resilience of communities. Poorly planned tourism, for instance, can overburden the natural environment, strain local resources, disrupt ecosystems, and introduce waste management crises, ultimately diminishing the quality of life for both residents and visitors alike.

Nature's response to human activity often manifests in the form of environmental crises: extreme weather events, rising sea levels, and the spread of zoonotic diseases, all of which underscore the urgent necessity for a more harmonious relationship with the environment. The COVID-19 pandemic serves as a stark reminder of the profound consequences of environmental neglect and the intricate links between ecological health and public health. It compels us to recognize that the well-being of society is inextricably bound to the health of the planet.

To address these pressing issues, it is essential to champion the development of R-T-Y cities that prioritize sustainability and environmental stewardship. This requires the implementation of thoughtful urban planning that respects natural ecosystems, promotes green infrastructure, and encourages responsible, sustainable tourism practices. By embedding ecological considerations into the very fabric of urban development, cities can bolster their resilience against the challenges of climate change while simultaneously improving the quality of life for their inhabitants.

Furthermore, fostering a deep sense of connection and responsibility toward the environment within communities can empower individuals to embrace sustainable practices. Educational initiatives, community involvement in environmental conservation, and the promotion of urban green spaces can inspire collective action and cultivate a shared commitment to safeguarding the planet.

Ultimately, by embracing the principles of R-T-Y city development, we can mitigate the harmful effects of unsustainable practices and create a future where human society exists in balanced harmony with nature. In doing so, we pave the way for resilient, sustainable urban landscapes where both humanity and the natural world can thrive together.

2.7.4 The Intrinsic Link Between Individuals and Society in HEARTY City

The relationship between individuals and society is a complex interplay, often characterized by a tension between the weight of institutional regulations and the desires of personal freedom. On one hand, societal norms and policies may impose restrictions on individual behavior, fostering frustration, alienation, and a sense of disempowerment. On the other, individuals may act in ways that defy these social expectations, creating friction that undermines the cohesion of the community and disrupts the collective harmony.

Social exclusion stands as a particularly urgent concern in the context of AI-driven and Resilient (A-R) cities, where the rapid acceleration of technological progress and urban expansion can unintentionally deepen pre-existing divides. Within such environments, marginalized groups may find themselves excluded from essential resources and opportunities, heightening feelings of isolation and disenfranchisement. This exclusion does more than fracture the social fabric—it fosters a broader sense of social insecurity, as individuals are left feeling disconnected from the decisions that shape their lives, stripped of agency in the very spaces they inhabit.

To address this dissonance, A-R cities should turn their focus toward inclusivity and accessibility, recognizing that a thriving city is one where all individuals are not only welcomed but empowered to participate in its unfolding. This can be accomplished by fostering policies and practices that engage diverse community voices in the urban planning process, ensuring that no one is left out of the conversation. Actively involving residents from all backgrounds—especially those who are too often overlooked—ensures that the needs of the entire community are considered, weaving a more inclusive social fabric.

Equally crucial is the enhancement of both physical and digital accessibility. In an increasingly interconnected world, the ability to engage with the public sphere cannot be limited by geographical or technological barriers. Urban spaces should be designed to be navigable and welcoming to all, while digital platforms should enable seamless access, ensuring that individuals can participate fully in the modern urban experience.

Education and outreach initiatives also play a pivotal role in bridging the gap between individuals and society. By fostering awareness of societal structures and encouraging active civic participation, communities can empower individuals to not only understand the forces shaping their lives but also assert their rights and advocate for meaningful change. Through these efforts, individuals become not passive recipients of societal norms but active agents in the shaping of their shared future.

Ultimately, the goal is to cultivate a sense of belonging and meaningful engagement within A-R cities, transforming what could be sources of division into opportunities for deeper collaboration and communal strength. By prioritizing inclusivity and accessibility, we can build urban environments that empower every resident, fostering resilience, social cohesion, and a society where all individuals, regardless of their background, are valued contributors to the collective whole.

2.7.5 Urbanimmunology *Bridging Individuals, Society, and Nature for Planetary Health*

Urbanimmunology, drawing inspiration from the intricate wisdom of the human immune system, proposes a novel paradigm for creating urban environments that are resilient, adaptive, and harmonious with both human needs and the health of the environment. This innovative framework emphasizes the profound interconnectedness of individuals, society, and nature, offering a holistic approach to urban living that nurtures community well-being while safeguarding ecological sustainability. The following explores how *Urbanimmunology* facilitates these vital connections.

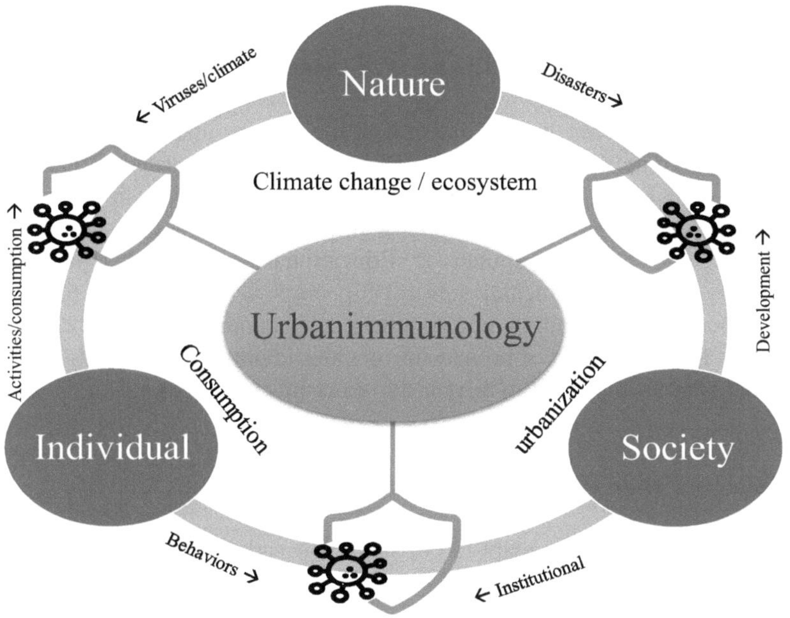

2.7.5.1 Creating Resilient Urban Ecosystems

Much like the immune system shields the body from external threats, *Urbanimmunology* advocates for the design of urban ecosystems resilient to environmental stressors. The integration of green infrastructure—such as parks, urban forests, and biodiversity corridors—into city planning is essential. By fostering biodiversity and enhancing ecosystem services, cities bolster their capacity to withstand the impacts of climate change, improve air quality, and provide sanctuaries for wildlife. This intricate web of natural systems not only supports individual well-being but also cultivates a flourishing, healthy environment for all urban inhabitants.

2.7.5.2 Fostering Social Immunity Through Community Networks

The immune system relies on a network of cells that communicate and collaborate to defend against threats. Similarly, *Urbanimmunology* emphasizes the importance of social networks within urban spaces. By nurturing community engagement and collaboration, cities can develop a form of social "immunity" that binds individuals together in mutual support. Neighborhood associations, local activism, and collective community events foster a sense of belonging and shared responsibility, empowering residents to address challenges collectively, combat isolation, and promote inclusivity. Such initiatives help weave a social fabric that strengthens the community's resilience.

2.7.5.3 Promoting Collective Health and Well-being

Urbanimmunology recognizes that individual health cannot be understood in isolation from the health of the community. The well-being of individuals is inseparably linked to the broader health of society. By prioritizing public health initiatives—such as access to healthcare, nutritious food, and clean environments—cities can enhance the resilience of their populations. Educational programs that promote healthy lifestyles and environmental stewardship empower individuals to take responsibility for their own health, while simultaneously contributing to the collective well-being. This collective approach mirrors the immune system's dynamic capacity to adapt and fortify itself in response to external challenges.

2.7.5.4 Adaptive Urban Planning for Environmental Change

The immune system's ability to adapt to ever-evolving pathogens serves as a profound model for urban planning. *Urbanimmunology* advocates for adaptive strategies that enable cities to respond effectively to environmental challenges such as climate change and resource depletion. Cities should design flexible infrastructure—such as permeable surfaces to manage stormwater or modular buildings that can be repurposed—to evolve over time. This adaptive capacity ensures that urban environments remain functional, livable, and sustainable, even as they are shaped by the unpredictable forces of the natural world.

2.7.5.5 Engaging Individuals in Sustainability Practices

Just as the immune system thrives when all of its components function in concert, *Urbanimmunology* envisions a sustainable urban future that requires the active participation of individuals. By engaging residents in initiatives like community gardens, recycling programs, and energy conservation efforts, cities foster a sense of ownership and shared responsibility for local resources. Empowering individuals to embrace sustainable practices nurtures a culture of environmental stewardship, reinforcing the critical connection between personal actions and their broader ecological consequences.

2.7.5.6 Leveraging Technology for Connectivity and Awareness

The immune system employs complex signaling mechanisms to coordinate its response, a function akin to how technology can enhance communication and connectivity in urban settings. *Urbanimmunology* proposes leveraging smart technologies to gather real-time data on environmental health, resource usage, and community needs. Such technologies empower residents to make informed decisions regarding their health and the environment, enhancing awareness and fostering collective action. Moreover, digital platforms can facilitate community feedback, ensuring that urban planning is attuned to the diverse needs and voices of all city dwellers.

2.7.5.7 Summary

Urbanimmunology offers a profound and compelling framework for bridging the connections between individuals, society, and nature in urban environments. Drawing inspiration from the principles of the human immune system, it fosters resilience, social cohesion, and ecological sustainability. By applying the insights of *Urbanimmunology*, cities can evolve into thriving ecosystems that not only support the well-being of their residents but also cultivate a deep, reciprocal relationship with the natural world. Ultimately, this paradigm fosters urban environments that nurture people, strengthen communities, and protect the planet—creating a harmonious interplay where all aspects of urban life flourish in tandem.

2.8 Navigating the Theoretical Challenges of *Urbanimmunology*

The emergence of *Urbanimmunology* as a new and transformative discipline invites us to confront a series of profound theoretical challenges, each of which should be carefully navigated to foster the evolution of next-generation urban and social development. The analogy between urban and social environments and the human immune system unveils rich insights into the nature of the resilience of cities and society, yet it also demands a rigorous engagement with the complexities inherent in such comparisons. As we embark upon the task of constructing this discipline, we are called to reflect deeply upon the following key theoretical challenges.

2.8.1 Interdisciplinary Integration

At its very core, *Urbanimmunology* necessitates the fusion of diverse intellectual landscapes, demanding a convergence of fields as varied as urban and social planning, ecology, public health, behavioral economics, sociology, and systems biology. Yet, this integration is fraught with challenges, for each domain carries its own lexicon, methods, and epistemological assumptions. These differences, while reflective of the richness of human inquiry, can also serve as barriers to effective communication and collaboration. To construct a unified theoretical framework for *Urbanimmunology*, it becomes imperative to cultivate a shared language—one that transcends disciplinary boundaries and fosters mutual understanding. In this endeavor, the creation of hybrid methodologies that synthesize both qualitative and quantitative approaches may prove essential, ensuring that the unique insights of each field coalesce into a more holistic comprehension of the resilience of cities and society. By fostering interdisciplinary partnerships, we unlock the potential for creative solutions that address the multifaceted challenges of urban and social life. Such collaboration, grounded in the mutual respect of diverse perspectives, not only deepens our collective wisdom but also lays the foundation for interventions that are attuned to the complex, evolving nature of urban and social environments. Through this synergy, we may cultivate cities and society that are not merely sustained, but thrive amidst the ever-shifting tides of change.

2.8.2 Dynamic and Adaptive Systems

Urban and social environments, by their very nature, are in a perpetual state of flux—shaped, molded, and reshaped by the ceaseless currents of technological advancement, demographic transformations, and the inexorable march of environmental change. Thus, the very notion of "immunity" within a city or society should extend beyond mere resilience to embrace the broader, more profound concepts of adaptability and evolution across time. In order to truly capture the intricacies of this dynamic reality, theoretical models should be crafted to accommodate the inherent fluidity of urban and social systems, incorporating principles such as adaptive capacity and feedback loops—concepts that reflect the deep interconnections across diverse components of the city or society. Just as the human immune system learns from prior encounters and refines its responses through memory, so too should urban and social systems evolve, learning from past disruptions to improve their future reactions. This calls for the integration of longitudinal studies and the application of complex adaptive systems theory—tools capable of unveiling the nuanced and often unpredictable behaviors that characterize urban and social life. In embracing this dynamic perspective, we may begin to understand how cities and society, as living, breathing entities, can not only survive but thrive amidst the constant ebb and flow of change, crafting urban and social environments that are not only resilient but also adaptive, fostering futures that are both sustainable and ever-evolving in the face of life's uncertainties.

2.8.3 *Measurement and Evaluation*

Defining and measuring the "immunity" of a city or society presents profound challenges, for the resilience of cities and society is a multifaceted, context-dependent concept—far more intricate and elusive than the biological metrics grounded in clear, standardized indicators. Unlike the immune system's measurable response to pathogens, the resilience of a city or society is woven from a complex tapestry of social, economic, and ecological threads, each with its own nuances and interdependencies. To truly assess urban and social immunity, we should cultivate a robust set of metrics that encompass both the qualitative and quantitative dimensions of resilience—ranging from social cohesion and economic stability to ecological health. Such an approach provides a more holistic and nuanced understanding of the capacity of a city or society to endure and evolve. The development of innovative tools for assessment, such as simulation models and participatory frameworks, offers an opportunity for communities to actively engage in the process, allowing residents to define resilience in their own terms. By involving the people who live in these environments, we ensure that the metrics reflect the unique realities and priorities of each city or society, fostering interventions that are not only more effective but more deeply aligned with local needs. In this way, measurement transcends a mere technical task, evolving into a philosophical practice—one that empowers communities, fosters collective ownership of resilience, and nurtures a shared vision for the future of cities and society, one that is both sustainable and adaptable to the uncertainties of the world ahead.

2.8.4 *Ethical Considerations and Equity*

Urbanimmunology should confront profound ethical challenges, particularly in relation to equity and access. The strategies designed to enhance the resilience of a city or society, while well-intentioned, may inadvertently privilege certain populations, thereby exacerbating existing inequalities and further marginalizing those already on the periphery. It is, therefore, essential that the theoretical frameworks guiding this discipline integrate explicit principles of social justice and equity. This requires a deep, reflective examination of how urban and social resilience strategies can be constructed in a way that is inherently inclusive, ensuring that the needs of the most vulnerable communities are not only acknowledged but prioritized in both planning and response efforts. True resilience is not merely the capacity to survive; it is the ability to ensure that no one is left behind in the process. To this end, actively engaging marginalized voices in the development of *Urbanimmunology* is not optional—it is an imperative. By fostering inclusivity and promoting a sense of belonging and agency among all community members, we can craft interventions that are both fair and effective, cultivating the resilience of a city or society in a manner that enhances not only its structural and ecological health but also the social and moral fabric of its inhabitants. When equity is placed at the center of these endeavors, the solutions we create not only will be ethically sound but will also foster more just, sustainable, and harmonious futures for cities and society.

2.8.5 Global Versus Local Contexts

The application of *Urbanimmunology* is inherently fraught with complexity, as it should contend with the profound variations that exist between global and local contexts, each shaped by its own geographical, cultural, and socio-economic dynamics. What proves effective in one urban and social environment may well be irrelevant, or even detrimental, in another, thus complicating the task of establishing universal theories or solutions. In response to this challenge, it is imperative that researchers embrace context-sensitive approaches—ones that account for the unique conditions, histories, and cultural narratives that define each city or society. Rather than attempting to impose a one-size-fits-all model, the discipline should cultivate flexible frameworks that adapt to the diverse and evolving landscapes of urban and social life, all the while preserving the core principles that underpin *Urbanimmunology*. This calls for a deep engagement with comparative studies across varied urban settings, which will not only illuminate the common threads of urban and social resilience but also reveal the distinctive struggles faced by different communities. Such a nuanced, thoughtful understanding opens the door to more tailored, contextually relevant solutions, ensuring that the principles of *Urbanimmunology* contribute meaningfully to the enhancement of urban and social resilience and public health in ways that resonate with the particularities of each place. Ultimately, it is through this careful balance of universal insight and local specificity that we can hope to forge urban and social environments capable of truly adapting and thriving in an ever-changing world.

2.8.6 Policy and Governance Frameworks

The translation of *Urbanimmunology* from theory into practice is fraught with formidable challenges, not the least of which are the entrenched governance structures, bureaucratic inertia, and political resistance that often stymie transformative change. To surmount these obstacles, it becomes essential to develop comprehensive policy and governance frameworks that can meaningfully bridge the gap between abstract principles and actionable strategies. This process demands more than technical expertise; it requires the active and collaborative engagement of policymakers, stakeholders, and community members—those who, together, can ensure that the wisdom of *Urbanimmunology* is woven into the very fabric of urban planning and governance. By advocating for policies that prioritize resilience and sustainability, we create the conditions for a broader acceptance of these ideas, laying the foundation for their eventual implementation in ways that meaningfully enhance both urban and social environments and public health. Yet, this approach is not merely about the efficient application of knowledge—it is a call to empower communities to participate in the process of transformation, allowing them to play an active, decisive role in shaping their urban and social landscapes. In doing so, we not only pave the way for more adaptive and healthier cities and society but also foster a deeper sense of collective responsibility, where governance is not a top-down imposition but a collaborative endeavor that draws upon the wisdom and agency of all.

2.8.7 Summary

The emergence of *Urbanimmunology* as a new discipline, while fraught with theoretical challenges, presents profound opportunities for innovation and interdisciplinary collaboration. By drawing on the rich insights of immunology and integrating wisdom from other fields, *Urbanimmunology* has the potential to offer a novel lens through which to understand urban and social resilience—one as dynamic and multifaceted as the cities and society it seeks to protect. However, addressing these challenges requires more than intellectual rigor; it demands a deep, thoughtful engagement with the complexities of urban and social life, acknowledging that resilience should be adaptive, equitable, and sustainable in the face of the disruptions that characterize our world. In this effort, the ongoing dialogue among researchers, practitioners, and communities is not merely a practical necessity, but a philosophical imperative—one that fosters a shared vision for cities and society that are not only structurally robust but also rich in their capacity to evolve. Through continued collaboration, we can refine both the theoretical foundations and practical applications of *Urbanimmunology*, ensuring it becomes a transformative force for cities and society that are not only resilient in the face of change but also alive to the possibilities of the future.

2.9 Shaping the Future: The Role of *Urbanimmunology* in Tomorrow's Cities and Society

As time unfolds and progress steadily moves forward, *Urbanimmunology* emerges as both a guiding light and a crucial field. It arises from the convergence of multiple systems—the human immune system, urban infrastructure, transportation, energy, tourism, health, social, and global systems—each offering its unique insights.

Together, these disciplines work toward a shared goal: to create cities and society that are not only resilient but adaptive, sustainable, and capable of thriving amidst change. As we face an uncertain future, the core principles of *Urbanimmunology* challenge us to imagine urban and social spaces that evolve in harmony with their environments, fostering growth and balance for generations to come.

2.9.1 The Weaving of Knowledge

At its core, *Urbanimmunology* is like a tapestry—a complex blend of diverse areas of knowledge, all interwoven by the invisible threads of various disciplines. From the microscopic world of biology to the broad systems of urban and social planning, from the intricate patterns of public health to the delicate balance of environmental science, this interdisciplinary approach creates a rich, layered framework for understanding not just resilience but the full spectrum of urban and social life. It's a framework that empowers us, offering practical tools to help cities navigate challenges, adapt to change, and thrive in an evolving world. In this process, collaboration flourishes, with city and social planners, environmentalists, health experts, and community leaders all contributing their unique perspectives to the shared goal of building stronger, more sustainable urban and social systems.

2.9.2 Vitality: The Essential Pulse of Cities and Society

At the heart of this vision lies a broader concept that goes beyond resilience—the dynamic and vital pulse of urban and social life. This idea encompasses the ability of a city or society to adapt, regenerate, and thrive in the face of challenges. Urban and social spaces should not only respond to change but also evolve in ways that strengthen their capacity to support diverse, thriving communities. In a world challenged by climate change, social inequality, and public health crises, this approach calls for a reimagining of urban and social spaces—not just as places of survival, but as interconnected systems that foster growth, well-being, and sustainability. Through this lens, cities and society can be seen as living organisms that not only withstand external pressures but also actively enhance the health and vitality of all who live within them.

2.9.3 Adaptive Cities and Society: Rethinking Design for Change

Just like the human immune system, cities and society can be designed to be dynamic, responsive, and always changing. *Urbanimmunology* promotes the idea that urban and social systems shouldn't be static or frozen in time but should be adaptive—constantly aware of potential threats and ready to respond, adjust, and evolve. This approach leads to innovative designs where flexibility, adaptability, and multifunctionality are core principles, not afterthoughts. For example, green infrastructure goes beyond being just an environmental feature; it becomes a key part of the resilience of a city or society. It helps manage flooding, supports biodiversity, transforms public spaces, and strengthens community ties, deepening the connection between people and their surroundings.

2.9.4 The Wisdom of Data

In today's world, data acts like a mirror—showing the health of the urban and social system, uncovering weaknesses, and offering the tools to strengthen its resilience. Just as the body's biomarkers reveal its internal health, data provides real-time insights into the state of cities and society. With this information, planners can act quickly and purposefully, ensuring that cities and society are proactive, not reactive—anticipating challenges and addressing them before they become serious problems. Predictive modeling offers a look into the future, helping cities and society plan, adapt, and respond with foresight, rather than scrambling after a crisis has already hit.

2.9.5 The Power of Community

Amid all the technology and planning, one of the strongest sources of strength is the human element—the heart of the city and society: its people. *Urbanimmunology* recognizes this. It understands that the power of a city or society doesn't just come from its infrastructure or systems, but from the collective spirit of its community. Much like the network of cells in the immune system, diverse social connections work together to protect and sustain the urban body. By involving communities in the planning and governance of their spaces, *Urbanimmunology* promotes a sense of shared responsibility. It empowers individuals to take action in times of crisis and builds a sense of unity that holds the city and society together when facing challenges.

2.9.6 Harmony: The Symphony of Systems

In the complex framework of *Urbanimmunology*, harmony is a key player. By understanding how urban and social systems are interconnected, cities and society can manage their resources carefully, making the most of water, energy, and waste while reducing vulnerabilities. This all-encompassing approach, based on the understanding of these connections, helps lower the environmental impact of cities and society, ensuring they can support not only the present but also future generations who will live in them.

2.9.7 A Holistic Vision of Well-being

Urbanimmunology calls for a broader view of urban and social well-being, one that includes physical, mental, and social aspects as interconnected. Through this perspective, cities and society can be reimagined as places that prioritize access to nature, clean air, social services, and lively public spaces. When these elements are thoughtfully integrated into daily life, they create environments where people can thrive—not just survive. It's a vision of cities and society as havens, where well-being is the norm, not the exception.

2.9.8 The Path Forward

In the end, *Urbanimmunology* has the potential to guide urban and social development in the years ahead. By viewing cities and society as living organisms with their own systems of strength and adaptability, this approach opens up new ways of designing and nurturing urban and social systems. It encourages us to think beyond just survival, imagining cities and society that not only withstand challenges but grow stronger and more vibrant with each one. In a world facing rapid change, *Urbanimmunology* offers hope, a vision, and a blueprint for the future—a future where cities and society can evolve, adapt, and thrive, no matter what storms come their way.

References

Abbas, A. K., Lichtman, A. H., & Pillai, S. (2023). *Basic immunology: Functions and disorders of the immune system* (7th ed.). Elsevier.

Alberti, M. (2023). Cities of the Anthropocene: Urban sustainability in an eco-evolutionary perspective. *Philosophical Transactions B, 379*, 20220264. https://doi.org/10.1098/rstb.2022.0264

Bogard, K., Murry, V. M., & Alexander, C. (2017). *Perspectives on health equity and social determinants of health*. National Academies Press. https://www.ncbi.nlm.nih.gov/books/NBK595256/

Bush, J., & Doyon, A. (2019). Building urban resilience with nature-based solutions: How can urban planning contribute? *Cities, 95*, 102483. https://doi.org/10.1016/j.cities.2019.102483

Castelli, K. R., Silva, A. M., & Dunning, J. B., Jr. (2021). Improving the biodiversity in urban green spaces: A nature based approach. *Ecological Engineering, 173*, 106398. https://doi.org/10.1016/j.ecoleng.2021.106398

Cheshmehzangi, A., Butters, C., Xie, L., & Dawodu, A. (2021). Green infrastructures for urban sustainability: Issues, implications, and solutions for underdeveloped areas. *Urban Forestry & Urban Greening, 59*, 127028. https://doi.org/10.1016/j.ufug.2021.127028

Clark, H. (2015). Governance for planetary health and sustainable development. *The Lancet, 386*(10007), E39–E41.

Datola, G. (2023). Implementing urban resilience in urban planning: A comprehensive framework for urban resilience evaluation. *Sustainable Cities and Society, 98*, 104821. https://doi.org/10.1016/j.scs.2023.104821

Delves, P. J., Martin, J. E., Burton, D. R., & Roitt, I. M. (2017). *Janeway's immunobiology* (9th ed.). Garland Science.

Ellena, M., Breil, M., & Soriani, S. (2020). The heat-health nexus in the urban context: A systematic literature review exploring the socio-economic vulnerabilities and built environment characteristics. *Urban Climate, 34*, 100676. https://doi.org/10.1016/j.uclim.2020.100676

Janeway, C. A., Travers, P., Walport, M., & Shlomchik, M. J. (2001). *Immunobiology: The immune system in health and disease* (5th ed.). Garland Science.

Murphy, K. M., Weaver, C., & Berg, L. J. (2022). *Janeway's immunobiology* (10th ed.). W. W. Norton & Company.

Myers, S., & Frumkin, H. (2021). *Planetary health: Protecting nature to protect ourselves*. MIT Press.

Steffen, W., Broadgate, W., Deutsch, L., Gaffney, O., & Ludwig, C. (2015). The trajectory of the Anthropocene: The great acceleration. *The Anthropocene Review, 2*(1), 81–98.

Steffen, W., Crutzen, J., & McNeill, J. R. (2007). The Anthropocene: Are humans now overwhelming the great forces of nature? *AMBIO: A Journal of the Human Environment, 36*(8), 614–621.

Whitmee, S., Haines, A., Beyrer, C., et al. (2015). Safeguarding human health in the Anthropocene epoch: Report of The Rockefeller Foundation–Lancet Commission on planetary health. *The Lancet, 386*(10007), 1973–2028.

Zhang, J. (2020). Transport policymaking that accounts for COVID-19 and future public health threats: A PASS approach. *Transport Policy, 99*, 405–418.

Zhang, J. (2021). Integrated carbon reduction policies on transportation in the era of post-COVID-19. *Urban Transport of China, 19*(5), 43–52. (in Chinese). https://doi.org/10.13813/j.cn11-5141/u.2021.0034

Zhang, J. (2022). Governance for post-COVID-19 carbon reduction: A case study of the transport sector. In J. Zhang & Y. Hayashi (Eds.), *Transportation amid pandemics: Lessons learned from COVID-19, chapter 34* (pp. 429–440). Elsevier.

Zhang, J. (2024). Interdisciplinary research on future cities and transportation systems. In *Selected speeches of 2023 world cities day events*. Shanghai Scientific and Technical Publishers.

Zhang, J. (2025). DIRECT approach. In J. Zhang, L. Zhang, & Y. Jiang (Eds.), *COVID-19 & pandemics, lifestyles, human mobilities, and cities: A perspective of planetary health*. Springer.

Zhang, J., Guan, K., & Jiang, Y. (2025). HEARTY city. In J. Zhang, L. Zhang, & Y. Jiang (Eds.), *COVID-19 & pandemics, lifestyles, human mobilities, and cities: A perspective of planetary health*. Springer.

Chapter 3
Energy-Disaster-Underground Integration Policymaking Informed by *Urbanimmunology*

3.1 The Pillars of Urban Resilience: Energy, Disaster Management, and Underground Infrastructure

Beneath the towering skyscrapers, within the labyrinth of streets, and beneath the very earth upon which the city pulses, lie three unseen yet essential pillars that hold the city together—energy, disaster management, and underground infrastructure. These elements, though each distinct in their role, are so intricately intertwined that they form the very skeleton of the urban environment. Each contributes to the vitality and survival of the city, working silently but steadfastly to ensure that urban life continues without interruption, even in the face of unforeseen trials.

3.1.1 Energy (E): The Lifeblood of the City

At the core of every urban landscape, energy flows like lifeblood through the city's veins. The buzz of electric vehicles, the movement of trains below ground, the glow of streetlights—all rely on energy. Cities, as hubs of human activity and innovation, need vast amounts of power to keep their economies running. From the tall office buildings to quiet residential areas, energy is crucial to the rhythm of daily life, powering everything from transportation and communication to industry and public services. As urban populations grow, so does the demand for energy, pushing cities to consume a large portion of the world's resources (IEA, 2024). However, this demand should be met in a sustainable way. Cities should lead the transition to renewable energy, improve energy efficiency, and develop smart grids to address future challenges. It's a delicate balance—meeting the current needs of the city while paving the way for a future where energy isn't just consumed, but regenerated, ensuring long-term security and sustainability (UNDP, 2024).

J. Zhang, *Urbanimmunology*, Urban Sustainability,
https://doi.org/10.1007/978-981-95-4878-1_3

3.1.2 *Disaster Management (D): The Shield Against the Unexpected*

The cities we live in are not immune to the forces of nature or the consequences of human mistakes. They are vulnerable to a wide range of disasters, from hurricanes and earthquakes to unseen threats like pandemics and terrorism (UN, 2018). In the face of such events, having a strong disaster management system is crucial. Cities should not only prepare for these challenges—they should be able to respond effectively. Buildings need to be designed to withstand earthquakes, infrastructure should be able to endure floods, and emergency systems should be ready to act quickly during a crisis. The impacts of climate change—such as rising seas and stronger storms—make these strategies even more important, as cities brace for unprecedented and devastating effects. With the rise of smart technologies, cities now have powerful tools to improve safety. Data analytics, artificial intelligence, and sensor networks enable urban planners to predict, prepare, and respond more accurately, reducing risks and ensuring the city is ready when disaster strikes (Liu & Ota, 2017).

3.1.3 *Underground Infrastructure (U): The Hidden Network*

Beneath the surface of our bustling cities lies an unseen world that supports the very foundation of urban life. Below the busy streets, homes, and offices, tunnels, pipes, and vast networks form the lifeblood of the city. Sewage systems, utility networks, and transportation tunnels make up the hidden infrastructure that keeps everything running smoothly, providing essential services like water, waste management, and public transit (Madryas et al., 2014). In dense urban areas, where space is limited, the underground offers more than just utilities. It provides room for growth—allowing for parks, housing, and commerce above ground while reducing congestion and boosting efficiency. These underground spaces also offer something invaluable—climate resilience. They provide shelter, storage, and systems for heating or cooling, helping cities cope with extreme temperatures and weather-related disasters. As urban areas continue to grow, the importance of underground infrastructure will only increase, meeting the demands of an ever-more urbanized world.

3.1.4 *The Interconnections of EDU: A Web of Resilience*

Energy, disaster management, and underground infrastructure (EDU) together form a complex network, with each element depending on the others to ensure a city's ability to withstand challenges. Energy is more than just a lifeline—it connects all systems, including underground networks. Cables, pipelines, and substations below the surface deliver the power that fuels homes, industries, and transportation. If these systems fail—especially during extreme weather—the consequences can be disastrous. But it's not just energy that supports disaster management; disaster management also relies on a steady energy supply. Communication systems, water treatment plants, hospitals, and emergency services all need power to operate, particularly in times of crisis. Without energy, even the best disaster response plans will fail, highlighting the vital link between energy and disaster management.

Underground infrastructure is equally essential in times of disaster. Beneath the city, underground spaces serve as shelters from floods, tornadoes, and heatwaves, providing protection when buildings collapse, streets flood, or extreme weather makes the air unbearable. The role of underground systems in recovery after a disaster is also crucial. They help manage debris, ensure access to utilities, and maintain the integrity of infrastructure, all of which are key to the city's ability to recover.

Ultimately, the smooth coordination between energy, disaster management, and underground infrastructure forms the foundation of a resilient city. Treating them separately is no longer enough—cities should see them as interconnected, each one supporting and enhancing the others. The future of urban resilience depends on this holistic, integrated approach, ensuring cities not only survive today's challenges but thrive in the face of future ones.

3.2 Immune-Inspired Urban Energy Policymaking

This section explores how insights from the human immune system (Janeway et al., 2001; Delves et al., 2017; Murphy et al., 2022; Abbas et al., 2023) can inform urban energy policymaking, particularly concerning multi-source renewable energy (e.g., solar, wind, heat) (Bristow & Mohareb, 2019). These insights are crucial for achieving carbon neutrality in cities through the development of complementary mechanisms among different energy sources. By understanding the roles of key elements within the immune system, the goal is not only to provide theoretical support for existing effective urban energy policies but also to derive more innovative and efficient policy recommendations inspired by the immune system's functions.

3.2.1 White Blood Cells: Defenders of Energy Systems

3.2.1.1 Resilience and Defense Mechanisms

Just as leukocytes protect the body from pathogens, urban energy systems need to be resilient against challenges such as climate change, energy shortages, and environmental degradation. Policymakers can learn the adaptive responses of leukocytes to enhance urban systems. Leveraging a variety of energy sources, such as solar, wind, and biomass, to generate electricity (Hassan et al., 2023) can serve as a "phagocyte" mechanism, capturing and converting diverse inputs to reduce the reliance on fossil fuels.

3.2.1.2 Diversity and Adaptation

Leukocytes demonstrate remarkable diversity, enabling the immune system to adapt to new and evolving threats. Similarly, urban energy policies should foster a diverse mix of renewable energy sources to enhance both adaptability and security. By responding to changing circumstances, technological advancements, market shifts, and emerging environmental and societal needs, such policies can ensure a reliable, stable, and sustainable energy supply for consumers, businesses, and industries. This approach can include incentivizing microgrids or community-based energy projects that utilize local resources, much like the various types of leukocytes that target specific pathogens. Such diversity in energy solutions enhances resilience and supports long-term sustainability (McGranaghan et al., 2016).

3.2.1.3 Monitoring and Feedback Loops

Lymphocytes are crucial for the adaptive immune response, providing long-term immunity and memory. Urban energy systems can benefit from similar feedback mechanisms by incorporating data analytics and smart technologies to monitor energy use and emissions. This allows for real-time adjustments in energy supply (UNECE, 2024), akin to how memory cells in the immune system retain information about past infections.

3.2.1.4 Community Engagement and Education

Just as leukocytes communicate and coordinate actions within the immune system, urban energy policymaking should foster collaboration among stakeholders—government, businesses, and the community (Suboticki et al., 2023). Public engagement campaigns can serve to educate citizens on the importance of renewable energy, much like the way leukocytes activate and inform other immune cells.

3.2.1.5 Health and Well-being Integration

Considering that leukocytes play a role in maintaining health, urban energy policies should also address public health outcomes. For example, promoting renewable energy can reduce air pollution, directly impacting health and community well-being. Integrating health impact assessments into energy policy decisions can ensure that energy strategies also contribute to a healthier urban environment (EC, 2019).

3.2.2 *Innate Immunity: Immediate Energy Response*

Drawing inspiration from innate immunity, cities should establish core functions that allow them to autonomously tackle fundamental energy challenges. In other words, these functions should serve as the essential foundation for urban resilience, enabling immediate energy responses without prior preparation.

3.2.2.1 Physical Barriers and Infrastructure Resilience

The skin serves as a protective barrier against pathogens, and similarly, urban infrastructure can be designed to shield against energy inefficiencies and vulnerabilities. Policymakers can prioritize the development of resilient energy systems that integrate renewable sources—such as solar, wind, and geothermal—into the existing grid (McGranaghan et al., 2016). This could involve promoting decentralized energy production, such as rooftop solar (Rathore et al., 2019), to reduce reliance on centralized fossil fuel sources, much like how the skin provides localized protection against external threats. Additionally, investing in infrastructure that can withstand environmental stresses, including extreme weather events, is crucial to ensure the continued functionality and resilience of urban energy systems, preventing potential systemic failures.

3.2.2.2 Chemical Barriers, Environmental Protection, and Energy Efficiency

Chemical barriers in innate immunity neutralize threats, and urban energy policy can adopt a comparable approach to protect environmental and public health. By promoting energy-efficient technologies, such as LED lighting and high-efficiency Heating, Ventilation and Air Conditioning (HVAC) systems (Jouhara & Yang, 2018), urban energy policies can act as a "chemical barrier" to excessive energy consumption and greenhouse gas emissions. These technologies, much like antimicrobial substances, reduce waste and minimize environmental impact. Additionally, pollution control measures—such as regulations limiting emissions from energy production—function similarly to how lysozyme breaks down bacterial cell walls. These regulations help neutralize harmful pollutants, promoting cleaner air and fostering healthier, more sustainable urban environments.

3.2.2.3 Cellular Components and Community Engagement

The immune system's cellular components, such as phagocytes and natural killer (NK) cells, respond quickly and adaptively to threats. Similarly, community engagement in urban energy policymaking can play a dynamic and responsive role in addressing energy challenges (Batidzirai et al., 2021). Encouraging community-led renewable energy projects allows residents to actively participate in both energy production and consumption, much like immune cells responding to localized infections. Additionally, developing educational programs and outreach initiatives can raise awareness about the benefits of renewable energy, fostering a collective response to energy challenges and enhancing overall community resilience.

3.2.2.4 Inflammatory Response and Adaptive Policymaking

The inflammatory response in innate immunity is crucial for mobilizing defenses against threats. In the same way, adaptive policymaking in urban energy systems is essential for addressing emerging challenges, enabling timely responses and ensuring that energy solutions remain resilient and effective as circumstances evolve (Strumińska-Kutra et al., 2023). Establishing responsive policy frameworks that can quickly adapt to shifting energy demands and environmental conditions—much like the immune system's rapid response to threats—is essential for enhancing resilience. For example, creating flexible regulatory structures that accommodate new technologies or evolving energy sources can help ensure long-term sustainability. Additionally, implementing real-time monitoring systems to track energy consumption and environmental impacts allows policymakers to adjust strategies as needed, akin to how cytokines signal and coordinate immune responses.

3.2.3 Adaptive Immunity: Evolving Energy Solutions

The concept of adaptive immunity teaches us how to learn from past experiences and integrate these lessons to improve future responses. Similarly, carbon-neutralization efforts in cities—through the deployment of diverse renewable energy sources—should focus on adopting proven best practices, while also adapting to local contexts and needs.

3.2.3.1 Specificity and Targeted Policy Development

Adaptive immunity is defined by its specificity and the ability to craft targeted responses. This concept translates to urban energy policy through the need for customized strategies that address the distinct energy demands and resources of individual communities. Just as B cells produce specific antibodies in response to particular pathogens, urban energy policies should support localized renewable energy solutions—such as solar, wind, or bioenergy—that best fit the geographic and socio-economic context of each neighborhood (Cloke et al., 2017). Moreover, targeted incentives and subsidies (Abbruzzese et al., 2024; Belyak et al., 2024;

Qadir yet al., 2024) should be developed based on local conditions, much like how the immune system tailors its response to specific threats.

3.2.3.2 Memory and Long-Term Planning

Adaptive immunity relies on memory, enabling faster and more effective responses upon re-exposure to the same pathogen. This principle applies to long-term energy planning and investments (Flores et al., 2014). By building frameworks that allow for ongoing learning from both successful and failed energy policies (Sokołowski and Heffron, 2022; Kirkegaard et al., 2023), cities can continually refine and improve their strategies. Maintaining comprehensive databases of past renewable energy projects and their lessons learned can provide valuable insights. Additionally, fostering long-term investments in renewable energy infrastructure will build resilience within the energy system, helping cities respond more effectively to future challenges—just as the immune system adapts to remember and combat past infections.

3.2.3.3 Collaboration and Enhanced Response

In urban energy systems, collaboration plays a key role, much like Helper T cells in the immune system, which activate and coordinate the broader immune response. To foster a successful energy transition, multi-stakeholder engagement—comprising governments, the private sector, NGOs, and local communities—is essential. Strong partnerships, especially public-private collaborations, can amplify the impact of energy initiatives, ensuring all actors work in synergy (Sanderink & Nasiritousi, 2020). These alliances can drive innovation, accelerate the adoption of renewable technologies, and enable cities to tap into new resources and solutions more efficiently.

3.2.3.4 Targeted Interventions and Monitoring

In urban energy systems, like the immune system's use of Helper T cells to coordinate and enhance responses, fostering collaboration between various stakeholders is crucial. Effective engagement among governments, private enterprises, NGOs, and communities ensures a unified approach to the energy transition (Sanderink & Nasiritousi, 2020). Just as Helper T cells stimulate immune responses, well-aligned partnerships can enhance the effectiveness of energy initiatives, accelerating the adoption of renewable energy technologies. Public-private partnerships, in particular, are instrumental in unlocking private investment and facilitating the swift deployment of cutting-edge energy solutions.

3.2.4 Antibodies: Targeting Energy Threats

Antibodies are highly effective at targeting specific antigens. Similarly, this function offers a fresh perspective for urban energy policymaking, highlighting the importance of identifying those who are unable to comply with or are facing challenges in adhering to multi-source renewable energy policies.

3.2.4.1 Targeting Energy Sources

Just as antibodies are engineered to selectively bind to specific antigens, urban energy policies can be designed to focus on particular energy sources and inefficiencies within cities. Instead of a one-size-fits-all approach, policies should prioritize renewable energy solutions that align with the unique characteristics of each area (Qudrat-Ullah, 2024). For example, cities with abundant sunlight should prioritize solar energy, while coastal regions may focus on harnessing wind power. To ensure the effectiveness of these strategies, a thorough assessment of local energy needs and carbon emissions is essential. By identifying areas with the greatest potential for impact, cities can craft tailored policies that directly address regional challenges, promoting energy transitions that are both efficient and sustainable.

3.2.4.2 Mitigation and Neutralization Strategies

Antibodies play a crucial role in neutralizing harmful pathogens, and this principle can inform urban energy policies aimed at reducing the environmental impact of conventional energy sources. Just as antibodies target and eliminate threats, cities can develop transition plans that gradually phase out high-emission energy sources—such as coal and oil—in favor of cleaner, renewable alternatives (Tian et al., 2022). This shift ensures that polluting practices are systematically replaced with more sustainable solutions. In addition, stringent pollution control measures, such as regulations to limit emissions from fossil fuel-based energy, can help alleviate their harmful effects on air quality and public health, much like antibodies protect the body from harmful agents.

3.2.4.3 Monitoring and Feedback Mechanisms

Effective urban energy strategies, like the immune system's response to pathogens, require continuous monitoring and adaptive feedback. To this end, real-time data systems that track energy consumption and emissions (Pang & Li, 2021) are essential for informing policy adjustments. These systems, akin to antibodies that monitor and respond to threats, provide invaluable insights that enable policymakers to make informed decisions and adapt energy strategies as new data emerges.

Moreover, adopting adaptive management techniques allows urban energy policies to remain flexible and responsive to shifting conditions, ensuring that efforts to transition to renewable energy remain dynamic and effective over time.

3.2.4.4 Building a Supportive Ecosystem

Just as antibodies function within a larger immune system, urban energy policies should nurture an ecosystem that fosters the development and integration of renewable energy sources. This can be achieved by creating incentives that support research, development, and deployment of clean energy technologies, paralleling the immune system's promotion of antibody production. Public engagement and education also play a pivotal role in raising awareness about the benefits of renewable energy and generating grassroots support for policies aimed at reducing carbon emissions. By cultivating an informed and active community, cities can build a collaborative environment where the transition to sustainable energy is collectively embraced.

3.2.4.5 Collaborative Networks for Change

Antibodies operate within a network of immune responses. Similarly, urban energy policy can benefit from the collaborative efforts of diverse sectors and stakeholders. Building partnerships between government agencies, private sector organizations, NGOs, and local communities fosters the exchange of knowledge, resources, and best practices in renewable energy deployment. Furthermore, creating multi-stakeholder platforms for open dialogue enables various groups to collectively address challenges, share solutions, and leverage their combined expertise in the transition to clean energy. Such collaborative networks are essential for driving impactful, long-term change in urban energy systems.

3.2.5 Antigens: Identifying Energy Vulnerabilities

The role of antigens is straightforward: they trigger immune responses when identified as harmful or foreign. In urban energy policymaking, particularly when advancing multi-source renewable energy, a similar approach is necessary: identifying key challenges and developing targeted, strategic solutions.

3.2.5.1 Identification of Key Challenges

Much like how antigens stimulate immune responses by being recognized as threats, identifying critical challenges or "antigens" in urban energy systems is essential for driving renewable energy adoption. Through vulnerability assessments (Gatto & Busato, 2020), cities can pinpoint major sources of carbon emissions and energy inefficiencies, which help prioritize interventions for improving energy resilience and sustainability. Moreover, engaging with local communities to understand their unique energy needs and concerns provides valuable insights—acting as "antigens" that guide the creation of tailored, context-specific policies. By recognizing and addressing these challenges, urban energy policies can be more effective and impactful.

3.2.5.2 Stimulating Policy Responses

As the immune system actively responds to antigens, urban energy policies should be equally responsive to the challenges identified within the energy landscape. Targeted policies should address the specific issues revealed through assessments. For example, if transportation is the primary source of emissions, policies might incentivize electric vehicle adoption or expand renewable-powered public transit. Encouraging innovation in energy technologies through research and development can also help overcome obstacles to renewable energy adoption, much like how the immune system generates antibodies to neutralize threats. This proactive approach ensures that energy policies remain dynamic, flexible, and capable of evolving with changing needs.

3.2.5.3 Creating Awareness and Engagement

Antigens act as signals that activate immune responses, and similarly, raising awareness and engaging the public is crucial for advancing energy transitions. Public education campaigns can highlight the harmful effects of traditional energy sources and the advantages of renewable alternatives, focusing attention on key issues such as fossil fuel dependence and pollution, which can generate widespread support for cleaner energy solutions. Involving community members in the policymaking process ensures that their voices are heard and their concerns addressed, fostering a sense of ownership and responsibility toward renewable energy projects. This inclusive approach strengthens the foundation for a more sustainable and equitable energy transition.

3.2.5.4 Adaptive Policymaking and Learning

Just as the immune system evolves to better respond to new antigens, urban energy policies should be designed with adaptability in mind, incorporating continuous feedback to remain responsive to emerging challenges. Establishing robust monitoring and evaluation systems to assess the effectiveness of renewable energy initiatives and their impact on emissions (Huisingh et al., 2015) allows for timely adjustments. Furthermore, creating feedback loops to gather community input on the successes and barriers of energy adoption enables policymakers to fine-tune strategies, much like how the immune system adjusts its response to new threats. This adaptive, iterative process is crucial for achieving long-term sustainability in urban energy transitions.

3.2.5.5 Collaborative Networks Against Threats

As antigens trigger a coordinated immune response, addressing energy challenges in urban environments requires collaboration across multiple sectors. Fostering cross-sector partnerships between government, academia, industry, and community organizations allows for the pooling of resources, knowledge, and innovative ideas, facilitating the transition to renewable energy. Additionally, establishing platforms for stakeholder engagement enables diverse groups to share insights, strategies, and best practices—similar to how the immune system relies on the collective action of different immune cells to combat threats. Such collaborative efforts are indispensable for driving meaningful progress toward sustainable energy solutions and carbon neutrality.

3.2.6 Cytokines: Signaling Energy Adaptation

Cytokines are crucial signaling proteins that enable communication and coordination within the immune system, offering valuable insights for urban energy policymaking. By drawing on their role in immune responses, cities can develop strategies to advance multi-source renewable energy and achieve carbon neutrality.

3.2.6.1 Facilitating Communication Among Stakeholders

Just as cytokines enable communication between immune cells, urban energy policies should foster effective engagement among diverse stakeholders, including government agencies, industry, NGOs, and community groups. Establishing platforms for continuous dialogue ensures alignment on shared goals, such as renewable energy adoption and carbon neutrality, akin to how cytokines coordinate immune cell actions. Centralized data-sharing systems on energy consumption, emissions, and renewable resources can further enhance synchronization, improve decision-making, and drive collective progress toward sustainable energy outcomes.

3.2.6.2 Coordinating Policy Responses

Cytokines coordinate immune responses, and urban energy policies should similarly align diverse initiatives within an integrated framework. A unified strategy that incorporates renewable energy efforts—such as solar, wind, and energy efficiency programs—ensures cohesive action across multiple sectors, much like how cytokines mobilize various immune cells. Cross-sector collaboration among energy producers, urban planners, and public health officials can further strengthen energy policies, addressing immediate needs while contributing to broader city objectives, including sustainability, public health, and resilience.

3.2.6.3 Promoting Activation and Engagement

Cytokines activate immune cells, and urban energy policies should similarly engage communities and stakeholders to take active roles. Outreach programs that inform and involve residents in renewable energy projects encourage public participation in energy conservation and clean energy adoption, akin to cytokines stimulating immune responses. Offering financial and technical incentives for community involvement can foster ownership, strengthen collaboration, and ensure the long-term success of energy policies through broad-based support for sustainable initiatives.

3.2.6.4 Feedback Mechanisms for Continuous Improvement

Cytokines regulate immune responses through ongoing feedback, and urban energy policies should incorporate continuous evaluation and adaptation. Comprehensive monitoring and evaluation systems that track the effectiveness of renewable energy initiatives, emissions reductions, and community engagement enable real-time adjustments, mirroring cytokines' role in fine-tuning immune responses. By leveraging these feedback mechanisms, policymakers can refine strategies to address new challenges and opportunities, ensuring flexibility and effectiveness in achieving long-term sustainability.

3.2.6.5 Enhancing Resilience Through Coordination

As cytokines help coordinate immune responses to build resilience, urban energy systems should integrate various components to stabilize energy supply and demand. Policies that promote resilient energy infrastructure—such as the integration of distributed energy resources (Rahman et al., 2015) and smart grid technologies (Tuballa & Abundo, 2016)—allow cities to respond effectively to supply fluctuations. Emergency preparedness plans incorporating coordinated energy management can further ensure reliable power supplies during crises, enhancing the system's capacity to withstand disruptions and maintain operational stability in challenging situations.

3.2.7 *Major Histocompatibility Complex (MHC): Matching Energy Needs with Solutions*

The Major Histocompatibility Complex (MHC) is essential in the immune response, presenting antigens to T cells and enabling the recognition of foreign substances. This concept provides valuable insights for urban energy policymaking, particularly in fostering a diverse portfolio of renewable energy sources to achieve carbon neutrality at the city level. The principles of MHC inform several key aspects of energy strategies.

3.2.7.1 Facilitating Recognition of Local Energy Needs

Just as MHC presents specific antigens to T cells, urban energy policies should identify and communicate local energy needs and challenges to relevant stakeholders. Comprehensive energy assessments that evaluate neighborhood-specific demands, resources, and obstacles are critical. Establishing localized data collection mechanisms—such as tracking energy consumption patterns and renewable resource availability—enables the development of targeted and effective solutions tailored to the unique needs of each area.

3.2.7.2 Encouraging Stakeholder Engagement

MHC facilitates communication within the immune system, and similarly, urban energy policymaking should foster active collaboration among government agencies, private sector actors, and local communities. Organizing workshops and forums to discuss energy challenges and solutions mirrors MHC's role in coordinating immune responses. Transparent platforms for sharing updates on energy policies, renewable opportunities, and local success stories build trust and cooperation, encouraging a coordinated and collective approach to achieving energy goals.

3.2.7.3 Integrating Diverse Energy Sources

MHC's ability to present both endogenous and exogenous antigens parallels the need for urban energy systems to integrate diverse renewable energy sources. Policies should promote the use of solar, wind, geothermal, and biomass energy to build a resilient and flexible infrastructure. Supporting research and investment in emerging energy technologies further complements existing systems, ensuring a robust and sustainable energy portfolio that meets the diverse needs of urban populations.

3.2.7.4 Adaptive Policy Development

MHC molecules adapt to present the most relevant antigens, offering a model for urban energy policies to remain flexible and responsive to evolving conditions. Feedback mechanisms that continuously evaluate the performance of energy initiatives and incorporate community input are essential for ongoing refinement. Responsive regulatory frameworks should also accommodate technological advancements and shifts in energy demand, ensuring policies remain effective and aligned with both present and future needs.

3.2.7.5 Enhancing Public Health and Resilience

MHC strengthens immune responses, and urban energy policies can similarly enhance public health and resilience by integrating sustainable practices. Health impact assessments embedded in energy policy development ensure that renewable energy initiatives not only reduce emissions but also improve community health outcomes. Additionally, prioritizing resilience planning prepares cities for extreme weather events and potential energy disruptions (Perera & Hong, 2023), ensuring a reliable and diverse renewable energy supply during crises. By fostering healthier communities and more resilient energy systems, these policies contribute to long-term sustainability and well-being.

By applying the principles of MHC, urban energy policies can become more comprehensive, adaptive, and collaborative, facilitating a successful transition to renewable energy systems that are efficient, resilient, and inclusive.

3.2.8 *Insights from Immunological Disorders: Lessons for Preventing Energy Failures*

Disorders of the immune system—such as autoimmune diseases, immunodeficiencies, and allergies—offer valuable lessons for urban energy policymaking, particularly in promoting diverse renewable energy sources and achieving carbon neutrality. These disorders illuminate systemic vulnerabilities, conflicts, and the need for adaptive strategies, insights that can inform the design and management of urban energy systems.

3.2.8.1 Addressing Autoimmune Responses: Systemic Reactions

Autoimmune diseases arise when the immune system mistakenly attacks its own tissues, offering a metaphor for internal conflicts within energy systems. Similar to resolving autoimmune disorders, urban energy policies should address tensions, such as those between fossil fuel reliance and renewable energy adoption. Policymakers should conduct comprehensive reviews to identify policies that inadvertently hinder clean energy goals. Integrated strategies that align government, industry, and community interests can create a unified vision for renewable energy, preventing systemic conflicts and ensuring a coordinated approach.

3.2.8.2 Enhancing Resilience Against Immunodeficiency: Building Capacity

Immunodeficiency disorders, which weaken the immune system's ability to respond to threats, parallel vulnerabilities in urban energy systems that lack diversity and resilience. To bolster energy security (Ang et al., 2015), cities should promote a diversified energy mix—including solar, wind, and geothermal sources—reducing dependence on any single resource. Workforce training programs focused on renewable energy initiatives can further enhance local capacity and resilience, akin to strengthening the immune system's ability to address diverse challenges.

3.2.8.3 Managing Allergies: Addressing Overreactions

Allergies result from excessive immune responses to harmless substances, reflecting how public resistance to new energy technologies (Biresselioglu et al., 2024) can derail progress. Policymakers should prioritize education campaigns to clarify the benefits of renewable energy and address misconceptions that fuel resistance. Like allergists managing allergic reactions, energy policies should remain flexible, allowing for adjustments based on public feedback. This adaptability fosters acceptance and ensures policies are responsive to community concerns.

3.2.8.4 Learning from Immune Disorders: Adaptive Management

Immune system disorders expose vulnerabilities that necessitate adaptive strategies, a lesson applicable to urban energy systems. Adaptive management practices—characterized by continuous evaluation and refinement—should guide energy policymaking. Feedback mechanisms from stakeholders can ensure that energy strategies remain responsive to changing conditions and community needs, mirroring how the immune system evolves to address new threats. This adaptability strengthens policy effectiveness over time.

3.2.8.5 Holistic Health and Energy System Integration

The complexity of immune system disorders underscores the importance of a holistic approach. Urban energy policies should integrate health and environmental considerations into energy systems. Health impact assessments (Nkyekyer & Dannenberg, 2018) should be embedded in policy development to ensure renewable energy initiatives improve public health outcomes, such as air quality. Moreover, aligning renewable energy projects with urban planning priorities—such as green spaces, sustainable transportation, and public health initiatives—can create synergies that enhance environmental sustainability and overall quality of life.

By drawing on lessons from immunological disorders, urban energy policies can become more robust, adaptive, and inclusive, fostering systems that are not only resilient and efficient but also aligned with broader health and environmental objectives.

3.2.9 Recommendations: Immunity-Inspired Urban Energy Systems for Renewable Advancements

Immunity-inspired urban energy systems provide an innovative and transformative framework for accelerating renewable energy development while achieving urban carbon neutrality. By drawing from the adaptive, resilient, and collaborative strategies of the human immune system, these systems guide policymakers toward establishing sustainable, robust, and community-centered energy infrastructures. This forward-thinking approach promises profound environmental, economic, and societal gains, empowering the global fight against climate change while fostering long-term energy resilience. The following recommendations offer policymakers a visionary roadmap to revolutionize urban energy systems, merging innovation with natural resilience to drive global sustainability.

3.2.9.1 Embrace Diverse and Complementary Energy Sources: Inspired by the Multitude of Leukocyte Functions

The immune system's diverse leukocytes address varied threats, and urban energy systems can emulate this by integrating a mix of renewable sources such as solar, wind, geothermal, and biomass. Policymakers should establish incentives for region-specific energy projects, such as solar farms in sunlit regions, wind farms in coastal areas, and geothermal plants in volcanic zones. A diversified energy mix reduces reliance on any single source, enhancing adaptability and resilience against disruptions like resource shortages or weather variability. This tailored approach ensures stability, efficiency, and optimized utilization of regional strengths.

3.2.9.2 Accelerate Decentralized Energy Systems: Inspired by Physical Barriers in Innate Immunity

The immune system's physical barriers provide localized protection, and urban energy systems can achieve similar resilience through decentralization. Promoting rooftop solar, microgrids, and community-based projects with incentives like tax benefits and grants ensures local energy security while reducing reliance on centralized grids prone to large-scale failures. Decentralized systems are inherently robust, capable of withstanding natural disasters or cyberattacks, and empower communities to produce and manage their energy independently, fostering resilience and sustainability.

3.2.9.3 Champion Energy Efficiency Technologies: Inspired by the Neutralizing Effects of Chemical Barriers

Just as chemical barriers in the immune system neutralize threats, energy efficiency technologies reduce waste and environmental harm. Policymakers should mandate the adoption of energy-efficient solutions, including LED lighting, high-efficiency HVAC systems, and smart energy management tools in all new constructions and retrofits. These measures act as preventive mechanisms against resource depletion and high emissions, optimizing energy use while cutting costs. By curbing unnecessary energy consumption, efficiency initiatives pave the way for sustainable urban development.

3.2.9.4 Advance Feedback-Driven Smart Energy Systems: Inspired by Monitoring and Feedback Loops in Adaptive Immunity

The immune system's feedback mechanisms enable swift, adaptive responses, and smart energy systems can replicate this using real-time data analytics. Policymakers should prioritize investments in smart grid infrastructure integrated with IoT technologies to monitor energy consumption, optimize distribution, and rapidly adapt to fluctuations. By leveraging data-driven insights, these systems reduce waste, enhance efficiency, and support the seamless incorporation of renewables, ensuring a flexible and future-ready urban energy landscape.

3.2.9.5 Forge Collaborative Energy Networks: Inspired by Helper T Cell Coordination

Helper T cells coordinate immune responses, demonstrating the power of collaboration. Policymakers should establish task forces uniting government agencies, private sectors, NGOs, and communities to co-develop and implement energy policies. Collaboration ensures diverse perspectives, shared ownership, and inclusive solutions tailored to local needs. This multi-stakeholder approach fosters trust, accelerates policy adoption, and strengthens the collective commitment to achieving transformative energy goals.

3.2.9.6 Deploy Targeted Renewable Solutions: Inspired by the Precision of Antibody Specificity

Antibodies target specific pathogens, and energy policies should similarly address regional and socio-economic needs. Policymakers should develop localized renewable energy strategies, prioritizing solar in sunny areas, wind in coastal regions, and bioenergy in agricultural zones. Tailored solutions maximize resource efficiency, ensure alignment with regional strengths, and promote equitable energy access. This targeted approach amplifies impact, delivering sustainable energy solutions where they are needed most.

3.2.9.7 Build Resilience Against Disruptions: Inspired by the Versatility of MHC Systems

MHC molecules enable the immune system to recognize and adapt to diverse threats, a principle that urban energy systems should adopt through resilience-focused design. Policymakers should establish regulatory frameworks that encourage integrating emerging technologies and renewable innovations into existing infrastructures. A flexible, multi-source energy system can swiftly adapt to challenges such as resource shortages, extreme weather events, or technological disruptions, ensuring uninterrupted energy supply and long-term stability.

3.2.9.8 Engage and Educate the Public: Inspired by Leukocyte Communication

The immune system thrives on coordination and signaling, emphasizing the value of public engagement in urban energy transitions. Policymakers should lead public education campaigns highlighting the benefits of renewable energy adoption, from economic savings to environmental gains. An informed citizenry is more likely to participate in and support sustainable initiatives, fostering a culture of shared responsibility and accelerating the renewable energy movement.

3.2.9.9 Prioritize Health and Environmental Synergies: Inspired by Lessons from Immune Disorders

The immune system underscores the connection between systemic health and individual components, a lesson urban energy policies should heed. Policymakers should integrate health impact assessments into energy project evaluations, prioritizing initiatives that reduce pollution and enhance community well-being. Cleaner energy solutions improve public health by mitigating air pollution, reducing respiratory illnesses, and fostering healthier, more vibrant communities while achieving environmental goals.

3.3 Immune-Inspired Urban Disaster Management Policymaking

Urban disasters arise from both global and local sources, each presenting unique challenges that demand tailored responses encompassing prevention, mitigation, and adaptation. Global threats are largely driven by climate change, which intensifies extreme weather events such as hurricanes, floods, and heatwaves. These events jeopardize urban infrastructure, strain essential services, disrupt economies, and threaten public health on a global scale.

Local disasters, by contrast, are highly varied and multifaceted. Man-made catastrophes—including fires, industrial accidents, and transportation incidents—can occur suddenly, often causing significant loss of life and property. Additionally, geological hazards such as landslides and wildlife-related incidents, including roadkill, contribute to urban risks. Many of these risks are exacerbated by human activities like construction, deforestation, and increased vehicular traffic, underscoring the need for proactive urban planning.

Addressing this spectrum of urban disasters requires a comprehensive and integrated approach. Key components include robust risk assessment, forward-thinking urban planning, and the implementation of preventive measures. Embedding disaster preparedness into urban development policies enables cities to build resilience against both global and local threats, ensuring a safer and more sustainable environment for their residents.

The subsequent section explores the parallels between urban disaster management and the human immune system, offering insights into how current frameworks can be reimagined for greater effectiveness. By leveraging immune-inspired principles, cities can develop adaptive and coordinated strategies to address complex disaster risks.

3.3.1 White Blood Cells: Defending Urban Resilience

The defensive role of white blood cells (leukocytes) in protecting the body against infections and injuries offers valuable analogies for urban disaster management policymaking. By mirroring the immune system's strategies, cities can enhance resilience to global and local hazards, ensuring safety and sustainability.

3.3.1.1 Rapid Response Mechanisms

Leukocytes respond swiftly to pathogens, emphasizing the need for agile urban disaster response systems. Real-time data monitoring and early warning networks are critical for enabling rapid action (Shah et al., 2019). Urban disaster policies should emulate the immune system's ability to identify and neutralize threats quickly by establishing protocols for prompt resource mobilization, efficient emergency services, and effective community alert systems.

3.3.1.2 Adaptive Resilience

The immune system's adaptive responses, particularly through lymphocytes, highlight the importance of flexibility in urban planning. Urban policies should allow dynamic adjustments to address evolving risks. For example, cities can promote green infrastructure that adapts to changing climatic conditions (Ramyar et al., 2021), similar to how the immune system evolves to counter emerging pathogens. This adaptability ensures that urban systems remain resilient in the face of uncertain future challenges.

3.3.1.3 Ecosystem Approach

Leukocytes' ability to address a broad spectrum of threats reflects the need for a holistic approach to disaster management. Urban policies should account for the interconnectedness of urban environments, infrastructure, and social systems. This includes assessing how urbanization affects local ecosystems and integrating biodiversity conservation into disaster planning. Acknowledging these interdependencies promotes long-term resilience by aligning disaster management with environmental sustainability.

3.3.1.4 Preventive Measures

Leukocytes' immunological memory underscores the importance of prevention in disaster management. Proactive measures, such as investing in resilient infrastructure, conducting regular risk assessments, and enforcing zoning regulations, can significantly reduce vulnerability to disasters (Choudhury & Haque, 2024). Furthermore, encouraging sustainable urban practices can help mitigate risks before they escalate into crises, ensuring that cities are prepared to address potential hazards effectively.

3.3.2 *Innate Immunity: Immediate Disaster Response*

Drawing on the principles of innate immunity provides a compelling framework for improving urban disaster management. Just as innate immunity offers the body its first line of defense against a wide range of threats, urban disaster management can adopt analogous strategies to build resilient cities that respond swiftly and effectively to diverse challenges.

3.3.2.1 Physical Barriers

The skin and mucous membranes in innate immunity serve as physical barriers to protect the body from harm. Similarly, urban planning can incorporate physical defenses against disasters. For instance, flood defenses such as levees, floodwalls, and permeable green spaces can mitigate the impact of excess rainwater (Gralepois et al., 2016). Building codes should emphasize resilient infrastructure capable of withstanding extreme weather events, ensuring both structural integrity and the safety of urban populations.

3.3.2.2 Chemical Barriers

Antimicrobial substances in innate immunity act as chemical barriers that neutralize potential threats. Analogously, urban disaster management can adopt environmental and public health policies to protect residents. Investment in advanced monitoring systems to detect pollutants or hazardous materials can reduce health risks during disasters (Young et al., 2004). Strengthening industrial emission regulations and promoting non-toxic materials in construction and manufacturing can further safeguard urban environments, much like how antimicrobial agents shield the body.

3.3.2.3 Cellular Components (Phagocytes and NK Cells)

Phagocytes and natural killer (NK) cells in innate immunity provide rapid, targeted responses to potential threats. Urban disaster management can draw on this model by creating specialized emergency response teams (Wex et al., 2014). These teams can quickly mobilize to address risks such as hazardous spills, fires, or structural collapses. Much like immune cells "engulfing" invaders, these units can swiftly assess and neutralize threats to ensure community safety.

3.3.2.4 Inflammatory Response

The immune system's inflammatory response emphasizes the need for efficient communication and coordination during crises. Urban disaster management policies should prioritize clear and effective communication strategies to facilitate collaboration among emergency services, local authorities, and residents (Marcillo-Delgado et al., 2022). This includes real-time updates, alerts, and information-sharing systems that enable communities to act collectively. Additionally, integrating community feedback into disaster planning ensures that local vulnerabilities are addressed, refining preparedness efforts.

3.3.2.5 Non-specific Response

The broad, non-specific nature of innate immunity, which responds to a variety of threats, provides guidance for flexible disaster management frameworks. Policymakers should develop multi-hazard risk assessment tools (Wang et al., 2020) to prepare for a wide range of potential disasters, from natural events to human-made incidents. Urban planning should incorporate adaptability into strategies and resource allocation, allowing for rapid shifts in focus depending on the type of disaster.

3.3.2.6 Community Engagement

Innate immunity depends on the coordinated actions of multiple components, much like the need for collective effort in urban disaster management. Fostering community engagement is essential (Geekiyanage et al., 2020). Policymakers should implement educational programs to empower residents in disaster preparedness. Initiatives such as training community volunteers in basic emergency response skills can strengthen neighborhoods' collective capacity to address crises, ensuring a cohesive and effective response.

By mirroring the principles of innate immunity, urban disaster management can be transformed into a resilient and adaptive system capable of protecting cities from a wide range of threats.

3.3.3 *Adaptive Immunity: Evolving Disaster Management*

Drawing on the principles of adaptive immunity can significantly enhance the specificity, effectiveness, and resilience of urban disaster management systems. Adaptive immunity is characterized by its ability to develop targeted responses over time, a concept that can guide urban planners and policymakers in creating more robust and adaptable disaster preparedness frameworks.

3.3.3.1 Targeted Risk Assessment

Much like how B cells produce specific antibodies in response to pathogens, urban disaster management can benefit from targeted risk assessments that address the unique vulnerabilities of different areas. By analyzing historical data and current threats, policymakers can identify localized risks—such as flood-prone areas or earthquake hotspots—allowing for customized disaster response strategies tailored to the specific needs of each region.

3.3.3.2 Data-Driven Decision-Making

The adaptive immune system relies on memory and learned responses to pathogens. Similarly, urban disaster management can make use of data analytics and historical reviews to refine disaster response policies. By collecting and analyzing data from previous disasters, cities can continually improve their preparedness and response strategies. After-action reports evaluating the effectiveness of past responses (Knox, 2021) can be key in informing these ongoing improvements.

3.3.3.3 Enhanced Training and Preparedness

Just as T cells are activated when exposed to specific antigens, urban emergency response teams should engage in regular, scenario-based training exercises (Niekerk et al., 2015). These drills help ensure that responders are prepared to act swiftly and effectively during real events. Training programs should be customized based on the risks facing each community, enhancing the readiness and capability of disaster teams to respond to specific threats.

3.3.3.4 Community Involvement and Education

The activation of helper T cells, which enhance the immune response, is a useful analogy for engaging communities in disaster preparedness. Policymakers can create educational programs that inform residents about local risks, emergency response strategies, and recovery processes. By fostering a culture of preparedness, communities can become active participants in disaster management, improving resilience and speeding up response times.

3.3.3.5 Cross-Sector Collaboration

The coordinated response of T cells serves as a model for fostering collaboration across multiple sectors in disaster management. Policymakers should encourage cooperation between government agencies, private sector organizations, non-profits, and local communities. Inter-agency task forces can streamline communication, share resources, and improve the coordination of disaster response efforts.

3.3.3.6 Feedback Mechanisms for Continuous Improvement

Just as adaptive immunity refines its response with each exposure, urban disaster management should include feedback loops to constantly improve disaster planning and response practices. Mechanisms for gathering community feedback post-disaster allow residents to share their experiences and suggestions, which can inform updates to disaster plans and ensure that they remain relevant and effective.

3.3.3.7 Investing in Long-Term Resilience

The enduring protection provided by adaptive immunity can guide long-term investments in resilient urban infrastructure. Policymakers should prioritize strengthening infrastructure to withstand specific hazards, such as floods or earthquakes. This might involve updating building codes, retrofitting public facilities, and ensuring that critical services—such as hospitals and emergency centers—are located in less vulnerable areas.

3.3.4 Antibodies: Targeting Disaster Risks

Drawing inspiration from the concept of antibodies, urban disaster management can adopt a focused, adaptive approach to improve resilience and responsiveness to a range of threats. Just as antibodies specifically target and neutralize pathogens, urban policies can be crafted to identify, address, and mitigate risks associated with both global and local disasters.

3.3.4.1 Targeted Risk Identification

Much like antibodies bind to specific antigens, urban disaster management should focus on identifying particular vulnerabilities within a community. Policymakers can conduct detailed assessments to pinpoint high-risk areas, such as flood zones, industrial districts, or aging infrastructure (Danladi et al., 2024). This tailored approach enables the development of disaster response plans that address the unique characteristics of each area.

3.3.4.2 Monitoring and Surveillance

Just as antibodies continually monitor the body for pathogens, urban disaster management needs real-time surveillance to detect emerging threats. By implementing advanced monitoring systems, such as sensors and data analytics, cities can track environmental risks or structural weaknesses, allowing for early detection and prompt intervention.

3.3.4.3 Community Engagement and Education

Antibodies require activation to neutralize threats, and similarly, community awareness and engagement are essential for effective disaster response. Policymakers should design educational programs to inform residents about local risks, preparedness strategies, and emergency protocols. Empowering communities to recognize vulnerabilities and actively participate in disaster preparedness fosters a form of "community immunity," enhancing collective resilience.

3.3.4.4 Strengthening Infrastructure

Antibodies mark pathogens for destruction, emphasizing the importance of reinforcing vital infrastructure. Urban planners should prioritize fortifying critical systems—such as roads, bridges, and utilities—to withstand potential disasters (Oh et al., 2012). Retrofitting buildings to endure earthquakes or floods ensures that essential services remain operational during emergencies.

3.3.4.5 Collaboration Among Stakeholders

Antibodies cooperate with other immune components to neutralize threats, illustrating the need for collaboration in urban disaster management. Policymakers should foster partnerships between government agencies, private sector entities, and community organizations. This cooperation can improve resource sharing, knowledge transfer, and the overall coordination of disaster response efforts.

3.3.4.6 Preparedness Drills and Simulations

Antibodies bolster the immune system's readiness, and similarly, urban disaster management benefits from regular preparedness drills and simulations. These exercises allow responders and communities to practice their roles, refine communication strategies, and improve coordination, ensuring that everyone knows how to act in an emergency.

3.3.4.7 Feedback Loops for Continuous Improvement

Antibodies evolve over time to enhance their effectiveness, and urban disaster management should adopt a similar adaptive approach. Policymakers can implement feedback loops, such as after-action reviews and community feedback sessions (Datta et al., 2024), to assess response effectiveness and adjust strategies as needed. Continuous evaluation helps refine disaster planning, ensuring it stays relevant and effective.

3.3.4.8 Resource Allocation and Prioritization

Just as antibodies neutralize specific pathogens, urban policies should prioritize resource allocation based on identified risks (Escudero et al., 2018). Funding and resources should be directed toward high-risk areas, with preventive measures concentrated where they are most needed. Strategic investments in infrastructure, emergency services, and community training can enhance overall urban resilience.

3.3.5 Antigens: Identifying Disaster Vulnerabilities

Applying the concept of antigens to urban disaster management provides a structured approach for identifying and addressing the unique risks posed by various hazards. Antigens, recognized by the immune system as foreign entities that provoke a response, serve as a useful analogy for how urban systems identify and react to potential disaster threats.

3.3.5.1 Identification of Threats

Just as antigens trigger immune responses by acting as identifiers, urban disaster management should focus on pinpointing specific threats to cities. Comprehensive risk assessments should examine both local and global disaster sources, such as climate change, industrial risks, and infrastructure vulnerabilities. By identifying these risks, policymakers can prioritize them in their planning and response strategies.

3.3.5.2 Development of Targeted Responses

The immune system tailors its responses to each antigen it encounters. Similarly, urban disaster management should develop customized response plans for different types of hazards. For example, cities can establish distinct protocols for flooding, wildfires, industrial accidents, or public health emergencies, ensuring that response teams are equipped to handle each scenario effectively.

3.3.5.3 Awareness and Education

Just as the immune system learns to recognize antigens over time, communities should be educated about the risks they face. Public awareness campaigns can help residents understand local hazards, recognize warning signs, and know how to respond. Empowering communities in this way builds resilience and enhances overall preparedness.

3.3.5.4 Monitoring and Early Warning Systems

Antigens provoke a swift immune response, and similarly, urban disaster management should incorporate monitoring and early warning systems to detect emerging threats. By leveraging technology to track environmental changes—such as shifts in weather patterns or seismic activity—cities can anticipate potential disasters and issue timely alerts to residents and emergency services.

3.3.5.5 Cross-Sector Collaboration

The immune system mobilizes various components to combat antigens, illustrating the value of collaboration in disaster management. Urban disaster planning can benefit from fostering cooperation among government agencies, non-profit organizations, businesses, and community groups. This multi-sector approach improves resource allocation, information sharing, and coordinated responses during disasters.

3.3.5.6 Feedback Mechanisms for Learning

Just as the immune system refines its response based on previous encounters with antigens, urban disaster management should integrate feedback loops to continuously improve response strategies. After-action reviews and community feedback sessions (Parker, 2020) can reveal what worked well and highlight areas for improvement, helping to adapt and refine future policies.

3.3.5.7 Resource Allocation Based on Threat Levels

The immune response varies depending on the threat level posed by the antigen. Similarly, urban disaster management should allocate resources according to the severity and likelihood of identified risks. By directing funding and support to areas most vulnerable to disasters, cities can ensure that preventative measures and emergency services are adequately equipped to handle potential crises.

3.3.5.8 Building Resilience Through Training

The immune system's capacity to respond to antigens underscores the importance of training in disaster preparedness. Regular drills and simulations for both emergency responders and community members enhance readiness, ensuring that individuals know their roles and responsibilities during a crisis. This practice builds resilience, allowing cities to react swiftly and effectively during real-world disaster situations.

3.3.6 Cytokines: Signaling Disaster Recovery

Drawing on the principles of cytokines can significantly enhance urban disaster management by improving coordination, communication, and the overall effectiveness of response strategies. Cytokines are signaling proteins that facilitate communication between immune cells, and this concept can be applied to urban disaster management to streamline collaboration and improve responsiveness.

3.3.6.1 Enhanced Communication Systems

Just as cytokines enable communication among immune cells, urban disaster management should prioritize robust communication channels between all stakeholders involved in disaster response, including government agencies, emergency services, NGOs, and community organizations. A centralized communication platform ensures rapid, efficient information exchange, facilitating real-time decision-making during emergencies.

3.3.6.2 Coordination of Resources

Cytokines regulate the activation and proliferation of immune cells, which can be mirrored in disaster management by coordinating the deployment of resources. Policymakers should establish inter-agency task forces (Shah et al., 2022) to streamline resource allocation, minimize duplication, and ensure that resources are deployed where they are needed most to address specific disaster scenarios.

3.3.6.3 Rapid Response Protocols

Just as cytokines trigger swift immune reactions to threats, urban disaster management should implement rapid response protocols that outline clear, actionable steps for various disaster types. These protocols should guide both emergency responders and community members, ensuring an organized and timely response that reduces confusion and increases efficiency during crises.

3.3.6.4 Training for Adaptive Responses

Cytokines help adapt the immune response to specific threats. Similarly, urban disaster management should regularly train emergency responders and community members to adapt to diverse disaster scenarios. Simulated exercises help practice roles, refine response strategies, and improve the ability to manage high-pressure situations effectively.

3.3.6.5 Feedback Mechanisms for Continuous Improvement

The immune system fine-tunes its responses through feedback from cytokine signaling. Similarly, urban disaster management should incorporate feedback loops to evaluate the success of each response. After-action reviews can identify strengths and weaknesses, allowing policymakers to refine strategies and improve future disaster management efforts.

3.3.6.6 Community Engagement and Empowerment

Just as cytokines mobilize immune cells, urban disaster management should engage and empower communities to actively participate in disaster preparedness. Policymakers can create educational programs that inform residents about local risks and encourage involvement in planning and response activities. Empowered communities become more resilient, fostering a sense of ownership and responsibility for disaster mitigation.

3.3.6.7 Multi-hazard Risk Assessment

Cytokines regulate responses to multiple threats, and urban disaster management should adopt a multi-hazard approach to risk assessment. This ensures that policies are flexible and can adapt to a range of potential disasters—whether natural, industrial, or public health-related. A broad, adaptive risk framework strengthens urban resilience by preparing cities for diverse challenges.

3.3.6.8 Promoting Resilience Through Support Networks

Just as cytokines help create a supportive environment for immune responses, urban disaster management should cultivate networks of support among various stakeholders. Policymakers can facilitate alliances between businesses, community groups, and emergency services, encouraging collaboration and resource-sharing both before and after disasters. This strengthens the community's collective capacity to respond.

3.3.7 Major Histocompatibility Complex (MHC): Coordinating Disaster Solutions

Applying the principles of the Major Histocompatibility Complex (MHC) to urban disaster management can enhance the ability to recognize, respond to, and mitigate various risks. MHC molecules are crucial in the immune system for identifying foreign substances. This concept can be adapted to urban planning by focusing on risk recognition, stakeholder collaboration, and targeted response strategies.

3.3.7.1 Enhanced Risk Recognition

Just as MHC molecules present antigens to T cells for identification, urban disaster management should establish systems to recognize and catalog potential hazards. A comprehensive database of risks—ranging from internal issues like aging infrastructure to external threats like climate-induced events—can help policymakers prioritize and address urban vulnerabilities more effectively.

3.3.7.2 Multi-layered Threat Assessment

MHC molecules have distinct classes that present different types of antigens, enabling the immune system to target specific threats. Similarly, urban disaster management should adopt a multi-tiered approach to risk assessment (Zhang et al., 2023). This involves distinguishing between natural disasters, technological risks, and public health crises, with tailored assessment protocols for each. Differentiating between threat types allows for more accurate resource allocation and planning.

3.3.7.3 Stakeholder Engagement and Communication

MHC molecules facilitate communication between immune cells, a principle that can enhance urban disaster management by fostering collaboration among stakeholders. Government agencies, emergency services, community groups, and residents should be connected through established communication channels. Regular forums and collaborative platforms will ensure smooth information flow, enhancing collective readiness and response capabilities.

3.3.7.4 Tailored Response Strategies

The specificity of MHC in presenting antigens enables a targeted immune response. Urban disaster management can mirror this approach by developing customized response strategies for various disaster scenarios. Policies should create detailed plans for specific emergencies—floods, wildfires, industrial accidents—ensuring that response teams are equipped and trained to handle each situation effectively.

3.3.7.5 Training and Capacity-Building

Just as MHC molecules prepare T cells to respond to specific threats, urban disaster management should prioritize training programs to equip responders and community members with the necessary skills to handle a range of hazards. Regular drills and simulations of real-world scenarios will build capacity and foster resilience, ensuring communities and responders are always prepared.

3.3.7.6 Feedback Mechanisms for Continuous Improvement

MHC molecules help refine immune responses over time through feedback. Similarly, urban disaster management should incorporate mechanisms for ongoing evaluation and improvement. After-action reviews, along with community feedback, can pinpoint areas for improvement, enabling policymakers to refine strategies and enhance future preparedness and response efforts.

3.3.7.7 Public Awareness Campaigns

Much like MHC molecules alert the immune system to potential threats, urban disaster management can launch public awareness campaigns that educate residents about local risks and appropriate response measures. By empowering communities with knowledge about hazards and how to act, policymakers can enhance collective resilience and foster a culture of preparedness.

3.3.7.8 Resource Allocation Based on Risk Profiles

The specificity of MHC allows for precise immune responses, which can inform how resources are allocated in urban disaster management. Policymakers should prioritize investments in high-risk areas—directing funding toward infrastructure, emergency services, and community education to mitigate vulnerabilities where they are most critical.

3.3.8 Insights from Immunological Disorders: Lessons for Preventing Disaster Management Failures

Applying insights from immunological disorders to urban disaster management can improve resilience, preparedness, and response strategies. Autoimmune diseases, immunodeficiency disorders, and allergies offer valuable metaphors for understanding vulnerabilities and strengthening urban systems to better protect communities during disasters.

3.3.8.1 Recognizing Vulnerabilities

Autoimmune diseases reveal how the immune system can mistakenly attack the body's own tissues. Similarly, urban disaster management should identify vulnerabilities within communities that may be exacerbated during a crisis (Ermagun et al., 2024). Populations such as those with chronic illnesses, disabilities, or those from marginalized backgrounds may be disproportionately affected by disasters. Policies should address these vulnerabilities by providing targeted resources and support tailored to their specific needs.

3.3.8.2 Building Resilience

Immunodeficiency disorders underscore the necessity of a robust immune system. In urban disaster management, this translates into strengthening community resilience through preparedness programs. Training residents in emergency response, developing local disaster plans, and ensuring access to vital resources can create a culture of readiness, empowering communities to better withstand disaster impacts.

3.3.8.3 Implementing Early Warning Systems

Allergic reactions illustrate the dangers of delayed or exaggerated responses to threats. Similarly, urban disaster management should focus on early detection and notification. Early warning systems that alert communities to impending risks—such as severe weather events or hazardous material spills—should be accessible and user-friendly, ensuring that all residents can respond promptly and effectively.

3.3.8.4 Customizing Emergency Response Plans

Just as autoimmune diseases require specific treatments tailored to individual conditions, urban disaster plans should be customized for different scenarios and populations. These plans should be flexible and adaptable to various types of disasters, ensuring that emergency responders can address the unique needs of each affected community effectively.

3.3.8.5 Collaborative Multi-agency Approaches

The immune system relies on the coordinated action of various cells to mount an effective response. Similarly, urban disaster management should foster collaboration between government agencies, non-profits, and community organizations. Multi-agency cooperation ensures efficient use of resources, enhances response efforts, and ensures a unified approach during crises.

3.3.8.6 Community Health and Safety Monitoring

Immunodeficiency disorders highlight the importance of continuous health monitoring. Similarly, urban disaster management can implement systems to monitor community health before, during, and after disasters. This includes tracking the well-being of vulnerable populations and providing necessary medical interventions to minimize the health impact of disasters.

3.3.8.7 Public Awareness and Education

Allergies often arise from a lack of awareness or understanding about certain triggers. Urban disaster management can improve public knowledge by implementing educational campaigns to inform residents about local risks, preparedness strategies, and response protocols. By increasing awareness, communities will be better prepared, reducing panic and confusion during actual emergencies.

3.3.8.8 Resource Allocation Based on Community Needs

Understanding the specific needs of individuals with immunological disorders can inform how resources should be allocated during a disaster. Urban disaster management should prioritize resource distribution based on the unique demographics and vulnerabilities within the community. This ensures that those most at risk—such as the elderly, the sick, or low-income groups—receive the necessary support, including medical care and emergency supplies.

3.3.9 Recommendations: Revolutionizing Urban Disaster Resilience with Immunity-Inspired Systems

Urban areas face unprecedented challenges from natural and man-made disasters, requiring disaster management systems to evolve into more resilient, adaptive, and proactive frameworks. By drawing inspiration from the human immune system's ability to respond, adapt, and recover, policymakers can ensure cities are prepared for the increasing frequency and intensity of these threats. These immunity-inspired disaster management recommendations provide an innovative, collaborative roadmap for building urban resilience, emphasizing the immune system's principles of adaptability, precision, and coordination. By applying these lessons, cities can develop systems that not only react swiftly to crises but also anticipate and mitigate risks before they escalate. This holistic, forward-thinking approach will strengthen urban infrastructure, empower communities, and create more sustainable, resilient cities in the face of future disasters.

3.3.9.1 Agile Response Systems: Drawing Inspiration from Leukocyte Activity

Disaster response should be as swift and decisive as the immune system's white blood cells (leukocytes) that rapidly identify and neutralize threats. Urban disaster management systems need to adopt real-time monitoring capabilities, implement centralized disaster command centers, and establish rapid deployment units with advanced logistics and communication tools. By integrating cutting-edge technologies like IoT sensors, AI analytics, and predictive algorithms, cities can detect potential threats early and ensure that resources are allocated efficiently. Training specialized first responders to act decisively in high-pressure scenarios enhances response times, minimizes damage, and saves lives. These proactive measures will reduce disaster escalation and improve coordination during emergencies, making cities more agile in their fight against unforeseen risks.

3.3.9.2 Unstoppable Adaptive Resilience: Learning from Lymphocyte Evolution

Just as lymphocytes evolve to recognize and fight new pathogens, urban disaster management systems should continuously adapt to emerging risks. By integrating dynamic urban planning that includes flood-resistant infrastructure, flexible transportation networks, and the use of green spaces like wetlands, cities can prepare for both known and unforeseen environmental challenges. Policymakers should promote adaptive strategies, including regular vulnerability assessments to identify weaknesses in urban systems and infrastructure. This approach will not only mitigate the immediate impacts of disasters but also enhance long-term resilience. As threats evolve, so too should our disaster management systems—ensuring they remain effective and responsive to changing circumstances.

3.3.9.3 Ecosystem-Based Resilience: Leveraging Nature's Defense Systems

The immune system functions as a vast network of interconnected components, and disaster management should take a similarly holistic, systems-based approach. Policymakers should prioritize ecosystem conservation and urban greening initiatives that protect natural buffers like mangroves, wetlands, and forests, which naturally mitigate the impacts of floods, storms, and other disasters. These ecosystems provide cost-effective solutions that reduce the need for artificial infrastructure while offering enhanced biodiversity, which helps strengthen urban resilience. Incorporating natural defenses into urban disaster management systems helps preserve the environment and creates sustainable, climate-resilient cities. By integrating nature-based solutions, we can mitigate disaster impacts while fostering environmental sustainability and protecting vital ecosystems.

3.3.9.4 Preventive Power: Anticipating Disasters Through Immunological Memory

The immune system's ability to remember past threats and respond proactively is a model for disaster management. Proactive risk assessment, preventive infrastructure investment, and strict zoning regulations are critical to reducing the likelihood of catastrophic disasters. Cities should implement regular risk assessments to identify high-risk areas for floods, fires, and other hazards, using this data to enforce policies that limit construction in vulnerable zones. Investment in flood defenses, firebreaks, seismic retrofits, and early warning systems should be prioritized to safeguard infrastructure and lives. Proactive disaster planning reduces recovery costs, minimizes fatalities, and ensures cities are built with resilience in mind, preventing major disasters before they even occur.

3.3.9.5 Empowering Communities—Building a Collective Immune System for Disaster Response

Just as leukocytes coordinate their efforts to respond to threats, urban disaster management relies heavily on the collaboration and engagement of local communities. Effective disaster management cannot be top-down alone; it requires a united effort from residents, government agencies, and local organizations. Policymakers should launch public awareness campaigns to educate citizens about disaster preparedness and provide training programs for community volunteers. Neighborhood-based preparedness networks can be formed, with regular drills and simulations to enhance local readiness. This grassroots involvement ensures that when disaster strikes, communities can respond quickly and effectively, reducing the burden on emergency services and fostering a culture of collective resilience.

3.3.9.6 Cutting-Edge Communication Systems: Real-Time Connectivity for Crisis Management

Cytokines enable rapid communication between immune cells to ensure swift responses to threats, and disaster management requires similarly efficient communication systems. Policymakers should invest in centralized platforms for real-time disaster alerts, providing the public with instant updates through mobile apps, social media, and public warning systems. These systems should be universally accessible, ensuring that vulnerable populations, including those without Internet access or in remote areas, receive timely notifications. Effective communication allows for coordinated, prompt action from citizens, emergency responders, and government officials. Clear and efficient messaging reduces confusion and ensures that all stakeholders are aligned and ready to respond at a moment's notice.

3.3.9.7 Comprehensive Multi-hazard Risk Assessment: Proactively Identifying Threats

The immune system's ability to recognize a wide array of threats through non-specific responses provides a model for urban disaster management. Policymakers should implement advanced multi-hazard risk assessment tools that use historical data, predictive modeling, and real-time analytics to identify and prioritize risks from both natural and man-made disasters. These tools should be integrated into urban planning processes to ensure that cities are prepared for a range of potential disasters, from earthquakes to chemical spills. By forecasting future risks, cities can allocate resources more effectively and develop tailored response strategies, improving disaster preparedness across all sectors.

3.3.9.8 Tailored Disaster Response Protocols: Precision in Action

The immune system's specificity in targeting pathogens through antibodies highlights the need for disaster management plans to be tailored to specific risks. Each type of disaster—whether floods, fires, earthquakes, or industrial accidents—requires a unique response strategy. Policymakers should develop and regularly update emergency response protocols for different disaster scenarios, ensuring that first responders and local authorities are well-prepared to act quickly and efficiently. Customizing these protocols, along with conducting scenario-specific drills, ensures that resources are directed where they are most needed and minimizes confusion during high-stress events. These targeted plans ensure a smoother, more effective response to disasters, ultimately saving lives and accelerating recovery efforts.

3.3.9.9 Continuous Learning and Improvement: Evolving Through Feedback Loops

Adaptive immunity improves through feedback, and disaster management should similarly evolve through constant evaluation and refinement. Policymakers should establish feedback loops that assess the effectiveness of disaster responses, conduct after-action reviews, and gather insights from affected communities. This continuous process of learning allows for real-time adjustments to policies, ensuring that strategies remain relevant and effective. Regular collaboration among stakeholders will also highlight areas for improvement, fostering a culture of innovation and resilience. By embracing feedback and learning from past events, cities can strengthen their disaster response capabilities and become more adaptable to new and unforeseen challenges.

3.3.9.10 Global Collaboration: Uniting Forces for Global Resilience

The immune system's ability to collaborate across its various components to address threats should be mirrored in disaster management through cross-sector partnerships. Policymakers should forge partnerships among governments, private sectors, NGOs, and local communities to streamline resource sharing and disaster response efforts. International collaboration is also crucial for exchanging best practices, sharing resources, and responding to transnational risks like pandemics and climate change. This collaborative approach ensures that cities have access to global expertise, technologies, and funding, enhancing their capacity to respond rapidly to disasters while also fostering a unified, global commitment to building resilience.

3.4 Immune-Inspired Policymaking of Underground Development in Cities

The development and management of underground systems in cities require innovative technologies and governance strategies that improve both resilience and functionality. Below, we explore how principles from the immune system can inform policymaking for underground urban development.

3.4.1 White Blood Cells: Defending Underground Infrastructure

The characteristics of white blood cells (leukocytes) offer a valuable metaphor for improving policymaking in underground urban development. Below are key ways this analogy can be applied.

3.4.1.1 Phagocytic Action: Proactive Monitoring and Maintenance

Just as phagocytes actively seek out and neutralize harmful pathogens to protect the body, urban policymakers can adopt a proactive approach to monitor and maintain underground infrastructure. This includes conducting regular inspections of critical systems, such as utility tunnels, sewage networks, and subway lines, to detect potential issues before they escalate. Additionally, leveraging real-time data from sensors and smart technologies can enable early detection of anomalies—such as leaks or blockages—functioning as an early warning system, much like the immune system's response to threats.

3.4.1.2 Lymphocyte Response: Adaptive Planning and Resilience

Lymphocytes play a crucial role in the immune system's ability to adapt to and counter new threats. In urban development, this concept calls for the adoption of resilient, adaptive strategies—such as dynamic urban planning frameworks (Nel et al., 2018)—that can evolve in response to changing population density, climate shifts, or other unpredictable challenges. Infrastructure should be designed with flexibility, allowing for future expansion or repurposing as needed. Moreover, engaging with the community—similar to how lymphocytes retain memory of past encounters—ensures that decisions about underground development are context-sensitive and responsive to local needs.

3.4.1.3 Immune System Collaboration: Interdisciplinary Coordination

The immune system's effectiveness hinges on the coordinated action of diverse cell types. Similarly, successful urban development requires strong interdisciplinary collaboration (Bibri, 2021). Policymakers should foster partnerships among urban planners, engineers, public health professionals, and local communities to develop comprehensive underground infrastructure solutions. A systems-thinking approach (Laspidou et al., 2020), which integrates transportation, utilities, and public health networks, can create a cohesive, adaptable infrastructure that responds efficiently to urban challenges—much like the immune system's coordinated defense against invaders.

3.4.1.4 Adaptive Immunity: Learning from Past Failures

The immune system's ability to learn from past infections is a powerful tool for refining responses to future threats. In urban planning, this concept can be applied through post-implementation reviews that assess the successes and shortcomings of infrastructure projects. These reviews offer valuable insights into performance, highlight areas for improvement, and refine future strategies. Creating a centralized repository of case studies—encompassing both successful and failed projects—can facilitate knowledge-sharing across cities and regions, enhancing the decision-making process for future underground infrastructure initiatives.

3.4.2 *Innate Immunity: Immediate Response to Infrastructure Threats*

By applying the principles of innate immunity to underground urban development, we can enhance the resilience and efficiency of infrastructure. Concepts such as physical barriers, chemical defenses, cellular responses, and the inflammatory response provide valuable insights for effective policymaking in the urban context.

3.4.2.1 Physical Barriers: Designing Resilient Infrastructure

Just as the skin acts as the body's first line of defense against harmful pathogens, urban infrastructure can be designed to protect against environmental threats. Rigorous construction standards (Athanasopoulo et al., 2019) should be enforced to ensure that underground facilities—such as subways and utility tunnels—are resilient to hazards like flooding, soil erosion, and seismic activity. Thoughtful site selection is equally crucial, with infrastructure placed in areas less prone to natural disasters or environmental degradation, much like how skin prevents harmful agents from entering the body. Together, these measures ensure that urban systems can withstand a range of external challenges.

3.4.2.2 Chemical Barriers: Preventive Measures and Maintenance

Just as the body relies on antimicrobial substances to ward off infections, urban infrastructure can benefit from proactive preventive strategies. Regular maintenance and system upgrades are essential to identify and address vulnerabilities before they result in failure or contamination. For example, cleaning sewage systems regularly to prevent blockages, and using durable, corrosion-resistant materials, can significantly enhance the longevity and resilience of underground infrastructure. These strategies ensure that systems remain strong and functional over time, much like the body's chemical defenses protect against harmful agents.

3.4.2.3 Cellular Components: Responsive Management Systems

The immune system's quick, targeted response to threats, through cells like phagocytes and natural killer cells, can be mirrored in urban management through rapid response teams and real-time monitoring systems. Specialized teams should be on standby to address emergencies such as utility disruptions or sewage leaks, similar to how immune cells quickly neutralize infections. Additionally, integrating sensors and smart technologies allows for continuous monitoring of underground systems, detecting anomalies and enabling immediate intervention (Shahrour et al., 2021).

This data-driven approach ensures that infrastructure remains resilient and responsive to emerging challenges.

3.4.2.4 Inflammatory Response: Proactive Community Engagement

The immune system's inflammatory response recruits additional cells to areas of injury or infection. A similar approach can be applied in urban development by fostering active community engagement. Involving local residents and stakeholders in the planning, monitoring, and maintenance of underground infrastructure ensures that potential issues are identified early and addressed effectively. Public awareness campaigns can further educate citizens on the importance of infrastructure upkeep, encouraging them to report early warning signs, such as unusual odors or water pooling. This collaborative approach not only mobilizes resources but also strengthens the community's role in maintaining a resilient urban environment.

3.4.3 Adaptive Immunity: Evolving Underground Systems

Drawing from the principles of adaptive immunity provides a more nuanced and resilient approach to urban policymaking, especially in the development and maintenance of underground infrastructure. The roles of B cells and T cells in the immune system offer valuable insights for creating targeted and sustainable urban strategies.

3.4.3.1 Specific Responses: Tailored Infrastructure Solutions

Just as adaptive immunity generates specific responses to individual pathogens, urban policies should be customized to meet the distinct needs of different neighborhoods. Urban planners can design underground infrastructure to address the unique challenges of each area, similar to how B cells produce antibodies for particular threats. For example, flood-prone areas may require advanced stormwater management systems, while other regions might need seismic reinforcements. Conducting detailed risk assessments allows for precise resource allocation, ensuring that infrastructure solutions are both effective and efficient, much like how T cells target specific infections. This approach ensures that each neighborhood is equipped with the right safeguards to bolster long-term resilience.

3.4.3.2 Long-Lasting Protection: Sustainable Practices

The lasting defense offered by adaptive immunity can be mirrored in urban development through sustainable practices that promote durability and long-term resilience. Underground infrastructure should be built with high-quality materials and advanced technologies to withstand environmental stressors. For example, using corrosion-resistant materials in utility tunnels (Kong, 2021) mirrors the enduring protection provided by adaptive immunity. Additionally, establishing mechanisms for regular reviews and updates to infrastructure policies ensures they remain adaptable to emerging challenges and data, just as the immune system refines its responses over time. This approach ensures that urban infrastructure remains robust and evolves to meet future demands.

3.4.3.3 Activation and Collaboration: Interdisciplinary Engagement

The activation of T cells underscores the importance of collaboration across diverse sectors, a principle that can enhance urban development through interdisciplinary engagement (Cook, 2015). By fostering cooperation among urban engineers, public health professionals, community members, and policymakers, urban planning can benefit from a range of perspectives, similar to how helper T cells activate other immune cells to mount a coordinated defense. Moreover, building local capacity through training programs for community organizations and governments strengthens their ability to identify and address infrastructure challenges in real time, ultimately improving the resilience of urban systems. This collaborative approach ensures that all stakeholders are well-prepared to respond effectively to emerging issues.

3.4.3.4 Learning from Experience: Data-Driven Decision-Making

Just as adaptive immunity improves over time through previous encounters, urban development can benefit from a data-driven approach that evolves through continuous learning. By establishing systems to collect and analyze data on infrastructure performance, failures, and successes, cities can gain valuable insights that inform future planning, just as immune cells rely on past experiences to refine their defenses. Implementing pilot projects for innovative infrastructure solutions (Hughes et al., 2018) enables testing, feedback, and iterative improvement, mirroring the immune system's ability to adapt and enhance its responses. This ensures that urban infrastructure evolves in line with both historical lessons and new innovations.

3.4.4 Antibodies: Protecting Critical Underground Networks

The concept of antibodies offers a powerful framework for improving the planning and management of underground urban infrastructure. Antibodies' unique characteristics—such as their specificity, ability to mark targets, neutralization functions, cooperative action, and capacity for learning—can provide actionable strategies for urban policymakers and planners.

3.4.4.1 Specificity: Targeted Interventions

The precision with which antibodies bind to specific antigens can serve as a model for urban infrastructure policies. Just as antibodies are tailored to target individual pathogens, urban strategies should be designed to address the unique challenges of each area. For example, flood-prone neighborhoods may require specialized stormwater management systems, while other regions may need infrastructure capable of withstanding seismic activity. Developing diagnostic tools to detect localized vulnerabilities—such as corrosion in utility tunnels—ensures that interventions are both focused and effective. This approach enables efficient resource allocation and guarantees that solutions are customized to meet the specific needs of each community.

3.4.4.2 Marking for Action: Prioritization of Issues

Much like antibodies signal the immune system to target pathogens, a structured approach to identifying and prioritizing infrastructure issues can ensure that the most pressing challenges are addressed first. By establishing a framework based on risk assessments, environmental impact, and community needs, policymakers can ensure that critical infrastructure concerns are prioritized. Additionally, implementing community reporting systems (Sadeghi-Bazargani et al., 2024) allows residents to flag emerging issues, providing real-time data that informs decision-making. These systems, like antibodies marking threats, direct attention and resources to areas requiring immediate intervention.

3.4.4.3 Neutralization: Mitigation Strategies

Just as antibodies neutralize pathogens to prevent harm, urban infrastructure policies can adopt similar strategies for risk mitigation. Implementing regular maintenance programs for underground systems—such as sewage and utility tunnels—helps identify and address potential issues before they escalate. This preventive approach mirrors the role of antibodies in neutralizing threats early in an immune response. Additionally, developing comprehensive risk mitigation plans, complete with contingency measures for system failures, ensures swift intervention to minimize the impact of disruptions. This proactive approach strengthens the resilience and reliability of urban infrastructure, safeguarding communities and services.

3.4.4.4 Collaborative Response: Building Partnerships

The cooperative action of antibodies with other immune components offers a valuable model for fostering collaboration in urban governance. Encouraging partnerships between public agencies, private companies, and community organizations can lead to a unified response to infrastructure challenges, similar to the coordinated efforts of immune cells. Public-private partnerships can also drive innovation in underground infrastructure technologies, benefiting from the diverse expertise and resources each sector brings. This collaborative approach ensures a comprehensive and adaptable response to urban infrastructure needs, ultimately strengthening resilience across the system.

3.4.4.5 Learning and Adaptation: Continuous Improvement

Just as antibodies evolve and improve through repeated encounters with pathogens, urban infrastructure can benefit from continuous learning and adaptation. Establishing feedback mechanisms to assess the performance of infrastructure systems enables policymakers to gather valuable insights that inform future planning and responses. Leveraging data analytics for evidence-based decision-making ensures that policies evolve in response to emerging challenges and shifting community needs. This cycle of learning and adaptation guarantees that urban infrastructure remains resilient and responsive, capable of meeting future demands with greater efficiency and effectiveness.

3.4.5 Antigens: Identifying Underground Infrastructure Risks

The concept of antigens provides a valuable framework for shaping urban infrastructure policies, particularly in the areas of risk identification, management, and adaptive improvement. By drawing parallels between how the immune system identifies and responds to foreign elements and how urban planners can address emerging risks, policymakers can develop more effective strategies for safeguarding underground systems.

3.4.5.1 Identifying Foreign Elements: Risk Assessment

Just as antigens prompt immune responses by being recognized as foreign, urban policies can use a similar approach to identify and assess risks within underground systems. Developing a comprehensive framework to pinpoint "antigenic" risks—such as aging infrastructure, environmental shifts, or disruptions from emerging technologies—enables policymakers to take proactive measures and prevent system failures. Additionally, creating community hazard maps that highlight areas prone to specific vulnerabilities (e.g., flood zones or outdated utility lines) helps focus interventions where they are most needed. This approach mirrors the immune system's ability to detect external threats and ensures resources are allocated effectively to safeguard urban infrastructure.

3.4.5.2 Targeted Responses: Tailored Infrastructure Solutions

Just as the immune system generates specific responses to each antigen, urban policies should be customized to address the unique needs of different neighborhoods and infrastructure challenges. For example, areas with high soil salinity may require specialized construction techniques for utility tunnels. Likewise, interventions such as maintenance and upgrades should be tailored to address the identified "antigenic" risks, ensuring solutions are both relevant and effective. This approach strengthens resilience by adapting to the distinct characteristics of each area and ensuring infrastructure remains robust and adaptable to evolving demands.

3.4.5.3 Monitoring and Surveillance: Proactive Systems

Similar to how the immune system continuously monitors for the presence of antigens, urban infrastructure management can benefit from constant vigilance through proactive monitoring systems. Implementing real-time surveillance using smart sensors and IoT technology provides ongoing assessments of underground infrastructure, enabling early detection of potential issues. This approach mirrors the immune system's continuous search for threats, allowing for swift responses to emerging risks. Additionally, establishing early warning systems ensures that city officials and the public are alerted to infrastructure problems before they escalate into crises, enabling timely interventions and minimizing disruptions.

3.4.5.4 Engaging Stakeholders: Collaborative Identification

The success of the immune system lies in the cooperation of various immune cells to identify and eliminate antigens—a principle that can be applied to urban governance through collaborative engagement. By fostering partnerships between community members, engineers, public health experts, and urban planners, infrastructure challenges can be identified from diverse perspectives. Furthermore, conducting workshops and training programs to raise awareness about potential risks within underground systems—similar to how the immune system trains its cells to recognize antigens—helps build a more informed and engaged public. This collaborative approach strengthens the resilience of urban infrastructure by tapping into a broad pool of knowledge and expertise, ensuring that responses are both comprehensive and effective.

3.4.5.5 Learning from Encounters: Adaptive Management

Just as the immune system adapts its responses based on prior encounters with antigens, urban infrastructure management can benefit from continuous learning and adaptation. Post-incident reviews following infrastructure failures or crises—such as floods or power outages—offer opportunities to refine strategies and improve future responses. Additionally, leveraging data analytics to evaluate past infrastructure performance helps guide policy adjustments, ensuring urban systems evolve in response to real-world challenges. This adaptive management approach fosters resilience, enabling infrastructure to better withstand future pressures by learning from past experiences, just as the immune system improves its defenses after exposure to new threats.

3.4.6 Cytokines: Signaling Adaptation for Underground Systems

Cytokines play a key role in the immune system by enabling communication and coordination between cells. Drawing inspiration from these signaling molecules, urban infrastructure policymaking can be enhanced through improved coordination, responsiveness, and resilience.

3.4.6.1 Facilitating Communication: Data Sharing and Integration

Just as cytokines facilitate communication among immune cells, urban infrastructure management can benefit from enhanced data integration and sharing. By developing centralized systems for real-time information exchange between stakeholders—such as city planners, utility providers, and emergency services—decisions can be made quickly, and responses can be coordinated more effectively (Lee et al., 2018). Open data initiatives (Neves et al., 2020) that make infrastructure health, performance metrics, and maintenance schedules publicly available increase transparency and foster collaboration. This enables all parties to engage more meaningfully, respond to emerging issues with agility, and ultimately improve the resilience of urban infrastructure.

3.4.6.2 Coordinated Responses: Multi-agency Collaboration

Similar to how cytokines orchestrate a unified immune response, urban infrastructure management relies on coordinating multiple stakeholders. Establishing formal collaboration protocols between city departments—such as transportation, environmental services, and public health—ensures diverse expertise is applied to complex infrastructure challenges. Regular inter-agency meetings can help align goals and share resources, strengthening collective responses to emerging issues. Public-private partnerships (Currie & Teague, 2015) also offer opportunities to leverage external expertise and innovation. Just as cytokines enhance cooperation within the immune system, such collaborations create a more integrated and responsive approach to managing urban infrastructure.

3.4.6.3 Adaptive Regulation: Responsive Policymaking

Much like cytokines regulate immune responses based on evolving needs, urban environments benefit from adaptive regulatory frameworks that remain flexible in the face of change. Implementing policies that can be adjusted in real time—such as updating construction standards or modifying maintenance schedules based on infrastructure performance—allows urban planning to stay responsive to emerging challenges. Continuous feedback mechanisms ensure that regulations evolve over time, improving their effectiveness. Just as cytokines fine-tune immune responses, dynamic regulatory systems help policymakers make more informed, agile decisions to meet the shifting demands of urban life.

3.4.6.4 Promoting Community Engagement: Local Empowerment

Cytokines activate immune responses by mobilizing cells—a process mirrored in urban policy through community empowerment. Establishing community action networks facilitates communication between residents and officials, encouraging local participation in identifying issues and contributing to planning efforts (Nik-Bakht & El-Diraby, 2020). Public awareness campaigns can further educate citizens about the importance of maintaining and monitoring underground infrastructure, much as cytokines enhance immune system awareness. This engagement boosts public understanding and fosters stronger community support for infrastructure initiatives, ensuring a more collaborative approach to urban resilience.

3.4.6.5 Crisis Response and Recovery: Rapid Mobilization

Just as cytokines enable rapid mobilization during immune crises, urban infrastructure management should be prepared for swift, coordinated responses to emergencies. Developing clear, regularly updated emergency response protocols for incidents such as power outages or sewage system failures ensures roles and communication channels are well-defined. Regular disaster preparedness drills and simulations help all stakeholders understand their responsibilities, facilitating a fast, coordinated reaction in the event of an infrastructure crisis. This proactive approach mirrors the immune system's ability to quickly mobilize cells to combat infection, ensuring urban systems can respond with similar efficiency during critical situations.

3.4.7 *Major Histocompatibility Complex (MHC): Integrating Underground Infrastructure Needs*

The Major Histocompatibility Complex (MHC) plays a crucial role in the immune system by enabling T cells to recognize and respond to foreign substances. This concept can provide valuable insights for urban policymaking, particularly in underground infrastructure, by fostering improved responsiveness, stakeholder collaboration, and adaptive planning.

3.4.7.1 Facilitating Recognition: Transparent Communication

Just as MHC molecules present antigens to immune cells for recognition, urban policymakers can promote transparency and raise awareness of infrastructure needs through clear, accessible communication. Establishing effective reporting mechanisms allows both engineers and the public to share insights about the condition of underground infrastructure, ensuring that issues are promptly identified and communicated to decision-makers. Additionally, creating public data portals (Lnenicka & Nikiforova, 2021) that provide up-to-date information on critical infrastructure—such as water quality or sewer system conditions—helps stakeholders recognize and prioritize areas in need of attention. This mirrors how MHC molecules present antigens for immune detection, ensuring essential data is readily available to guide timely, informed responses.

3.4.7.2 Classifying Needs: Tailored Responses

Much like MHC molecules categorize antigens as either endogenous or exogenous, urban planning can benefit from classifying infrastructure needs to facilitate more targeted interventions. Developing a classification system for various underground systems—such as utility tunnels, sewage networks, and stormwater management—ensures more strategic allocation of resources and customized planning. By recognizing the distinct challenges of each infrastructure type, policymakers can ensure that interventions are both appropriate and effective. For example, tailored maintenance strategies can be developed for aging utility tunnels versus newer stormwater systems, similar to how MHC molecules distinguish between different types of antigens, optimizing resource allocation based on the unique characteristics of each system.

3.4.7.3 Engaging Stakeholders: Collaborative Decision-Making

Just as MHC molecules present antigens to T cells to trigger a coordinated immune response, effective urban infrastructure management requires engaging multiple stakeholders in decision-making processes. Organizing workshops and consultations that bring together engineers, public health experts, community members, and urban planners ensures that a variety of perspectives are considered in infrastructure planning. Promoting partnerships between public agencies and private companies fosters knowledge exchange and the sharing of best practices in underground infrastructure management. This collaborative network strengthens the recognition of infrastructure needs and enhances overall resilience, ensuring that planning and responses are comprehensive and inclusive.

3.4.7.4 Adaptive Management: Responsive Policy Frameworks

Similar to how MHC molecules enable the immune system to adapt to new and evolving threats, urban policy can benefit from flexible frameworks that respond dynamically to emerging challenges. By implementing policies that can be adjusted based on real-time data about infrastructure performance, urban systems can remain resilient and responsive to change. Just as MHC molecules present new antigens, policies should be updated to address emerging issues effectively. Additionally, establishing feedback mechanisms to capture insights from infrastructure users and maintenance teams creates a continuous flow of information, enabling policymakers to refine strategies based on current needs and experiences. This adaptive approach ensures that urban policies remain relevant and effective in the face of evolving challenges.

3.4.7.5 Crisis Preparedness: Rapid Response Planning

Just as MHC molecules play a critical role in the immune system's swift recognition of pathogens, urban infrastructure should be prepared for rapid identification and response during crises. Developing clear, well-defined emergency response protocols for infrastructure failures—such as major leaks, sewer blockages, or power outages—ensures that assessment procedures are in place for immediate action. Regularly conducting simulations and drills prepares all stakeholders to respond efficiently when emergencies arise, fostering a coordinated and rapid mobilization of resources. This proactive approach mirrors the immune system's ability to quickly react to threats, ensuring that infrastructure systems recover swiftly and effectively from crises, thereby strengthening overall urban resilience.

3.4.8 Insights from Immunological Disorders: Lessons for Preventing Infrastructure Failures

Insights from immune system disorders offer valuable lessons for urban policymakers, particularly when managing underground infrastructure. Challenges posed by autoimmune diseases, immunodeficiency disorders, and allergies can inspire strategies to enhance resilience, public health, and safety in urban environments.

3.4.8.1 Autoimmune Diseases: Identifying Systemic Risks

Autoimmune disorders reveal the dangers of misidentification, where the immune system mistakenly attacks the body's own tissues. Similarly, urban infrastructure can suffer from misidentifying risks, leading to severe consequences if left unchecked. To prevent this, comprehensive risk assessment frameworks should be implemented to identify systemic vulnerabilities within underground systems. This involves analyzing internal factors—such as mismanagement or unchecked wear—that could lead to failures, much like how the immune system may mistakenly target its own cells. Additionally, preventive maintenance programs, including routine inspections of utility tunnels, can help identify issues early, ensuring that risks like leaks or structural collapses are mitigated before they compromise public safety or infrastructure integrity.

3.4.8.2 Immunodeficiency Disorders: Enhancing Resilience

Immunodeficiency disorders highlight the need for systems that can effectively respond to external threats. This concept can guide efforts to strengthen the resilience of urban infrastructure. To protect underground systems from external shocks, such as natural disasters or power outages, infrastructure should be designed with redundancies and backup systems—similar to enhancing the immune system's ability to defend against pathogens. Furthermore, establishing comprehensive emergency response plans, supported by regular training and simulations, ensures that city agencies are prepared to respond quickly and efficiently to failures or emergencies. This proactive approach mirrors the immune system's enhanced capacity to respond to infections, ensuring that urban infrastructure remains adaptable and resilient in the face of unexpected challenges.

3.4.8.3 Allergies: Managing Overreactions

Allergic reactions occur when the immune system overreacts to harmless substances. This principle can be applied to urban planning by managing overreactions to environmental or social changes (Peters et al., 2017). To prevent public backlash

(Voeten, 2022) or resistance to environmental changes, thorough environmental impact assessments should be conducted before launching new underground projects. This helps address and mitigate potential concerns, preventing exaggerated reactions, much like managing allergies. Additionally, fostering open communication with the community about planned developments can help manage expectations and reduce opposition. By educating the public about the benefits of underground infrastructure and addressing concerns upfront, urban planners can smooth project implementation and prevent misunderstandings, similar to preventing allergic overreactions by managing exposure.

3.4.8.4 Holistic Health Approaches: Integrating Public Health and Infrastructure

Immunological disorders emphasize the interconnectedness of health and environmental factors—an idea that can be mirrored in urban policymaking through a holistic approach. Encouraging collaboration between urban planners, public health officials, and environmental scientists is essential for evaluating how underground infrastructure, such as sewage and stormwater systems, affects public health (Kondo et al., 2015). This integrated approach ensures that health considerations are embedded in infrastructure design and management from the outset. Moreover, promoting sustainable construction and maintenance practices for underground systems minimizes environmental impact, safeguarding public health and enhancing community resilience against health risks associated with poor environmental conditions. This strategy fosters a forward-thinking urban development approach that prioritizes both infrastructure and public well-being.

3.4.8.5 Monitoring and Evaluation: Adaptive Management

Just as monitoring immune disorders provides crucial insights for adapting treatments, continuous monitoring and evaluation of urban infrastructure can drive resilience and efficiency. Real-time monitoring systems for critical underground networks (Zhang et al., 2020) allow for the quick detection of issues, preventing minor faults from escalating into significant failures—similar to how healthcare systems monitor patients with immunological disorders. Additionally, establishing feedback mechanisms that enable stakeholders to report concerns ensures ongoing dialogue between infrastructure users and policymakers. This approach facilitates timely interventions, ensuring urban systems remain responsive, adaptable, and effective over time, fostering continuous improvement and better long-term management.

3.4.9 *Recommendations: Immunity-Inspired Urban Underground Infrastructure Systems*

Underground infrastructure systems are the lifelines of urban environments, enabling essential services such as transportation, utilities, and waste management. Yet, as cities grapple with mounting environmental, technological, and societal challenges, these systems face unprecedented risks that demand bold, transformative action. Inspired by the unparalleled efficiency and adaptability of the human immune system, this proposal presents a groundbreaking framework to revolutionize underground infrastructure management. By integrating proactive monitoring, adaptive planning, interdisciplinary collaboration, and data-driven decision-making, policymakers can forge a new era of urban resilience. These immunity-inspired strategies empower cities to not only withstand current pressures but also anticipate and adapt to future challenges, ensuring the enduring safety, functionality, and sustainability of urban life.

3.4.9.1 Proactive Infrastructure Vigilance: Phagocytic Action in Real-Time Monitoring and Maintenance

Underground infrastructure systems require continuous monitoring and proactive maintenance to prevent failures. Drawing inspiration from the immune system's ability to detect and neutralize threats, underground systems should be equipped with real-time monitoring systems and sensors that detect anomalies such as leaks, corrosion, or structural weaknesses in utilities, subways, and sewage networks. Rapid-response maintenance teams should be established to address these issues promptly. Early detection and intervention can prevent minor issues from escalating into costly failures, ensuring that infrastructure remains reliable and reducing risks to urban populations.

3.4.9.2 Resilient Urban Evolution: Lymphocyte-Like Adaptation in Designing Future-Ready Underground Systems

Urban underground infrastructure should be designed to evolve with changing needs, such as population growth or climate shifts. Just as lymphocytes adapt to new threats, urban planning frameworks should be flexible and allow for modular designs that can be expanded or repurposed over time. Adaptive systems should incorporate continuous assessments of vulnerabilities and stakeholder feedback, ensuring the infrastructure remains functional and relevant despite evolving challenges. This approach ensures that cities are prepared for future demands, safeguarding long-term resilience.

3.4.9.3 Unifying Forces: Immune System Collaboration in Fostering Interdisciplinary Coordination for Robust Infrastructure

Effective underground infrastructure management requires collaboration among diverse disciplines. Just as the immune system's efficiency relies on coordinated actions among various cells, urban planners, engineers, public health professionals, and environmental scientists should work together to design and manage integrated infrastructure systems. By fostering interdisciplinary partnerships, cities can create cohesive strategies that address transportation, utilities, and public health needs while maximizing efficiency and resilience, ensuring a systems-thinking approach that addresses complex urban challenges.

3.4.9.4 Fortifying the Foundations: Physical Barrier-Like Resilience in Building Underground Systems for Tomorrow's Challenges

To withstand environmental hazards like flooding, soil erosion, and seismic activity, underground infrastructure systems should be designed with the same resilience as the body's physical barriers. This means enforcing rigorous construction standards and utilizing advanced materials, such as corrosion-resistant alloys, to ensure the longevity of underground systems. Well-designed systems will reduce vulnerability to environmental stresses, ensuring the long-term safety and functionality of critical infrastructure, thus enhancing urban resilience in the face of unpredictable natural events.

3.4.9.5 Preventive Power: Chemical Barrier-Like Maintenance to Safeguard Urban Infrastructure

Routine maintenance and preventive measures are essential to the longevity and reliability of underground infrastructure. Inspired by the immune system's chemical barriers, urban systems should benefit from scheduled inspections and upgrades to critical systems such as sewage and utility tunnels. Integrating predictive maintenance technologies can also address issues before they lead to failures. Proactive upkeep reduces long-term costs, enhances system reliability, and prevents disruptions that could compromise urban populations' safety and well-being.

3.4.9.6 Rapid Response Mastery: Cellular Component-Like Agility in Crisis Management

In emergency situations, swift responses are essential. Just as immune cells like phagocytes and natural killer cells respond quickly to emerging threats, underground infrastructure management should develop specialized, rapid-response teams to handle crises such as utility failures, flooding, or hazardous leaks. These teams should be equipped with advanced diagnostic tools and real-time data access, ensuring quick interventions that minimize disruptions, prevent cascading failures, and safeguard urban systems and communities from long-term damage.

3.4.9.7 Community-Powered Infrastructure: Inflammatory Response-Like Engagement for Smarter Cities

Active community engagement is key to effective underground infrastructure management. Just as the immune system recruits resources to areas of infection, urban infrastructure systems should involve local communities in monitoring and maintaining systems. Public awareness campaigns should educate residents about the importance of underground infrastructure, and platforms should be created for citizens to report issues such as unusual odors, water pooling, or structural changes. Fostering a culture of shared responsibility will enhance early detection of problems and improve overall urban resilience.

3.4.9.8 Precision Solutions: Adaptive Immunity-Like Customization for Unique Urban Challenges

Just as the immune system's adaptive immunity generates targeted responses to specific threats, underground infrastructure should be customized to meet the unique needs of each urban area. Detailed risk assessments should be conducted to identify localized challenges, such as flood risks or seismic vulnerabilities, and tailored solutions should be designed to address them. This might include advanced stormwater management systems or reinforced utility tunnels. Customized approaches ensure efficient resource allocation and effective infrastructure solutions that meet the specific needs of urban communities.

3.4.9.9 Continuous Evolution: Learning from Adaptive Immunity to Build Ongoing Infrastructure Resilience

Urban infrastructure systems should continuously evolve to meet new challenges. Just as the immune system refines its responses over time, cities should establish review processes to evaluate the effectiveness of completed projects. Centralized databases of infrastructure performance data should be maintained to inform future planning and decision-making, ensuring that lessons learned from past experiences lead to ongoing improvements. A culture of continuous improvement ensures infrastructure systems remain robust, adaptable, and resilient in the face of future challenges.

3.4.9.10 Data-Driven Mastery—Monitoring and Feedback-Like Insights for Smarter Infrastructure Management

Data analytics can optimize underground infrastructure management, just as the immune system relies on constant monitoring to detect and respond to threats. By implementing IoT-based monitoring networks that collect real-time data on infrastructure conditions, cities can leverage predictive analytics to identify potential issues before they escalate. Data-driven decision-making improves resource allocation, enhances operational efficiency, and ensures timely maintenance, enabling cities to adapt quickly to emerging issues while maintaining a high standard of infrastructure resilience.

3.5 Using *Urbanimmunology* to Inform EDU Policymaking

Energy, Disaster, and Underground (EDU) systems are crucial to the functioning of cities. Policymaking that addresses the challenges arising from their interdependencies can be significantly strengthened by drawing lessons from the human immune system. The key is to approach these interdependent EDU systems holistically. Just as the immune system operates as a unified, adaptive response network—constantly monitoring, responding, and adapting to threats while maintaining balance—urban EDU systems can also be integrated into a cohesive policy framework. This framework would ensure that energy, disaster management, and underground infrastructure complement each other, fostering a resilient urban environment. Below are key lessons from the immune system that can inform EDU policymaking.

3.5.1 Holistic Detection and Early Warning Systems: Surveillance for Energy, Disaster, and Underground Networks

The immune system continuously monitors the body for potential threats, with specialized cells such as dendritic cells and macrophages detecting foreign invaders and initiating tailored responses. Similarly, effective urban resilience depends on integrated surveillance mechanisms that monitor disruptions across energy, disaster, and underground systems. These systems should detect vulnerabilities early, allowing for proactive intervention before issues escalate into larger crises. Just as the immune system identifies and addresses threats before they fully develop, EDU systems should be equipped to identify risks across all sectors in a coordinated manner, enabling swift, targeted responses.

To achieve this, integrated monitoring systems should track vulnerabilities across the three EDU sectors simultaneously. These systems could include energy grid sensors to detect faults, overloads, or failures; flood and seismic sensors to predict or monitor natural disasters; and smart infrastructure capable of assessing the health of underground utilities like water, sewage, and communications networks. By utilizing data-sharing platforms, these systems can aggregate and analyze data from across all sectors, providing a holistic view of potential risks. This approach ensures that urban planners and policymakers can anticipate issues in a timely manner, enabling a coordinated and proactive response, much like the immune system's ability to detect and address threats before they fully manifest.

To further enhance this capability, a centralized urban monitoring hub should be established where data from energy, disaster, and underground systems are continuously assessed. This hub would enable real-time responses and efficient resource allocation based on the severity and interconnections between system vulnerabilities. For example, if a hurricane is detected, the monitoring system could issue early warnings assessing risks within both the energy grid and underground infrastructure, such as potential flooding in subway systems. This unified, adaptive system would ensure that urban environments can respond effectively to emerging threats, just as the immune system adapts with precision and speed.

3.5.2 Coordinated and Adaptive Response Mechanisms: Activation for Resilience

Once a threat is identified, the immune system activates a coordinated response, deploying white blood cells like T cells and B cells to neutralize the threat. Inflammatory responses also work to repair and protect the body. Similarly, when disruptions occur in urban systems—whether in energy, underground infrastructure, or due to natural disasters—coordinated, adaptive responses are required to restore functionality and prevent further damage. Just as the immune system mobilizes and adapts to respond to immediate threats, urban systems should exhibit flexibility and speed during crises to minimize disruption and ensure continued operation.

Urban policies should establish adaptive response protocols to ensure smooth collaboration between energy providers, disaster agencies, and underground infrastructure managers. During a disaster, for example, energy systems should work with emergency response teams to maintain power and quickly deploy necessary services. Likewise, underground systems like subways and water utilities should be rapidly repaired or rerouted to support affected infrastructure, while energy grids should be adjusted or fortified to maintain city functions. This cross-sector coordination ensures that urban systems work together, enhancing resilience during crises.

To support these efforts, policies should prioritize the development of integrated management frameworks that unite energy, underground infrastructure, and disaster management authorities under a single command during crises. Specialized response teams should be trained to handle interconnected failures, ensuring a swift, coordinated response to multiple challenges. Emergency coordination platforms should link utility companies, municipal agencies, and other relevant stakeholders, enabling real-time sharing of information about system failures—such as power outages, flooding, or transport disruptions—ensuring an efficient and effective response.

3.5.3 Resilience Through Redundancy, Flexibility, and Recovery: Immunological Memory in Urban Systems

The immune system adapts to new threats by forming immunological memory, enabling quicker and more effective responses to future attacks. It also builds resilience through redundancy, ensuring that multiple immune cells are ready to act if one system is compromised. This ability to adapt and recover rapidly is key to the immune system's efficiency. Similarly, urban systems should prioritize redundancy and flexibility to ensure continuous operation during and after a crisis.

Integrated EDU policies should focus on ensuring that each sector has backup mechanisms and alternative pathways to maintain urban functionality. For energy systems, this could mean combining centralized and decentralized sources, such as renewable energy and storage solutions, to activate if the primary grid fails. Underground systems should have multiple delivery networks and failover mechanisms to reroute services in case of damage. Disaster recovery plans should allow for the restoration of each sector individually, preventing cascading disruptions across systems.

Policy recommendations should emphasize developing redundant energy grids, incorporating alternative power sources like solar, wind, and backup battery systems alongside the main grid. Underground utilities should be redesigned with redundant pathways for essential services such as water, sewage, and electricity. Post-disaster recovery plans should ensure that energy restoration efforts are synchronized with repairs to underground infrastructure, ensuring recovery processes are efficient and cohesive across all sectors.

3.5.4 Adaptive Learning and Long-Term Healing: Memory and Adaptation in Urban Resilience

After an infection, the immune system strengthens its defenses through immunological memory, adapting to prevent reinfection. This ability to learn and adapt is crucial for long-term resilience. Similarly, urban policies should incorporate adaptive learning to strengthen future resilience. Policymakers should consistently evaluate past crises to improve future responses, ensuring that EDU systems are better prepared for emerging challenges.

Urban policies should incorporate continuous feedback loops to learn from past failures, such as energy grid disruptions, underground infrastructure damage, and shortcomings in disaster responses. Post-event assessments should be mandatory, allowing policymakers to analyze the causes and impacts of system failures. This process of adaptive learning will inform future urban planning and enhance the interconnectedness and resilience of EDU systems. By incorporating lessons learned from past disasters, cities can better withstand future challenges, just as the immune system strengthens its defenses over time.

To facilitate this, policy recommendations should include mandatory after-action reviews for each disaster or significant event involving EDU system failures. These reviews will ensure that lessons are documented and used to inform future policies. Additionally, a framework for adaptive learning should be developed, integrating real-time data into the design of future energy, disaster management, and underground infrastructure systems. Cities should prioritize the continuous upgrading of EDU infrastructure based on emerging technologies and insights gained from past experiences, ensuring urban environments remain resilient and prepared for evolving risks.

3.5.5 Recommendations of Integrated Policymaking

Drawing inspiration from the human immune system, with its interconnected components working as a cohesive whole, integrating energy systems, disaster management, and underground infrastructure is supported by both rational and theoretical foundations. In light of the growing complexity and interdependence of urban systems, it is imperative for policymakers to adopt a rigorous, forward-thinking approach to ensure the resilience and continuity of cities. As urban environments face increasingly frequent and severe disruptions—from energy failures to natural disasters and infrastructure damage—traditional, siloed approaches are no longer sufficient. The following policy recommendations outline a comprehensive, integrated framework to fortify the critical systems that underpin modern cities: energy, disaster management, and underground infrastructure. These recommendations call for decisive action to enhance system monitoring, crisis response coordination, infrastructure resilience, continuous adaptation, and long-term, systems-wide integration. Only through such a unified, proactive approach can cities effectively safeguard against current and future challenges, ensuring they remain resilient, adaptive, and operational in the face of unpredictable events.

3.5.5.1 Unyielding Vigilance: Strategic Risk Detection and Proactive Response Framework

To safeguard urban functionality, a comprehensive surveillance and early detection framework is essential, monitoring vulnerabilities across energy systems, disaster management, and underground infrastructure simultaneously. These interconnected sectors require integrated, real-time monitoring to detect emerging risks before they escalate into major crises, ensuring timely interventions and proactive management. A centralized citywide monitoring system should be mandated, continuously assessing all three sectors and sharing real-time data to enable comprehensive situational awareness. This system should include early warning protocols that span across sectors, triggering coordinated actions to prevent widespread disruptions. Additionally, cross-sector data-sharing between energy providers, emergency services, and infrastructure managers is crucial to offer a holistic view of urban vulnerabilities and facilitate prompt, effective responses.

3.5.5.2 Decisive Crisis Leadership: Unified Command and Cross-Sector Response System

During a crisis, an uncoordinated or delayed response can significantly worsen the damage, making it crucial to establish a robust, centralized command structure that integrates response teams from energy, disaster management, and underground infrastructure into one unified operational framework. This centralized approach ensures that critical actions across all sectors are synchronized, enabling swift mitigation of impacts, rapid recovery, and the restoration of city functions with precision. To support this system, policy imperatives include institutionalizing a centralized crisis coordination hub that consolidates all sectors—energy, disaster management, and underground systems—under one unified authority. Additionally, mandatory joint exercises and crisis simulations should be implemented to ensure seamless collaboration between energy, emergency, and infrastructure teams during real-world crises, while developing real-time communication platforms that guarantee immediate, synchronized decision-making across all sectors, ensuring efficient resource allocation and response.

3.5.5.3 Uninterrupted Continuity: Guaranteed Infrastructure Redundancy and Operational Resilience

Urban systems should be designed for resilience, with infrastructure redundancy and flexibility embedded across energy, disaster management, and underground sectors to ensure operational continuity during crises. Backup systems and alternative pathways are crucial to prevent a single point of failure from compromising the city's critical systems. Energy grids, underground infrastructure, and emergency services should all incorporate fail-safe mechanisms that enable continued service delivery even when one sector faces disruption. Policy imperatives to support this include mandating the development of fail-safe systems with redundant pathways and backup systems for energy, underground utilities (such as water and sewage), and emergency services. Additionally, decentralized energy solutions, including renewable sources, storage systems, and backup grids, should be integrated to complement centralized power systems and ensure energy resilience during grid failures. Finally, underground utilities should be designed with multiple, independent service delivery routes to minimize disruptions in the event of damage to primary pathways.

3.5.5.4 Relentless Evolution: Continuous System Evaluation and Adaptive Policy Innovation

Urban resilience is a dynamic, ongoing process that requires continuous evaluation and adaptive policy innovation. Post-crisis evaluations should assess the performance of energy, disaster, and underground systems to extract actionable insights, enabling the ongoing improvement of infrastructure and response strategies. A feedback loop should be institutionalized across all sectors to adapt policies and systems based on the latest lessons learned from past events, ensuring that urban systems evolve and become more resilient over time. To support this, policy imperatives include enforcing mandatory after-action reviews for every significant urban disruption, focusing on how energy, underground infrastructure, and disaster management systems performed, and identifying areas for improvement. Additionally, adaptive feedback loops should be institutionalized in urban policy to integrate insights from past failures into the next generation of designs for energy, disaster management, and infrastructure. Finally, continuous investment in emerging technologies should be prioritized to enhance the flexibility and adaptability of energy grids, underground infrastructure, and disaster response strategies.

3.5.5.5 Holistic Urban Dominance: Comprehensive Systems Integration for Long-Term Resilience

Urban environments should function as interconnected systems, where energy, disaster management, and underground infrastructure are not treated in isolation. A holistic, systems-level planning framework should be developed to ensure these three critical sectors are designed, planned, and managed as interdependent components, minimizing vulnerabilities, enhancing operational efficiency, and preventing cascading failures from disruptions in any one sector. To support this, policy imperatives include mandating the adoption of systems-thinking in urban planning policies to account for the interdependencies of energy, disaster management, and underground infrastructure. Collaborative governance frameworks should also be required to unite energy providers, infrastructure managers, and emergency response agencies, ensuring long-term alignment and coordination. Additionally, prioritizing whole-system risk assessments is essential to address vulnerabilities across these sectors and ensure that policies reflect the interconnected nature of urban systems.

3.6 Conclusion: A New Blueprint for Urban Resilience

In the quiet, unseen intricacies of the human immune system, one can find a remarkable blueprint—a model for how urban environments might adapt, thrive, and endure in the face of ever-evolving challenges. By drawing upon the immune system's remarkable ability to defend against, respond to, and recover from threats, we can envision a fresh, innovative approach to urban policymaking, particularly within the realms of energy, disaster resilience, and underground infrastructure (EDU). These sectors, often treated in isolation, are fundamentally interwoven, and like the body's defense mechanisms, they should function in harmony to safeguard the future of cities.

At the heart of this vision lies the principle of adaptability, a lesson drawn from the immune system's fluid response to emerging threats. Just as the immune system continuously learns from and adapts to the ever-changing landscape of pathogens, urban energy systems should evolve to meet the complex challenges of a world in flux. Cities, now more than ever, should design energy infrastructure capable of detecting and responding to disruptions in real time. Policies should prioritize renewable energy sources, decentralized grids, and demand-side management, much like the immune system's ability to self-regulate and adapt to changing conditions. With the integration of microgrids, battery storage, and other innovations, cities can build energy systems that not only meet the demands of today but also weather the storms of tomorrow—be they natural disasters, economic upheavals, or technological disruptions.

But energy is only one part of the equation. In the domain of disaster management, the immune system's role in identifying threats, mobilizing defenses, and promoting recovery offers profound insights for urban planning. The shift toward a more proactive and responsive disaster management model is essential, where early detection systems act like immune sensors, rapidly deployed interventions mirror immune response cells, and long-term recovery processes resemble the body's healing mechanisms. Urban disaster policies, therefore, should evolve from reactive strategies to proactive, data-driven approaches—ones that anticipate threats before they emerge and engage communities in building resilience against a multitude of risks, from the rise of extreme weather events to the looming threats of social and political unrest.

As the immune system operates through a vast network of cells, proteins, and organs, each in constant communication and coordination to maintain balance, so too should the various layers of urban infrastructure work in concert. Underground networks—water, sewage, gas, electricity—are the lifeblood of cities, yet they often go unnoticed until failure strikes. These hidden systems, much like the immune system's interconnected components, rely on seamless integration to function properly. When crafting policies for underground infrastructure, we should consider the interdependencies between energy, disaster resilience, and infrastructure systems, for a breakdown in one area can send ripples of failure across the entire urban ecosystem. A power outage, for example, could cripple water and sewage services, precipitating a public health crisis, while a disaster-related failure in underground systems could undermine the city's ability to respond effectively to emergencies.

Thus, the integration of these vital elements—energy, disaster management, and underground infrastructure—becomes not just a policy priority but a necessity. Urban systems, much like biological ones, do not exist in isolation. To strengthen resilience, urban governance should adopt a holistic, interconnected approach that embraces the complex interdependencies between these systems. The energy policies that shape a city's future should anticipate the impacts of climate-related disasters. Disaster response plans should incorporate the condition and capacity of underground infrastructure, recognizing its potential to either mitigate or exacerbate the impacts of crises.

This shift from siloed thinking to integrated policymaking is fundamental to crafting cities that are both resilient and sustainable. Drawing inspiration from the immune system, we can envision a new kind of urban governance—one that is adaptive, proactive, and flexible enough to respond to the challenges of a constantly changing world. Effective policies will not only address the needs of each sector in isolation but will foster synergies that reinforce one another, creating a cohesive system of resilience.

To achieve this vision, collaboration across sectors is essential. Interdisciplinary research, adaptive governance, and a deep understanding of urban systems will empower policymakers to craft solutions that are not only reactive but anticipatory—solutions that embrace innovation while acknowledging the complexities of an interconnected world. By understanding the interdependencies of energy, disaster resilience, and underground infrastructure, cities can better navigate the uncertainties of the future, ensuring that they are not only livable but capable of flourishing in an increasingly unpredictable world.

In the end, the shift toward informed, interconnected policymaking is not just a matter of improving urban systems. It is about creating cities that can endure, adapt, and thrive in the face of the challenges to come. By drawing on the lessons of the immune system—its resilience, adaptability, and proactive nature—we can craft cities that are not only prepared for the threats of today but also poised to flourish in the uncertain world of tomorrow.

References

Abbas, A. K., Lichtman, A. H., & Pillai, S. (2023). *Basic immunology: Functions and disorders of the immune system* (7th ed.). Elsevier.

Abbruzzese, M., Infante, D., & Smirnova, J. (2024). The diffusion of renewable energy production in European countries: The role of incentives. *Economics of Innovation and New Technology, 34*(1), 70–93.

Ang, B. W., Choong, W. L., & Ng, T. S. (2015). Energy security: Definitions, dimensions and indexes. *Renewable and Sustainable Energy Reviews, 42*, 1077–1093.

Athanasopoulo, A., Bezuijen, A., Bogusz, W., et al. (2019). *Standardisation needs for the design of underground structures*. European Commission – JRC Technical Reports. https://doi.org/10.2760/615209

Batidzirai, B., Trotter, P. A., Brophy, A., Stritzke, S., Moyo, A., Twesigye, P., Puranasamriddhi, A., & Madhlopa, A. (2021). Towards people-private-public partnerships: An integrated community engagement model for capturing energy access needs. *Energy Research & Social Science, 74*, 101975.

Belyak, N., Gabriel, S. A., Khabarov, N., & Oliveira, F. (2024). Renewable energy expansion under taxes and subsidies: A transmission operator's perspective. *Journal of Cleaner Production, 451*, 141955. https://doi.org/10.1016/j.jclepro.2024.141955

Bibri, S. E. (2021). The core academic and scientific disciplines underlying data-driven smart sustainable urbanism: An interdisciplinary and transdisciplinary framework. *Computational Urban Science, 1*(1), 1. https://doi.org/10.1007/s43762-021-00001-2

Biresselioglu, M. E., Solak, B., & Savas, Z. F. (2024). Unveiling resistance and opposition against low-carbon energy transitions: A comprehensive review. *Energy Research & Social Science, 107*, 103354.

Bristow, D. N., & Mohareb, E. A. (2019). From the urban metabolism to the urban immune system. *Journal of Industrial Ecology, 24*(2), 300–312.

Choudhury, M., & Haque, C. E. (2024). Disaster management policy changes in Bangladesh: Drivers and factors of a shift from reactive to proactive approach. *Environmental Policy and Governance, 34*(5), 445–462. First published: 07 January 2024. https://doi.org/10.1002/eet.2094

Cloke, J., Mohr, A., & Brown, E. (2017). Imagining renewable energy: Towards a social energy systems approach to community renewable energy projects in the global south. *Energy Research & Social Science, 31*, 263–272.

Cook, I. R. (2015). Policy mobilities and interdisciplinary engagement. *International Journal of Urban and Regional Research, 39*(4), 835–837.

Currie, D., & Teague, P. (2015). Conflict management in public–private partnerships: The case of the London underground. *Negotiation Journal, 31*(3), 237–266.

Danladi, A. J., Ludovico, D. D., Eugeni, F., Dusserre, G., & Yusuf, M. K. (2024). Disaster management on a regional scale: Knowledge platform toward overall hotspot identification. *International Journal of Research Publications and Reviews, 5*(8), 2106–2121.

Datta, N., Frank, S. C., Grewe, M. E., Yang, C., Lawal, K.-K., & Bilheimer, A. (2024). Enhancing the impact of a community feedback session service through ongoing evaluation. *Journal of Clinical and Translational Science, 8*(s1), 68–68.

Delves, P. J., Martin, J. E., Burton, D. R., & Roitt, I. M. (2017). *Janeway's Immunobiology* (9th ed.). Garland Science.

EC. (2019). Energy and Climate Policy (ECP) Scenario: Impact assessment of policies and measures. Annex 2 to the national energy and climate plan for 2021–2030. https://energy.ec.europa.eu/system/files/2020-08/pl_final_necp_part_5_en_0.pdf

Ermagun, A., Smith, V., & Janatabadi, F. (2024). High urban flood risk and no shelter access disproportionally impacts vulnerable communities in the USA. *Communications Earth & Environment, 5*, 2. https://doi.org/10.1038/s43247-023-01165-x

Escudero, L. F., Garín, M. A., Monge, J. F., & Unzueta, A. (2018). On preparedness resource allocation planning for natural disaster relief under endogenous uncertainty with time-consistent risk-averse management. *Computers & Operations Research, 98*, 84–102.

Flores, J. R., Montagna, J. M., & Vecchietti, A. (2014). An optimization approach for long term investments planning in energy. *Applied Energy, 122*, 162–178.

Gatto, A., & Busato, F. (2020). Energy vulnerability around the world: The global energy vulnerability index (GEVI). *Journal of Cleaner Production, 253*, 118691.

Geekiyanage, D., Fernando, T., & Keraminiyage, K. (2020). Assessing the state of the art in community engagement for participatory decision-making in disaster risk-sensitive urban development. *International Journal of Disaster Risk Reduction, 51*, 101847.

Gralepois, M., Larrue, C., Wiering, M., Crabbé, A., Tapsell, S., Mees, H., Ek, K., & Szwed, M. (2016). Is flood defense changing in nature? Shifts in the flood defense strategy in six European countries. *Ecology and Society, 21*(4), 37. https://doi.org/10.5751/ES-08907-210437

Hassan, Q., Algburi, S., Sameen, A. Z., Salman, H. M., & Jaszczur, M. (2023). A review of hybrid renewable energy systems: Solar and wind-powered solutions: Challenges, opportunities, and policy implications. *Results in Engineering, 20*, 101621.

Hughes, S., Yordi, S., & Besco, L. (2018). The role of pilot projects in urban climate change policy innovation. *The Policy Studies Journal, 48*(2), 271–297.

Huisingh, D., Zhang, Z., Moore, J. C., Qiao, Q., & Li, Q. (2015). Recent advances in carbon emissions reduction: Policies, technologies, monitoring, assessment and modeling. *Journal of Cleaner Production, 103*, 1–12.

IEA. (2024). *World energy outlook 2024*. The International Energy Agency.

Janeway, C. A., Travers, P., Walport, M., & Shlomchik, M. J. (2001). *Immunobiology: The immune system in health and disease* (5th ed.). Garland Science.

Jouhara, H., & Yang, J. (2018). Energy efficient HVAC systems. *Energy and Buildings, 179*, 83–85.

Kirkegaard, J. K., Nyborg, S., Georg, S., & Horst, M. (2023). Towards failed renewable energy communities? Activist attempts to change market conditions in the Danish wind energy market. *Energy Research & Social Science, 102*(2023), 103152. https://doi.org/10.1016/j.erss.2023.103152

Knox, C. C. (2021). Systematic analysis of after-action reports: A plan evaluation methodological approach. *Natural Hazards Review, 22*(4), 06021004. https://doi.org/10.1061/(ASCE)NH.1527-6996.0000519

Kondo, M. C., Low, S. C., Henning, J., & Branas, C. C. (2015). The impact of green stormwater infrastructure installation on surrounding health and safety. *American Journal of Public Health, 105*(3), e114–e121.

Kong, K. W. K. (2021). Double corrosion protection of sewage tunnel linings. *Australian Journal of Civil Engineering, 20*(1), 156–173.

Laspidou, C. S., Mellios, N. K., Spyropoulou, A. E., Kofinas, D. T., & Papadopoulou, M. P. (2020). Systems thinking on the resource nexus: Modeling and visualisation tools to identify critical interlinkages for resilient and sustainable societies and institutions. *Science of the Total Environment, 717*, 137264.

Lee, P.-C., Wang, Y., Lo, T.-P., & Long, D. (2018). An integrated system framework of building information modelling and geographical information system for utility tunnel maintenance management. *Tunnelling and Underground Space Technology, 79*, 263–273.

Liu, Z., & Ota, K. (2017). *Smart technologies for emergency response and disaster management (Advances in public policy and administration)*. Information Science Reference.

Lnenicka, M., & Nikiforova, A. (2021). Transparency-by-design: What is the role of open data portals? *Telematics and Informatics, 61*, 101605.

Madryas, C., Kolonko, A., Szot, A., & Nienartowicz, B. (2014). *Underground infrastructure of urban areas 3*. CRC Press.

Marcillo-Delgado, J. C., Alvarez-Garcia, A., & García-Carrillo, A. (2022). Communication strategies on risk and disaster management in South American countries. *International Journal of Disaster Risk Reduction, 76*, 102982.

McGranaghan, M., Houseman, D., Schmitt, L., Cleveland, F., & Lambert, E. (2016). Enabling the integrated grid: Leveraging data to integrate distributed resources and customers. *IEEE Power and Energy Magazine, 14*(1), 83–93.

Murphy, K. M., Weaver, C., & Berg, L. J. (2022). *Janeway's Immunobiology* (10th ed.). W. W. Norton & Company.

Nel, D., Plessis, C. D., & Landman, K. (2018). Planning for dynamic cities: Introducing a framework to understand urban change from a complex adaptive systems approach. *International Planning Studies, 23*(3), 250–263.

Neves, F. T., de Castro Neto, M., & Aparicio, M. (2020). The impacts of open data initiatives on smart cities: A framework for evaluation and monitoring. *Cities, 106*, 102860.

Niekerk, V., Coetzee, D., Botha, C., et al. (2015). Planning and executing scenario based simulation exercises: Methodological lessons. *Journal of Homeland Security and Emergency Management, 12*(1), 193–210.

Nik-Bakht, M., & El-Diraby, T. E. (2020). Beyond chatter: Profiling community discussion networks in urban infrastructure projects. *Journal of Infrastructure Systems, 26*(3), 05020006. https://doi.org/10.1061/(ASCE)IS.1943-555X.0000555

Nkyekyer, E. W., & Dannenberg, A. L. (2018). Use and effectiveness of health impact assessment in the energy and natural resources sector in the United States, 2007–2016. *Impact Assessment and Project Appraisal, 37*(1), 17–32.

Oh, E. H., Deshmukh, A., & Hastak, M. (2012). Criticality assessment of lifeline infrastructure for enhancing disaster response. *Natural Hazards Review, 14*(2), 98–107. https://doi.org/10.1061/(ASCE)NH.1527-6996.0000084

Pang, B., & Li, H. (2021). A review of real-time energy consumption and carbon dioxide emission monitoring system in ports. In *2021 6th International Conference on Transportation Information and Safety (ICTIS), Wuhan, China* (pp. 273–277). IEEE. https://doi.org/10.1109/ICTIS54573.2021.9798560

Parker, G. W. (2020). Best practices for after-action review: Turning lessons observed into lessons learned for preparedness policy. *Revue Scientifique et Technique (International Office of Epizootics), 39*(2), 579–590. https://doi.org/10.20506/rst.39.2.3108

Perera, A. T. D., & Hong, T. (2023). Vulnerability and resilience of urban energy ecosystems to extreme climate events: A systematic review and perspectives. *Renewable and Sustainable Energy Reviews, 173*, 113038.

Peters, B. G., Jordan, A., & Tosun, J. (2017). Over-reaction and under-reaction in climate policy: An institutional analysis. *Journal of Environmental Policy & Planning, 19*(6), 612–624.

Qadir, S. A., Ahmad, F., Al-Motairi, H., et al. (2024). Renewable energy incentives for project owners representing communities with heavily subsidised fossil fuel-based electricity. *Sustainable Cities and Society, 101*, 105152. https://doi.org/10.1016/j.scs.2023.105152

Qudrat-Ullah, H. (2024). A framework for developing and implementing FIT policies for renewable energy based on local economic conditions. *Sustainable Futures, 7*, 100170.

Rahman, H. A., Majid, M. S., Jordehi, A. R., Gan, C. K., Hassan, M. Y., & Fadhl, S. O. (2015). Operation and control strategies of integrated distributed energy resources: A review. *Renewable and Sustainable Energy Reviews, 51*, 1412–1420.

Ramyar, R., Ackerman, A., & Johnston, D. M. (2021). Adapting cities for climate change through urban green infrastructure planning. *Cities, 117*, 103316.

Rathore, P. K. S., Chauhan, D. S., & Singh, R. P. (2019). Decentralized solar rooftop photovoltaic in India: On the path of sustainable energy security. *Renewable Energy, 131*, 297–307.

Sadeghi-Bazargani, H., Golestani, M., Saadati, M., Samadirad, B., Azami-Aghdash, S., & Jafari-Khounigh, A. (2024). A community-based safety reporting system; development, implementation, and evaluation: An experience of safe communities in Iran. *Medical Journal of the Islamic Republic of Iran, 38*(1), 81–87.

Sanderink, L., & Nasiritousi, N. (2020). How institutional interactions can strengthen effectiveness: The case of multi-stakeholder partnerships for renewable energy. *Energy Policy, 141*, 111447.

Shah, I., Mahmood, T., Khan, S. A., Elahi, N., Shahnawaz, M., Dogar, A. A., Subhan, F., & Begum, K. (2022). Inter-agency collaboration and disaster management: A case study of the 2005 earthquake disaster in Pakistan. *Journal of Disaster Risk Studies, 14*(1), 1088. https://journals.co.za/doi/epdf/10.4102/jamba.v14i1.1088

Shah, S. A., Seker, D. Z., Hameed, S., & Draheim, D. (2019). The rising role of big data analytics and IoT in disaster management: Recent advances, taxonomy and prospects. *IEEE Access, 7*, 54595–54614.

Shahrour, I., Bian, H., Xie, X., & Zhang, Z. (2021). Smart technology applications for the optimal management of underground facilities. *Underground Space, 6*(5), 551–559.

Sokołowski, M. M., & Heffron, R. J. (2022). Defining and conceptualising energy policy failure: The when, where, why, and how. *Energy Policy, 161*, 112745. https://doi.org/10.1016/j.enpol.2021.112745

Strumińska-Kutra, M., Dembek, A., Hielscher, S., & Stadler, M. (2023). Innovating urban governance for sustainable energy transitions: Between institutional design and institutional adaptation. *Environmental Innovation and Societal Transitions, 48*, 100751.

Suboticki, I., Heidenreich, S., Ryghaug, M., Skjølsvold, T.M. (2023). Fostering justice through engagement: A literature review of public engagement in energy transitions. Energy Research & Social Science, 99, 103053. https://doi.org/10.1016/j.erss.2023.103053.

Tian, J., Yu, L., Xue, R., Zhuang, S., & Shan, Y. (2022). Global low-carbon energy transition in the post-COVID-19 era. *Applied Energy, 307*, 118205.

Tuballa, M. L., & Abundo, M. L. (2016). A review of the development of Smart Grid technologies. *Renewable and Sustainable Energy Reviews, 59*, 710–725.

UN. (2018). Majority of the world's cities highly exposed to disasters, UN DESA warns on World Cities Day. https://www.un.org/tr/desa/majority-world%E2%80%99s-cities-highly-exposed-disasters-un-desa-warns-world-cities-day

UNDP. (2024). Cities have a key role to play in tackling climate change—Here's why. https://climatepromise.undp.org/news-and-stories/cities-have-key-role-play-tackling-climate-change-heres-why

UNECE. (2024). Case study: Balancing the electricity supply and demand with Artificial Intelligence. Task force on digitalization in energy, group of experts on energy efficiency, United Nations Economic Commission for Europe. https://unece.org/sites/default/files/2024-10/electricity.balancing.AI_case.study_.2024.pdf

Voeten, E. (2022). Is the public backlash against globalization a backlash against legalization and judicialization? *International Studies Review, 24*(2), viac015. https://doi.org/10.1093/isr/viac015

Wang, J., He, Z., & Weng, W. (2020). A review of the research into the relations between hazards in multi-hazard risk analysis. *Natural Hazards, 104*, 2003–2026.

Wex, F., Schryen, G., Feuerriegel, S., & Neumann, D. (2014). Emergency response in natural disaster management: Allocation and scheduling of rescue units. *European Journal of Operational Research, 235*(3), 697–708.

Young, S., Balluz, L., & Malilay, J. (2004). Natural and technologic hazardous material releases during and after natural disasters: A review. *Science of the Total Environment, 322*(1–3), 3–20.

Zhang, S., Shang, C., Fang, X., He, S., Yu, L., Wang, C., & Yan, L. (2020). Wireless monitoring–based real-time analysis and early-warning safety system for deep and large underground caverns. *Journal of Performance of Constructed Facilities, 35*(2), 04020147. https://doi.org/10.1061/(ASCE)CF.1943-5509.0001534

Zhang, X., Yin, L., Chen, J., et al. (2023). A study of the multi-tiered risk assessment method of site groundwater contamination considering transport-induced risk. *Hydrogeology & Engineering Geology, 50*(2), 160–170.

Chapter 4
Driving *Urbanimmunology* Forward with Integrated Perspectives

4.1 Deriving Core Principles of *Urbanimmunology*

In the previous chapters, human immune intelligence has been widely applied to address various urban and social challenges, demonstrating its potential to guide future developments and yielding a wealth of novel insights. While these applications span diverse contexts, common patterns and underlying mechanisms frequently emerge. These recurring observations suggest that, despite the differences in urban and social environments, there are shared principles that can inform sustainable development of cities and society.

To bridge these insights with a more systematic framework, we now turn to the task of deriving core principles for establishing *Urbanimmunology* as a robust scientific system. By synthesizing these common observations, we identify nine key principles: Critical Dependency, Avoidance, Tolerance, Adaptation, Regulation, Coordination, Diversity, Robustilient Regeneration, and Coherence.

Though these principles are still in development, they serve as a foundation for understanding how immune system concepts and mechanisms can inspire urban and social development. Together, they form a framework for shaping urban and social systems that not only tackle present challenges but also remain adaptable and responsive to future demands. By incorporating these principles, cities and society can better manage a range of pressures—from environmental to social—while maintaining a balance between growth and sustainability. Ultimately, these principles provide a path forward for ensuring that urban and social systems are capable of preventing, managing, and recovering from disruptions, fostering long-term vitality and resilience in an ever-evolving world.

J. Zhang, *Urbanimmunology*, Urban Sustainability,
https://doi.org/10.1007/978-981-95-4878-1_4

4.1.1 Critical Dependency Principle—Building Essential Urban and Social Systems for Sustained Existence

In immunology, the proper functioning of the immune system relies on two critical components: the bone marrow, which produces immune cells, and the thymus, which matures T cells and ensures their functionality. Without either, the immune system would collapse, leaving the body defenseless against threats. Similarly, urban and social systems depend on their most fundamental structures to support all other activities and responses. These foundational systems should work in harmony to enable cities and society to adapt, respond, and thrive. The *Critical Dependency Principle* underscores that just as the immune system relies on the bone marrow and thymus for survival and function, cities and society depend on their core systems and governance structures. Without these dependencies, urban and social systems would fail to maintain resilience, adaptability, and long-term prosperity.

4.1.1.1 Essential Production (Urban and Social "Bone Marrow")

Just as the bone marrow is responsible for producing immune cells, urban and social systems require core infrastructure—such as transportation, energy, water, and waste management systems—to generate the resources and services that sustain urban and social life. Without these essential systems, a city or society cannot function.

4.1.1.2 Functional Maturation (Urban and Social "Thymus")

In parallel to the thymus, cities and society need mechanisms to refine and coordinate their foundational systems. Governance structures, planning institutions, and regulatory bodies ensure that the core systems are optimized, calibrated, and prepared to handle challenges. These mechanisms mature the responses of both the city and society, ensuring that its fundamental systems function cohesively and effectively.

4.1.2 Avoidance Principle—Arresting Development

In immunology, avoidance mechanisms are designed to prevent exposure to pathogens, thereby reducing the likelihood of infection. These mechanisms include both anatomic barriers—physical structures that prevent pathogen entry—and behavioral modifications—actions that reduce exposure to harmful environments or situations. Similarly, in urban and social systems, avoidance mechanisms can be applied to minimize risks from various disruptions, such as environmental hazards, disease outbreaks, or social disturbances, by integrating both structural and systemic strategies.

4.1.2.1 Physical Avoidance

In urban and social systems, physical avoidance is achieved through infrastructure and design that limit exposure to potential threats. This includes creating protective environments that reduce the impact of environmental or systemic disruptions and developing resilient urban and social infrastructure that can withstand or minimize harm from external risks.

4.1.2.2 Behavioral Avoidance

On a broader scale, cities and society can implement behavioral modifications that guide the actions of individuals or systems to avoid high-risk situations. This involves fostering proactive strategies, data-driven decision-making, and adaptive policies that help communities avoid exposure to harmful events or conditions. By influencing behavior through information, planning, and response mechanisms, urban and social systems can effectively reduce vulnerability to disruptions.

4.1.3 Tolerance Principle—For Navigating Dilemmas

Tolerance, as a strategy in immunology, refers to mechanisms that enhance a tissue's ability to withstand and limit damage caused by microbial infections, rather than directly targeting the pathogens themselves. This strategy focuses on minimizing the negative impact of infection on the host, even when pathogens are present. In urban and social systems, tolerance can serve as a vital principle for maintaining functionality and resilience in the face of disruptions or dilemmas. Rather than focusing solely on preventing or resisting challenges, cities and society can adopt strategies that enable them to absorb and adapt to impacts, minimizing harm and ensuring continuity. This approach mirrors the immune system's focus on preserving tissue integrity and overall function despite the presence of stressors.

4.1.3.1 Physical Tolerance

Urban and social systems can enhance their capacity to withstand damage by adopting designs and materials that prioritize durability and adaptability. For example, resilient infrastructure, such as flood-tolerant buildings, permeable pavements, and flexible utility networks, allows cities and society to maintain functionality during environmental stressors like extreme weather or natural disasters. By integrating repairable and modular designs, cities and society can recover quickly from damage and reduce long-term vulnerabilities.

4.1.3.2 Systemic Tolerance

Cities and society can build systemic tolerance by fostering redundancy and flexibility within their networks. For instance, decentralized energy systems, diverse transportation options, and distributed water supply networks ensure that critical services remain operational even when parts of the system are disrupted. These adaptive measures enable cities and society to sustain essential functions and avoid catastrophic failures during crises.

4.1.3.3 Community Tolerance

Just as tissues adapt to minimize damage, communities can develop tolerance to social and economic disruptions. Initiatives like affordable housing, universal healthcare, and robust social safety nets help communities absorb shocks such as economic downturns or health crises without significant harm to their well-being. Strengthening community resilience enhances societal stability and reduces the overall impact of disruptions.

4.1.3.4 Environmental Tolerance

Urban and social systems can tolerate environmental stressors by implementing strategies that minimize harm to ecosystems and enhance their capacity to recover. Green infrastructure, urban forests, and nature-based solutions improve air and water quality, reduce heat islands, and mitigate the effects of pollution and climate change. These measures allow cities and society to coexist sustainably with their environment, even under stress.

4.1.4 Adaptation Principle—Flexibility Triumphs Over Rigidity

In immunology, adaptation refers to the immune system's ability to learn from prior experiences and enhance its response to future threats. This dynamic process involves recognizing patterns, refining responses, and improving efficiency over time. Similarly, in urban and social systems, adaptation can serve as a principle for fostering resilience and responsiveness to evolving challenges by incorporating mechanisms for learning, innovation, and continuous improvement.

4.1.4.1 Systemic Adaptation

Urban and social systems can implement adaptive strategies by continuously analyzing data, learning from past disruptions, and refining operations to improve future outcomes. This includes integrating real-time feedback loops, predictive analytics, and scenario modeling to adjust policies, resource allocation, and operational protocols in response to emerging challenges.

4.1.4.2 Structural Adaptation

Cities *and society* can embed flexibility into their physical and organizational structures to enable dynamic adjustments to changing conditions. This includes designing infrastructure with modular and scalable components, as well as fostering cross-sector collaboration that enables collective problem-solving and resource-sharing during crises.

4.1.4.3 Cultural Adaptation

At a societal level, fostering a culture of adaptability enables urban and social communities to embrace innovation and remain agile in the face of uncertainty. This involves promoting continuous learning, cultivating diverse skillsets, and encouraging proactive engagement with emerging trends and technologies.

4.1.5 Regulation Principle—Strengthening Cities and Society Through Self-Discipline

In immunology, regulation mechanisms ensure that immune responses are precisely calibrated to the challenges at hand, *upregulating* activity to fight infections and *downregulating* it to prevent harmful overreactions, such as autoimmune diseases. This balance protects the host while minimizing collateral damage. Similarly, in urban and social systems, regulation plays a vital role in maintaining balance, adaptability, and stability amid disruptions. By fine-tuning responses to challenges, cities and society can optimize resource use, ensure equity, and enhance resilience across various domains. As a foundational principle of *Urbanimmunology*, regulation emphasizes the importance of precision, adaptability, and systemic balance, enabling cities and society to remain functional and responsive in the face of complexity and uncertainty.

4.1.5.1 Infrastructure Regulation

Urban and social infrastructure regulation mirrors the immune system's dynamic adjustment mechanisms by employing adaptive and scalable designs that respond to changing demands or disruptions. For example, smart grids and energy systems adjust power supply in real-time to prevent overloads or shortages. Flood management systems dynamically control water flow through dams, retention basins, or permeable surfaces. These adaptive strategies ensure cities *and society* can balance growth and safety while maintaining functionality.

4.1.5.2 Behavioral and Policy Regulation

Cities and society can regulate community behaviors and systemic functions through targeted policies and interventions. Adaptive traffic management systems that control congestion during peak hours or emergencies demonstrate precision in addressing localized challenges. Similarly, zoning regulations that balance residential, commercial, and industrial areas prevent conflicts and inefficiencies. Evidence-based, flexible policymaking enables cities and society to adjust activity levels dynamically, reducing disruptions and fostering societal well-being.

4.1.5.3 Economic and Social Regulation

The immune system's balance between vigilance and restraint mirrors the efforts of cities and societies to regulate economic and social dynamics. Mechanisms like progressive taxation, social safety nets, and inclusive development initiatives act as regulatory tools to maintain cohesion and prevent destabilizing inequalities. By distributing resources equitably and reducing systemic stressors, cities and society ensure no part of the urban system is overstressed or vulnerable, fostering stability and adaptability over the long term.

4.1.5.4 Environmental Regulation

The immune system's self-regulation to prevent overreaction aligns with urban and social strategies for environmental management. Measures such as carbon emission controls, efficient waste management systems, and green building standards enable cities and society to operate within environmental thresholds while adapting to climate challenges. These practices mitigate harm and enhance urban and social sustainability, much like the immune system protects its host by preventing excessive damage during its responses.

4.1.6 Coordination Principle—For a Tacit and Trustful Future

In immunology, coordination is critical for mounting an effective defense, as various immune cells should work together in a synchronized manner to identify, respond to, and eliminate threats. This involves the seamless integration of diverse functions, signaling pathways, and responses. Similarly, in urban and social systems, coordination can be applied as a principle to enhance collective resilience and ensure an effective response to challenges by fostering synergy among different components, systems, and stakeholders as cohesive, adaptive entities, ensuring timely, efficient, and integrated responses to both immediate challenges and long-term changes.

4.1.6.1 Systemic Coordination

Urban and social systems can enhance coordination by integrating communication and operational frameworks across sectors, such as transportation, public health, energy, and emergency services. This ensures that resources, information, and actions are aligned during routine operations and crises, improving the overall efficiency and effectiveness of responses.

4.1.6.2 Infrastructural Coordination

Cities and society can design interconnected infrastructure systems that work in harmony to address challenges. For example, integrating energy grids, transportation networks, and communication systems allows for mutual support and quick recovery in the face of disruptions, ensuring that critical services remain operational.

4.1.6.3 Stakeholder Coordination

Collaboration among diverse urban and social stakeholders, including government agencies, private organizations, and local communities, is essential for creating a unified approach to problem-solving. Clear communication channels, shared goals, and coordinated efforts can ensure that actions taken by different entities complement rather than conflict with one another.

4.1.7 Diversity Principle—For a Risk-Averse Growing Future

In immunology, the principle of diversity is exemplified by the wide variety of leukocytes, each specialized to recognize and respond to different pathogens and threats. This diversity equips the immune system to adapt to a continuously evolving landscape of challenges, ensuring robust defense mechanisms. Similarly, in urban and social systems, diversity fosters adaptability, resilience, and innovation, enabling cities and society to address dynamic and complex challenges while enhancing their functional capacities.

4.1.7.1 Functional Diversity

Just as leukocytes include T cells, B cells, and macrophages, each with distinct roles in the immune response, urban and social systems thrive when they integrate diverse functions and services. A city or society that incorporates multi-functional infrastructure, such as green spaces doubling as flood control and public gathering spaces, mirrors the versatility of the immune system's components. This functional diversity ensures redundancy and adaptive capacity in urban and social operations.

4.1.7.2 Ecological and Demographic Diversity

Leukocytes evolve and diversify to counter new pathogens, emphasizing the importance of genetic variation in immunity. In urban and social contexts, ecological and demographic diversity strengthens cities and society by fostering inclusive and heterogeneous communities. These communities support innovation, creativity, and robust social networks, enabling cities and society to respond effectively to new challenges, such as economic shifts or cultural integration.

4.1.7.3 Spatial and Systemic Diversity

The immune system's ability to deploy specialized cells to targeted sites demonstrates the value of spatial diversity. Cities and society can emulate this by distributing critical services and infrastructure across different locations to avoid overreliance on centralized systems. Similarly, systemic diversity ensures that cities and society do not depend solely on a single mode of transportation, energy source, or economic sector. Diversified systems increase redundancy and flexibility, reducing vulnerabilities to targeted disruptions.

4.1.7.4 Evolutionary Diversity

The immune system continuously evolves, adapting to emerging threats through mutation and memory. Urban and social systems can adopt a similar approach by embracing adaptive planning and dynamic policies. Encouraging diverse perspectives in decision-making processes enhances the ability to anticipate and adapt to future challenges, ensuring sustained growth and innovation.

4.1.8 Robustilient Regeneration Principle—Ensuring Adaptive Recovery and Long-Term Vitality

In *Urbanimmunology*, the robustilient regeneration principle underscores the dynamic process by which urban and social systems not only recover from disruptions but also grow stronger and more adaptable in response to these challenges. It combines the immediate restorative actions required after damage with a forward-thinking approach to enhance the city's and society's ability to withstand and evolve amid future stresses.

4.1.8.1 Adaptive Recovery

Urban and social systems should be capable of recovering from physical, social, and economic shocks while improving their capacity for adaptation. This includes rebuilding and restoring damaged infrastructure, rejuvenating ecosystems, and revitalizing communities. However, recovery goes beyond simply "bouncing back"—it involves rethinking how cities and society can be rebuilt with enhanced resilience, ensuring that they are more robust, sustainable, and prepared for future disruptions.

4.1.8.2 Multidimensional Regeneration

Robustilient regeneration operates across multiple levels: physical (e.g., urban infrastructure and buildings), social (e.g., community networks, cultural dynamics), and environmental (e.g., ecosystems, biodiversity). By integrating these layers into the recovery process, cities and society can ensure a holistic regeneration that fosters long-term stability and interconnection between various urban and social functions.

4.1.8.3 Continuous Evolution

The regeneration process in urban and social systems is not static. It is an ongoing cycle of growth, learning, and adaptation. As cities and society recover, they do so with the intent of becoming more resilient to future challenges, incorporating innovations, sustainable practices, and proactive policies that anticipate future needs. This process fosters cities and society that continuously evolve and thrive, adapting not just to recover but to improve their capacity for handling future crises.

4.1.8.4 Empowering Urban Communities

Robustilient regeneration emphasizes the importance of community involvement in recovery and adaptation. Communities should be equipped with the knowledge, resources, and networks needed to support their recovery efforts and shape the evolution of cities and society. This includes fostering local leadership, building social cohesion, and ensuring inclusive and participatory governance.

The robustilient regeneration principle reflects the city's and society's ability to recover, adapt, and thrive, ensuring that urban and social systems can continuously evolve in response to both anticipated and unforeseen challenges. This integrated process promotes long-term urban vitality, empowering cities and society to grow in resilience while reinforcing their foundational strength for the future.

4.1.9 Coherence Principle—For the Harmonious Functioning of Cities and Society

Inspired by the immune system's ability to coordinate and adapt effectively, this principle highlights that the proper functioning of urban and social systems relies on their capacity to operate as an integrated, harmonious whole. It ensures alignment and mutual reinforcement among foundational components, emphasizing the following core aspects:

4.1.9.1 Systemic Interconnection

Urban and social systems should maintain structural and functional linkages, enabling seamless integration across their components. This interconnectedness ensures that each element is responsive to the others, fostering unity and preventing isolated disruptions. A well-connected system can more efficiently identify and address potential issues before they escalate, ensuring a smoother operation overall.

4.1.9.2 Communication and Feedback

Effective communication and feedback mechanisms are essential for ensuring that urban and social responses are timely, adaptive, and aligned with overarching objectives. Feedback loops allow systems to respond to changes dynamically, ensuring adjustments are made in real-time to address emerging needs. Transparent communication enhances collaboration among stakeholders, promoting shared understanding and decision-making that aligns with collective goals.

4.1.9.3 Balance

Urban and social systems require regulatory frameworks to ensure proportional and balanced responses to disruptions, maintaining systemic harmony. These regulations must be flexible enough to adapt to varying conditions while ensuring consistent and equitable outcomes. Well-designed frameworks help to mitigate extreme fluctuations, ensuring that the system remains stable even in the face of challenges.

4.1.9.4 Redundancy

Redundancy in critical urban and social functions ensures robustness, allowing systems to continue operating despite failures in individual components. This safety net minimizes vulnerabilities, making it easier to recover from disruptions without major consequences. Additionally, resilience planning prepares systems for unforeseen challenges, ensuring that even under stress, the system can maintain essential functions.

4.1.9.5 Learning

Urban and social systems should continuously learn from past experiences to refine their processes and improve their capacity to handle recurring or new challenges. By embracing a culture of learning, systems can evolve to meet changing needs and environments, reducing the risk of stagnation. Responses based on continual learning ensure that urban and social systems remain relevant and effective in the long term.

The *Coherence Principle* provides the foundation for *Urbanimmunology* to function as a scientific discipline by defining the criteria for integrated, adaptive, and resilient urban and social systems, ensuring a unified and systematic approach to urban and social development. This principle guides cities and society toward fostering sustainable growth while prioritizing the well-being of their inhabitants, creating environments where both urban infrastructure and communities can thrive together.

4.1.10 *Toward a Scientifically Sound and Practice-Ready* Urbanimmunology

Urbanimmunology offers a holistic approach to urban and social systems, drawing inspiration from the adaptive, self-regulating qualities of the human immune system. At its core, it emphasizes Critical Dependency, highlighting the need for essential urban and systems that ensure the city's and society's long-term survival, while Avoidance stresses the importance of halting unsustainable or hazardous growth to protect urban and social integrity. Through Tolerance, cities and society can navigate complex dilemmas, fostering an environment that embraces compromise and diverse perspectives. Adaptation calls for flexibility, encouraging cities and society to embrace change and innovation rather than be constrained by rigid structures. Regulation ensures that urban and social systems maintain balance—adjusting through upregulation or downregulation—to preserve public order and well-being. Coordination fosters seamless collaboration across sectors, creating a cohesive urban and social landscape built on trust and mutual understanding. The principle of Diversity encourages resilience in the face of uncertainty, nurturing varied approaches to growth, culture, and resources. Robustilient Regeneration emphasizes that cities and society should not only recover from disruptions but evolve to become more adaptable, strengthening their ability to withstand future challenges. Finally, Coherence ensures that all urban and social functions work harmoniously toward shared goals. Together, these principles present a model where cities and society act as dynamic, self-organizing entities—thriving in complexity, learning from challenges, and continually adapting to both internal and external shifts, much like living organisms.

4.2 Features of Truly Scientific Systems

J. R. Kantor (1958), a prominent psychologist, outlined a framework for what constitutes a robust scientific system, emphasizing three critical attributes: validity, significance, and comprehensiveness (Fryling & Hayes, 2010). These attributes collectively define the quality and utility of a scientific discipline, shaping its capacity to contribute meaningfully to knowledge and practice.

Validity refers to the internal consistency of a scientific system, emphasizing the absence of contradictions within its theoretical framework and methodologies. A valid scientific system ensures that its core principles, assumptions, and derived propositions align logically and cohesively. This internal coherence allows the system to withstand critical scrutiny, forming a stable foundation for further theoretical and practical advancements. Without validity, a scientific system risks internal disarray, undermining its reliability and the trust of the broader scientific community.

Significance describes the relational value of a scientific system within the broader scientific landscape. A system achieves significance when its meta-assumptions—fundamental principles shared by multiple systems—resonate with those of other scientific disciplines. This shared foundation facilitates productive interactions, cross-disciplinary integration, and the mutual enrichment of knowledge. However, significance is contingent on validity; a scientific system should first ensure internal consistency before engaging effectively with others. While validity lays the groundwork, significance positions a scientific system within the interconnected web of scientific inquiry, enhancing its influence and applicability.

Comprehensiveness pertains to the system's capacity to account for the full range of phenomena within its defined scope. A comprehensive scientific system offers explanations that encompass all relevant events, variables, and dynamics related to its domain of study. This attribute ensures that the system is not overly narrow or exclusionary, allowing it to provide a holistic understanding of complex phenomena. However, comprehensiveness is meaningful only when grounded in a system that is both valid and significant. A comprehensive approach without validity risks incoherence, while one lacking significance may isolate the system from broader scientific discourse.

Kantor's criteria collectively highlight the interconnectedness of these attributes. Validity establishes the foundation, significance situates the system within a wider intellectual context, and comprehensiveness ensures that the system provides a thorough and meaningful account of its subject matter. Together, these qualities define the essence of a robust scientific system, guiding its development and ensuring its relevance in advancing knowledge and addressing real-world challenges.

4.3 Relevance of *Urbanimmunology*'s Core Principles to Kantor Criteria

To help *Urbanimmunology* become a robust scientific system, we can examine how its nine core principles align with J. R. Kantor's (1958) three criteria for a scientific system: **validity**, **significance**, and **comprehensiveness**.

4.3.1 Relevance of Critical Dependency Principle to Kantor Criteria

Validity: This principle ensures that urban systems rely on foundational structures to function effectively, thus creating internal coherence. It establishes a reliable basis for further exploration by identifying the critical components necessary for resilience and adaptation, ensuring that *Urbanimmunology* is grounded in solid, valid assumptions about system interdependencies.

Significance: By drawing parallels with the immune system's critical components, it aligns *Urbanimmunology* with broader biological and ecological theories, establishing significance across fields like ecology, sociology, and systems theory. The application of this principle can influence various scientific disciplines interested in complex adaptive systems.

Comprehensiveness: This principle is comprehensive in that it acknowledges the complex interrelations between core urban and social systems, ensuring the discipline addresses a wide range of phenomena related to resilience, adaptation, and governance.

4.3.2 Relevance of Avoidance Principle to Kantor Criteria

Validity: The avoidance of risks is internally consistent with *Urbanimmunology*, as it mirrors immune responses to pathogens by creating urban and social systems that anticipate and mitigate potential disruptions. This principle ensures logical coherence by prioritizing proactive measures over reactive ones.

Significance: This principle can resonate with a variety of disciplines concerned with risk management, urban planning, public health, and environmental science. Its application positions *Urbanimmunology* within a broader, interdisciplinary context, enhancing its significance.

Comprehensiveness: It addresses the need to consider a wide array of potential disruptions—social, environmental, and health-related—making it comprehensive in its scope. The principle suggests strategies that prevent risk while maintaining flexibility and resilience.

4.3.3 Relevance of Tolerance Principle to Kantor Criteria

Validity: This principle's consistency with biological systems, which adapt to threats without overreacting, supports the internal logic of *Urbanimmunology*. It underscores the importance of adaptation, with urban systems learning to absorb and minimize disruptions instead of overextending their resources.

Significance: It enhances significance by linking social and urban systems' ability to cope with challenges to biological resilience, making it relevant to the broader scientific community, especially in disciplines focused on resilience and sustainability.

Comprehensiveness: This principle broadens the scope of *Urbanimmunology* by recognizing that not all challenges require elimination; some may be absorbed or mitigated. This approach makes it applicable to a range of phenomena, from natural disasters to social unrest.

4.3.4 Relevance of Adaptation Principle to Kantor Criteria

Validity: Adaptation is foundational to both the immune system and urban resilience, ensuring internal coherence within *Urbanimmunology*. This principle's emphasis on learning and evolving over time fits well with the discipline's focus on dynamic systems and their ability to adjust to changing conditions.

Significance: The concept of adaptation is significant because it links *Urbanimmunology* with theories of evolution, systems theory, and ecological resilience. It helps position *Urbanimmunology* as a key player in understanding the long-term survival of urban environments.

Comprehensiveness: By encouraging continuous learning and improvement, this principle expands the range of phenomena *Urbanimmunology* can address, from climate change to socio-economic shifts, ensuring it has broad applicability.

4.3.5 Relevance of Regulation Principle to Kantor Criteria

Validity: The regulation principle ensures balance within urban systems, avoiding both overreaction and neglect. Its logical consistency supports the validity of *Urbanimmunology*, emphasizing the importance of maintaining equilibrium within complex systems.

Significance: This principle aligns *Urbanimmunology* with other systems-focused disciplines (such as ecology, economics, and public health), enhancing its significance by contributing to interdisciplinary dialogues about balance and sustainability.

Comprehensiveness: Regulation is a comprehensive principle because it deals with the fine-tuning of responses across various domains—social, environmental, and economic—ensuring the urban system is neither too rigid nor too chaotic.

4.3.6 Relevance of Coordination Principle to Kantor Criteria

Validity: Coordination within urban systems mirrors immune responses, where various cells and mechanisms work together for a common purpose. The internal coherence of *Urbanimmunology* is enhanced by this principle, which ensures that diverse system components work in harmony.

Significance: The principle of coordination positions *Urbanimmunology* within a broader context, encouraging interdisciplinary collaboration, particularly in fields concerned with complex system integration, such as urban planning, sociology, and disaster management.

Comprehensiveness: By emphasizing collaboration and mutual reinforcement, this principle ensures that the discipline addresses not only individual components but the complex interdependencies of urban and social systems, fostering a holistic view.

4.3.7 Relevance of Diversity Principle to Kantor Criteria

Validity: Diversity, as an essential component of both biological and urban systems, maintains internal consistency by ensuring that urban systems are equipped to handle a wide range of challenges. This principle supports the resilience and adaptability of the system, which is crucial for *Urbanimmunology*'s validity.

Significance: The diversity principle connects *Urbanimmunology* to broader discussions on biodiversity, social diversity, and innovation, making it significant in terms of how cities can grow and evolve in a complex, interconnected world.

Comprehensiveness: Diversity ensures that *Urbanimmunology* comprehensively addresses the various components of urban systems, including social, economic, and environmental factors, which need to work together to mitigate risks and foster growth.

4.3.8 Relevance of Robustilient Regeneration Principle to Kantor Criteria

Validity: This principle's emphasis on both recovery and forward-thinking aligns with the internal consistency of *Urbanimmunology*, providing a well-rounded approach to system resilience. It ensures that cities are not only able to recover but also grow stronger after disruptions.

Significance: It enhances the significance of *Urbanimmunology* by linking it to larger themes of regeneration and sustainability, which are central to environmental science, economics, and urban studies.

Comprehensiveness: Robustilient regeneration ensures that *Urbanimmunology* addresses both the immediate recovery and the long-term vitality of urban systems, making it a comprehensive approach to urban resilience.

4.3.9 Relevance of Coherence Principle to Kantor Criteria

Validity: The coherence principle ensures that all aspects of urban and social systems are working in harmony, which is essential for maintaining validity within the scientific system of *Urbanimmunology*. Without coherence, the system would lack the internal consistency necessary for sound conclusions and predictions.

Significance: This principle underscores the importance of systemic integration, making *Urbanimmunology* relevant across disciplines concerned with complex system functioning, including systems theory, sociology, and public policy.

Comprehensiveness: By emphasizing the harmonious integration of all components, the coherence principle ensures that *Urbanimmunology* addresses the full spectrum of challenges that cities and societies face, from governance to infrastructure to public health.

4.4 Shortcomings of Existing Urban Theories

The core tenets of existing urban theories (e.g., Jacobs, 1992; Fainstein, 2005; Storper & Manville, 2006; Fainstein & Campbell, 2011; Edensor & Jayne, 2012; Lin & Mele, 2013; Harding & Bloklan, 2014; Scott & Storper, 2014; Short, 2014; Derickson, 2015; Robinson & Roy, 2015; Jayne & Ward, 2017; Jazeel, 2018; Chakrabarti, 2023) are briefly summarized, with a discussion of their shortcomings in relation to Kantor's criteria—a perspective that is less explored in the existing literature. Existing urban theories are not static; they evolve over time. For example, in Urban Ecology, theories and disciplines from other fields—such as urban design, architecture, economics, sociology, anthropology, climate science, environmental education, complexity science, resilience theory, and sustainability science—have been integrated to enhance the field (e.g., McPhearson et al., 2016). However, this raises the question: does such integration justify renaming the theory? This analysis reflects the author's current understanding and expertise, and invites critique from experts in the field. Nevertheless, the author recognizes the value in each theory, which will be leveraged to strengthen Kantor's criteria for *Urbanimmunology*, as shown in the next subsection. These insights collectively contribute to the advancement of more scientifically rigorous urban theories. Here, 21 existing urban theories are selected.

4.4.1 Validity-Related Shortcomings

Urban Ecology (Ramalho & Hobbs, 2012) naturalizes social inequalities like gentrification as inevitable outcomes of "ecological succession," disregarding underlying power imbalances. *Central Place Theory* (Mulligan et al., 2012) relies on economic determinism, neglecting cultural, political, and historical factors that shape urban development. Similarly, *Neo-Marxist Urban Theory* (Jaret, 1983) focuses too heavily on economic determinism, overlooking the role of cultural, environmental, and technological forces. These theories fail to integrate multiple dimensions, compromising their validity in explaining urban environments.

Urban Political Economy (Swanstrom, 1993) and *New Urban Economics* (Mills & MacKinnon, 1973), also fall short by overemphasizing macroeconomic factors.

Urban Political Economy neglects local and cultural influences, while *New Urban Economics* assumes rational markets without considering discrimination or power imbalances. *World-Systems Theory* (Straussfogel, 1997) focuses too much on global capitalism, ignoring local agency, and *Global City Theory* (Ancien, 2011) portrays cities as passive in global networks, neglecting the active role of local agencies. These theories lack sufficient consideration of local, human, and contextual factors, limiting their scientific validity.

Critical Urban Theory (Brenner et al., 2009), *Right to the City* (Merrifield, 2011), and *Actor-Network Theory* (de Munck, 2017) are criticized for lacking actionable frameworks. *Critical Urban Theory* offers insights into urban inequalities but lacks clear solutions. *Right to the City* has no operational frameworks, and *Actor-Network Theory* obscures power hierarchies by treating all actors as equal. These theories fail Kantor's validity criterion due to their lack of practical, actionable components essential for scientific inquiry.

Feminist Urban Theory (Peake et al., 2021) and *Postcolonial Urban Theory* (Roy, 2014) focus on specific issues but have limited scope. *Feminist Urban Theory* narrowly addresses gender, sidelining class and race, while *Postcolonial Urban Theory* offers context-specific insights but lacks generalizability. *New Urbanism* (Ellis, 2002) lacks empirical support for addressing inequality, and *Landscape Urbanism* (Thompson, 2011) prioritizes ecology over equity, potentially leading to green gentrification. *Creative Class Theory* (Nathan, 2008) promotes exclusionary growth, exacerbating urban inequality, highlighting its failure to integrate essential social dimensions.

Theories like *Urban Resilience Theory* (Kong et al., 2022), *Everyday Urbanism* (Kelbaugh, 2000), and *Informal Urbanism* (Acuto et al., 2019) also struggle to scale or address systemic challenges. *Urban Resilience Theory* depoliticizes vulnerabilities, *Everyday Urbanism* lacks scalability, and *Informal Urbanism* romanticizes informality while downplaying structural issues. *Post Urbanism* (Kelbaugh, 2000) presents abstract critiques without practical frameworks, and *Theory of Urban Sustainability* (Childers et al., 2014) faces challenges in practical implementation. These theories fail Kantor's criterion by not addressing larger systemic structures or providing scalable, real-world solutions.

4.4.2 Significance-Related Shortcomings

Urban Ecology overlooks class struggle, racial discrimination, and political influence in urban development. *Central Place Theory* marginalizes socio-cultural influences like community ties, and *Neo-Marxist Urban Theory* fails to engage with non-economic perspectives, such as gender and postcolonialism. These theories lack the complexity needed to fully understand urban dynamics, diminishing their significance.

Other theories focus too narrowly on economic factors and fail to integrate broader social forces. *Urban Political Economy* neglects cultural dimensions like community resilience, and *New Urban Economics* prioritizes market efficiency over equity, overlooking displacement. *World-Systems Theory* is disconnected from urban design, neglecting the human-scale needs of urban spaces. These limitations reduce the significance of these theories in capturing the complexities of urban systems.

Some theories also suffer from weak interdisciplinary integration. *Critical Urban Theory* has limited engagement with technological innovations and policymaking, while *Urban Morphology* ignores social sciences, such as governance and inequality. *Global City Theory* overlooks the contributions of informal economies. Without connecting with these crucial areas, these theories fail to provide a comprehensive understanding of urban processes.

Theories grounded in social justice or critical perspectives also face limitations. *Feminist Urban Theory* lacks connections to economic and ecological frameworks, and *Postcolonial Urban Theory* remains marginalized in policy discussions. *Actor-Network Theory* lacks ethical and political frameworks for justice, reducing its ability to address inequality. These theories often lack the broad applicability needed for significance.

Theories on governance and resilience also fall short. *Right to the City* is disconnected from technological and ecological innovations, and *Urban Resilience Theory* has weak integration with political economy. *Landscape Urbanism* overlooks political-economic drivers in favor of ecological concerns. These gaps prevent these theories from fully addressing contemporary urban challenges.

Finally, theories like *Everyday Urbanism* and *Creative Class Theory* are often marginalized or overlook key urban factors. *Everyday Urbanism* is not discussed in policy circles, and *Creative Class Theory* overlooks non-creative labor. *Post Urbanism* offers limited actionable policy insights, and *Informal Urbanism* is disconnected from formal planning systems. *Theory of Urban Sustainability* prioritizes global goals over local needs, undermining its practical significance. These theories fail to meet Kantor's significance criterion due to their lack of actionable, locally relevant solutions.

4.4.3 Comprehensiveness-Related Shortcomings

Urban Ecology overlooks socio-political drivers like colonialism and capitalism, while *Central Place Theory* excludes informal economies like grassroots markets. *Neo-Marxist Urban Theory* ignores informal economies and non-capitalist practices in the Global South. These exclusions weaken the ability of these theories to offer a comprehensive view of urban systems.

Several theories also fail to account for grassroots initiatives and marginalized perspectives. *Urban Political Economy* neglects these voices, while *New Urban Economics* excludes informal economies. *World-Systems Theory* ignores micro-level adaptations, particularly informal economies. These gaps limit the ability of these theories to capture the complexity of urban processes.

Other theories lack concrete tools for measurement, limiting their practical applicability. *Critical Urban Theory* lacks metrics for assessing progress, and *Urban Morphology* neglects socio-economic drivers like gentrification. *Global City Theory* focuses on elites and global networks while neglecting urban peripheries and informal economies. These limitations prevent these theories from fully engaging with the forces shaping urban environments.

Theories focused on social justice and cultural critique also show shortcomings. *Feminist Urban Theory* overlooks intersectionality, particularly the experiences of low-income women of color, while *Postcolonial Urban Theory* ignores synergies between formal and informal urban practices. These gaps prevent these theories from providing a fully integrated understanding of urban systems.

Several theories fail to account for structural inequities and broader economic and ecological interdependencies. *Actor-Network Theory* neglects structural inequities like racism and colonialism, and *Right to the City* overlooks the economic-ecological interdependencies in urban systems. These theories don't provide a full understanding of the broader forces at play in urban environments.

Theories related to urban design and sustainability also face challenges in comprehensiveness. *New Urbanism* excludes marginalized populations from design considerations, and *Urban Resilience Theory* neglects historical and cultural contexts. *Landscape Urbanism* focuses on ecology while ignoring market forces, and *Everyday Urbanism* neglects global economic forces. These omissions restrict the scope of these theories in capturing urban dynamics.

Theories on informality and sustainability, such as *Informal Urbanism* and *Theory of Urban Sustainability*, also struggle with comprehensiveness. *Informal Urbanism* neglects the influence of global market dynamics, while *Theory of Urban Sustainability* overlooks grassroots solutions. These gaps limit their applicability to real-world urban issues.

4.5 Enhancing Theoretical Foundations of *Urbanimmunology* by Integrating with Advantages of Existing Urban Theories

Although *Urbanimmunology* is proposed to address both urban and social issues, many social challenges are closely linked to urbanization and cities. It is important to emphasize that urban theories themselves often tackle a range of social issues. As such, this subsection focuses solely on existing urban theories. While these theories have their limitations, they also provide valuable insights for addressing specific urban and social challenges. These strengths merit further exploration to enhance the validity, significance, and comprehensiveness of *Urbanimmunology*, transforming it into a robust scientific framework capable of systematically addressing urban and social issues. The following section highlights selected components from existing urban theories, demonstrating how these examples can contribute to the advancement of *Urbanimmunology* by addressing the interconnected physical, social, economic, and environmental factors that shape urban and social systems.

4.5.1 Enhancing the Validity of Urbanimmunology

To enhance the validity of *Urbanimmunology* as a truly scientific system based on Kantor's criteria, it is essential to integrate theories that provide empirical frameworks, robust methodologies, and analytical tools for accurately assessing urban and social systems across multiple dimensions.

A framework grounded in ecological science allows for a deeper understanding of urban ecosystems, thereby enhancing the accuracy of urban resilience assessments (Urban Ecology). By offering a spatial framework to understand the distribution of services and resources, resilience planning becomes more precise and better aligned with urban realities (Central Place Theory). Linking capitalism to urban inequality and vulnerabilities such as gentrification provides a valid analysis of systemic issues in urban health, making it a crucial element in assessing resilience (Neo-Marxist Urban Theory). Similarly, examining how political and economic structures impact urban resilience helps to understand the complex relationship between governance and recovery, ensuring the inclusion of all relevant factors (Urban Political Economy).

Using market equilibrium models to explain urban resource distribution enhances validity by offering analytical tools for understanding urban economic dynamics and their influence on resilience outcomes (New Urban Economics). The global-local dynamics of urban resilience are better understood when the impact of global economic systems on local urban structures is considered, validating the broader forces at play (World-Systems Theory). The analysis of power structures and their impact on urban vulnerability strengthens the validity of resilience strategies by incorporating socio-political underpinnings into the framework (Critical Urban Theory). Additionally, a systematic approach to urban layouts and their effects on resilience adds to the validity by focusing on the built environment and how its organization influences response capabilities (Urban Morphology).

By examining how cities and society participate in global economic networks, urban resilience becomes more comprehensively understood in the context of their international roles, making global connections crucial for resilience planning (Global City Theory). Incorporating gendered exclusions and recognizing the need for inclusive design enhances the validity of resilience strategies by accounting for marginalized groups and ensuring equitable outcomes (Feminist Urban Theory). The focus on context-specific urban strategies and local vulnerabilities challenges universalist assumptions, contributing to the validity of urban resilience models across diverse global contexts (Postcolonial Urban Theory). Mapping complex interrelations between human and non-human actors within urban systems enhances the validity by expanding resilience analysis beyond traditional models (Actor-Network Theory).

A normative framework rooted in social justice principles ensures the validity of urban resilience strategies by prioritizing fair and just outcomes for all residents, particularly the most vulnerable (Right to the City). Concrete design principles, such as mixed-use development, directly link urban design to health outcomes and resilience, reinforcing the practical validity of urban planning strategies (New Urbanism). Using scientific methodologies to measure urban resilience and assess a city's capacity to withstand and recover from disruptions further ensures the credibility of resilience evaluations (Urban Resilience Theory). Integrating ecological principles into urban design, with a focus on green infrastructure, enhances validity through evidence-based strategies for maintaining urban health (Landscape Urbanism).

Validating the role of micro-level practices, such as community gardening and mutual aid, in resilience strategies emphasizes the importance of everyday urban activities, ensuring that bottom-up approaches are considered (Everyday Urbanism). The link between creativity, innovation, and urban resilience highlights how cultural and economic dimensions shape a city's capacity to adapt, thus supporting the overall validity of resilience planning (Creative Class Theory). The incorporation of technological advancements and decentralized planning fosters flexible, adaptable urban systems, which strengthens the validity of modern resilience models (Post Urbanism). A more complete understanding of urban function is achieved by recognizing the importance of informal practices, further enriching the validity of resilience assessments beyond formal institutional frameworks (Informal Urbanism).

Finally, providing evidence-based strategies for sustainable urban development ensures that resilience is viewed through the lens of long-term urban health, emphasizing sustainability as a key pillar of valid resilience planning (Theory of Urban Sustainability).

4.5.2 Enhancing the Significance of Urbanimmunology

To enhance the significance of *Urbanimmunology* as a truly scientific system based on Kantor's criteria, it is crucial to integrate various theoretical perspectives that provide insights into urban and societal resilience from multiple dimensions—ecological, spatial, social, political, and economic. These theories contribute to a comprehensive understanding of the complex interrelations within urban and social systems, ensuring the relevance and robustness of *Urbanimmunology*.

Urban Ecology addresses critical interactions between urban areas and natural environments, emphasizing the ecological impacts of urbanization (Urban Ecology). Central Place Theory demonstrates how the spatial organization of resources impacts residents' access to essential services, underscoring its significance for urban resilience (Central Place Theory). Neo-Marxist Urban Theory connects *Urbanimmunology* to critiques of neoliberalism, highlighting economic forces that drive urban health crises (Neo-Marxist Urban Theory). Urban Political Economy highlights how wealth, power, and governance influence urban vulnerability, making it significant in understanding differential resilience within cities (Urban Political Economy). New Urban Economics links neoliberal policies to urban development, underscoring the significance of market-driven urbanization for resilience outcomes (New Urban Economics). World-Systems Theory shows how global economic trends create urban vulnerabilities, making it significant for understanding urban resilience in a global context (World-Systems Theory).

Critical Urban Theory advocates for equitable urban development and challenges power structures, making it significant for ensuring resilience strategies are just and inclusive (Critical Urban Theory). Urban Morphology provides insights into how physical design impacts urban resilience, emphasizing the significance of the built environment (Urban Morphology). Global City Theory highlights the importance of global cities in shaping global trends, underlining their significance for international collaboration in resilience efforts (Global City Theory). Feminist Urban Theory aligns *Urbanimmunology* with intersectional justice movements, making it significant in addressing gender, race, and class inequalities in urban health (Feminist Urban Theory). Postcolonial Urban Theory connects *Urbanimmunology* to decolonial scholarship, highlighting the significance of local and context-specific resilience strategies (Postcolonial Urban Theory).

Actor-Network Theory (ANT) demonstrates how diverse urban networks (human and non-human) affect resilience, enhancing the significance of interconnected urban systems (Actor-Network Theory). The Right to the City stresses the importance of inclusivity and equitable access, making it highly significant for developing just urban resilience strategies (Right to the City). New Urbanism focuses on urban design that fosters community cohesion and sustainability, contributing significantly to neighborhood-level resilience (New Urbanism). Urban Resilience Theory emphasizes adaptive capacity, ensuring that resilience strategies are designed for long-term stability, enhancing the significance of urban planning (Urban Resilience Theory). Landscape Urbanism advocates for nature-based solutions, highlighting the significance of green infrastructure for maintaining urban health (Landscape Urbanism).

Everyday Urbanism emphasizes the importance of small, everyday practices in building resilience, enhancing its significance at the community level (Everyday Urbanism). Creative Class Theory demonstrates how creativity and innovation drive resilience, making it significant in fostering a dynamic, adaptive urban environment (Creative Class Theory). Post Urbanism recognizes the importance of non-hierarchical planning and adaptive reuse, emphasizing the significance of flexible, decentralized urban systems (Post Urbanism). Informal Urbanism acknowledges the adaptive capacity of informal systems, making it significant in understanding urban resilience outside formal governance structures (Informal Urbanism). Finally, the Theory of Urban Sustainability, grounded in long-term sustainability, emphasizes the significance of maintaining urban resilience for future generations (Theory of Urban Sustainability).

4.5.3 Enhancing the Comprehensiveness of Urbanimmunology

To enhance the comprehensiveness of *Urbanimmunology* as a truly scientific system based on Kantor's criteria, a holistic and multidimensional approach is required. This approach integrates diverse theoretical insights that address ecological, economic, political, social, and cultural factors that shape urban and societal resilience.

Urban resilience can be broadened by integrating humans, ecosystems, and the built environment into a cohesive framework (Urban Ecology). By incorporating geographic and economic factors in the distribution of services, resilience thinking expands and improves comprehensive urban resilience models (Central Place Theory). This ensures that marginalized groups are included in resilience planning, addressing economic disparities and systemic vulnerabilities (Neo-Marxist Urban Theory). Political and economic considerations are also integral to understanding urban resilience, providing a more comprehensive view of the factors that shape a city's vulnerability and capacity for recovery (Urban Political Economy). Furthermore, balancing market efficiency with equity concerns ensures that issues such as housing affordability and economic fairness are considered in resilience planning (New Urban Economics).

Urban resilience is further contextualized within the global economic system, highlighting how global trends influence local urban dynamics (World-Systems Theory). A critique of power structures is essential for integrating equitable urban development and ensuring resilience strategies are just and inclusive (Critical Urban Theory). The physical layout of cities also plays a significant role in shaping their capacity to respond to crises, contributing to a comprehensive understanding of urban resilience (Urban Morphology). In parallel, considering the global economic roles of cities offers a broader perspective on urban resilience, particularly in fostering international collaboration (Global City Theory).

Addressing social inequalities is key to comprehensive resilience planning, prioritizing the needs of marginalized groups and promoting social equity in urban environments (Feminist Urban Theory). Resilience models also need to be inclusive of informal economies and diverse global urban experiences, enriching urban resilience strategies through context-specific approaches (Postcolonial Urban Theory). Expanding the resilience framework to include both human and non-human actors enhances the analysis of interconnected urban systems and their responses to disruptions (Actor-Network Theory).

A comprehensive urban resilience system should integrate social justice and human rights, ensuring that all residents' needs are addressed, particularly those who are most vulnerable (Right to the City). Design, social interaction, and ecological sustainability should be incorporated into urban resilience strategies, ensuring that multiple urban dimensions are considered and fostered (New Urbanism). Recognizing the multiple interconnected systems—social, environmental, and economic—that shape urban resilience ensures a more comprehensive understanding of urban health (Urban Resilience Theory). Additionally, emphasizing the role of green infrastructure through nature-based solutions reinforces a comprehensive approach to urban health and resilience (Landscape Urbanism).

Integrating micro-level activities such as community gardens or mutual aid into resilience strategies ensures that everyday urban practices are considered in broader resilience efforts (Everyday Urbanism). Creativity and innovation contribute to resilience by fostering a dynamic, adaptive urban environment, emphasizing the role of cultural and economic factors in resilience planning (Creative Class Theory). Moreover, flexibility in urban planning—especially through non-hierarchical and adaptive spaces—enhances the capacity of cities to evolve and respond to shifting societal needs (Post Urbanism). Understanding the role of informal systems further enriches our view of urban resilience by acknowledging the adaptive capacity of informal practices alongside formal governance structures (Informal Urbanism).

Finally, resilience planning should consider long-term sustainability, addressing the environmental, social, and economic pillars to ensure urban systems are able to endure and thrive across generations (Theory of Urban Sustainability).

4.6 Establishing Theoretical Foundations for *Urbanimmunology* as a Robust Scientific System

By aligning with Kantor's criteria of validity, significance, and comprehensiveness, the nine core principles of *Urbanimmunology* have been evaluated for their collective potential to establish the discipline as a robust scientific system. These principles not only ensure internal coherence but also position *Urbanimmunology* within broader scientific discussions while addressing the diverse phenomena linked to urban resilience and sustainability. Each principle contributes unique depth and flexibility, enabling *Urbanimmunology* to serve as an integrative, adaptable, and comprehensive framework for understanding and enhancing the resilience of cities and societies.

Recognizing that no theory is perfect, it is important to acknowledge the limitations of existing urban theories alongside their valuable insights for addressing specific urban and social challenges. As discussed previously, these strengths provide a foundation for further exploration, aimed at enhancing the validity, significance, and comprehensiveness of *Urbanimmunology*.

This section combines discussions on the relevance of the nine core principles to Kantor's three criteria with the advancement of *Urbanimmunology* through the strategic use of existing urban theories. The goal is to establish *Urbanimmunology* as a robust scientific system, emphasizing how each element—the principles and the theories—works together to achieve validity, significance, and comprehensiveness.

4.6.1 Validity of Urbanimmunology

To enhance the validity of *Urbanimmunology* as a scientific system according to Kantor's criteria, it is crucial to incorporate a range of theoretical frameworks that offer empirical grounding, robust methodologies, and analytical tools to address the complexities of urban and social systems. The nine principles of *Urbanimmunology*, in alignment with Kantor's validity criterion, ensure internal coherence and logical consistency within the framework. When integrated with the strengths of existing urban theories, these principles foster synergies and complementary effects that strengthen the overall validity of *Urbanimmunology*.

4.6.1.1 Synergetic Effects

Synergies arise when the principles of *Urbanimmunology* and the strengths of existing urban theories complement one another, reinforcing *Urbanimmunology*'s internal coherence and logical structure. These synergies ensure that *Urbanimmunology* remains both scientifically rigorous and practically applicable, thereby enhancing its validity.

Critical Dependency Principle and Urban Ecology The *Critical Dependency Principle* in *Urbanimmunology* ensures the identification of foundational components essential for resilience and adaptation, while *Urban Ecology* provides an empirical framework for understanding urban ecosystems. By integrating these, *Urbanimmunology* gains a solid foundation for comprehending how urban systems depend on their ecological environment. This synergy strengthens *Urbanimmunology* by grounding it in empirical ecological science, enhancing the validity of resilience assessments and ensuring that urban systems are understood in relation to their environmental context, which is crucial for accurate predictions of resilience.

Avoidance Principle and Urban Political Economy The *Avoidance Principle* in *Urbanimmunology*, which focuses on proactively preventing risks, aligns with *Urban Political Economy*, a theory that examines the relationship between political-economic systems and urban resilience. Urban Political Economy emphasizes understanding governance structures and their role in preventing urban vulnerabilities, making this combination essential for effective urban resilience planning. The synergy between these concepts strengthens *Urbanimmunology* by integrating political-economic dimensions into resilience planning, ensuring that *Urbanimmunology* not only addresses immediate risks but also accounts for broader systemic vulnerabilities that influence urban resilience.

Tolerance Principle and New Urban Economics The *Tolerance Principle* in *Urbanimmunology*, which emphasizes absorbing disruptions without overextending resources, aligns with *New Urban Economics*, which uses market equilibrium models to explain urban resource distribution. New Urban Economics provides analytical tools for understanding the economic dynamics of urban systems, including their ability to absorb and mitigate economic disruptions. The synergy between these two principles ensures that *Urbanimmunology* incorporates both economic resilience and adaptive capacity, enhancing the validity of resilience assessments by integrating economic factors into the framework.

Adaptation Principle and Urban Resilience Theory The *Adaptation Principle*, a core concept in both *Urbanimmunology* and *Urban Resilience Theory*, focuses on how cities adjust to shocks and stresses. Urban Resilience Theory provides the scientific foundation for assessing the ability of urban systems to adapt and recover over time. The synergy between these two ensures the internal coherence of *Urbanimmunology* by linking the adaptive capacity of cities to scientifically validated resilience models, thereby strengthening the validity of predictions about how cities can respond to various challenges.

Regulation Principle and Landscape Urbanism The *Regulation Principle* in *Urbanimmunology* emphasizes maintaining balance within urban systems, aligning with *Landscape Urbanism*, which integrates ecological principles into urban design to ensure sustainability. Landscape Urbanism provides evidence-based strategies for creating resilient urban landscapes that promote long-term ecological and social health. By combining regulatory strategies with landscape-oriented urban design, this synergy ensures that *Urbanimmunology* remains grounded in sustainable, evidence-based practices, enhancing its validity as a scientific framework for urban resilience.

4.6.1.2 Complementary Effects

Complementary effects arise when the core principles of *Urbanimmunology* and the strengths of existing urban theories offer distinct perspectives that enrich *Urbanimmunology*'s overall validity. These complementary theories add depth and context, enhancing the scientific rigor of *Urbanimmunology* and strengthening its validity.

Coordination Principle and Actor-Network Theory (ANT) The *Coordination Principle* in *Urbanimmunology* ensures that different components of urban systems work together in harmony, while *Actor-Network Theory (ANT)* emphasizes the role of both human and non-human actors in shaping urban systems. ANT expands the scope of analysis beyond traditional models by mapping the interconnectedness of actors within urban systems. The complementary effect between these concepts ensures that while the Coordination Principle fosters harmony in urban systems, ANT provides the methodology to explore the complex relationships between diverse actors. This synergy strengthens *Urbanimmunology*'s capacity to assess and coordinate the various elements of urban resilience, ultimately enhancing the framework's validity.

Diversity Principle and Feminist Urban Theory The *Diversity Principle* in *Urbanimmunology* ensures that a broad range of urban challenges is considered, while *Feminist Urban Theory* advocates for inclusive and equitable urban development, emphasizing gendered exclusions and social justice. The intersection of these principles ensures that *Urbanimmunology* is both comprehensive and just, addressing the diversity of urban systems and the need for inclusivity. The complementary effect between these concepts broadens the scope of resilience strategies, while Feminist Urban Theory ensures that this diversity includes marginalized voices and experiences, reinforcing the validity of *Urbanimmunology* by grounding it in equity and inclusivity.

Robustilient Regeneration Principle and New Urbanism The *Robustilient Regeneration Principle* in *Urbanimmunology* focuses on both recovery and forward-thinking resilience, while *New Urbanism* emphasizes urban design that promotes social cohesion, sustainability, and accessibility. By integrating these perspectives, *Urbanimmunology* ensures that both immediate recovery and long-term urban vitality are central to resilience planning. The complementary effect between these principles allows New Urbanism to provide practical design solutions that support robustilient regeneration, ensuring that urban resilience is not only theoretically sound but also operationalized through design principles that foster community cohesion and sustainability.

Coherence Principle and Right to the City The *Coherence Principle* in *Urbanimmunology* emphasizes the importance of integration and harmony within urban systems, while the *Right to the City* advocates for equitable access to urban resources and services. This combination ensures that resilience strategies in *Urbanimmunology* are not only internally consistent but also socially just and inclusive. The complementary effect between these principles guarantees that, while the Coherence Principle maintains the system's integrity, the Right to the City ensures that the benefits of resilience are equitably distributed, enhancing both the scientific and ethical validity of *Urbanimmunology*.

Tolerance Principle and Post Urbanism The *Tolerance Principle* in *Urbanimmunology* emphasizes the ability of urban systems to absorb and adapt to disruptions, while *Post Urbanism* promotes flexibility and adaptive urban planning. Together, these principles ensure that *Urbanimmunology* embraces non-hierarchical, flexible urban systems capable of responding to changing needs. The complementary effect of Post Urbanism enhances the adaptability of *Urbanimmunology*, reinforcing the idea that resilience is not static but evolves with changing social and environmental conditions, thus strengthening the validity of the framework in the face of uncertainty.

4.6.1.3 Summary

The integration of the nine principles of *Urbanimmunology* with existing urban theories creates both synergies and complementary effects that enhance the validity of *Urbanimmunology* as a scientific system. The synergies ensure that *Urbanimmunology* remains scientifically rigorous, internally coherent, and logically consistent by reinforcing its foundational principles. Meanwhile, the complementary theories deepen the framework's understanding of urban resilience by adding perspectives from ecological, economic, political, and social dimensions. Together, these elements strengthen the scientific credibility and applicability of *Urbanimmunology*, making it a robust framework for addressing urban resilience in a complex and rapidly changing world.

4.6.2 Significance of Urbanimmunology

The integration of Kantor's significance criterion with the nine principles of *Urbanimmunology*, combined with the strengths of existing urban theories, provides a nuanced enhancement of *Urbanimmunology*'s significance in addressing urban resilience. The following is a detailed discussion of the synergistic and complementary effects of these connections.

4.6.2.1 Synergetic Effects

Synergy arises when the principles of *Urbanimmunology* and the strengths of existing urban theories intersect and reinforce each other, thereby deepening the overall significance of *Urbanimmunology*.

Critical Dependency Principle and Urban Ecology The *Critical Dependency Principle*, which emphasizes interconnectedness within the urban system, aligns with *Urban Ecology*, which focuses on the interactions between urban areas and the environment. Together, they highlight the vital role of urban environments' resilience to ecological changes. Through these lenses, *Urbanimmunology* not only acknowledges dependencies within urban systems but also emphasizes the need for adaptive and sustainable urban planning in response to ecological challenges. The synergistic effect of this combination enhances the significance of *Urbanimmunology* by grounding it in ecological reality, emphasizing urban environments' resilience to natural forces like climate change, and positioning it as a key component of environmental sustainability.

Tolerance Principle and Critical Urban Theory The *Tolerance Principle*, which focuses on the ability of urban systems to cope with challenges, aligns with *Critical Urban Theory*'s advocacy for equitable urban development. Urban systems require tolerance not only for environmental or economic shocks but also for social inequalities. Critical Urban Theory challenges existing power structures and advocates for resilience that is both just and inclusive. The synergistic effect of this pairing enhances *Urbanimmunology*'s significance by ensuring that resilience strategies are not only efficient but also equitable, addressing inequalities that may undermine urban resilience.

Adaptation Principle and Urban Resilience Theory The *Adaptation Principle* in *Urbanimmunology* aligns with *Urban Resilience Theory*, which focuses on the adaptive capacity of urban environments. Both theories emphasize the need for cities to adjust to changing conditions, with *Urbanimmunology*'s adaptation principle complementing the long-term focus of resilience theory, ensuring that urban environments can evolve in response to challenges over time. The synergistic effect of integrating these principles enhances the significance of *Urbanimmunology* by providing comprehensive, adaptable strategies for long-term urban sustainability, linking it to global resilience frameworks.

Coordination Principle and Actor-Network Theory (ANT) The *Coordination Principle*, which focuses on the integration and alignment of urban systems, works in synergy with *Actor-Network Theory (ANT)*, which examines how various human and non-human actors shape urban resilience. By emphasizing coordination across different elements—such as systems, stakeholders, and policies—*Urbanimmunology* strengthens the concept of interconnected networks within urban systems. The synergistic effect of this integration enhances the significance of *Urbanimmunology* by highlighting the complexity of urban systems and the importance of collaboration among diverse actors, ensuring a more integrated and cohesive approach to urban resilience.

Diversity Principle and Feminist Urban Theory The *Diversity Principle*, which highlights the importance of diversity in urban systems for their resilience, aligns with *Feminist Urban Theory*, which emphasizes gender, race, and class inequalities. Urban resilience must account for diverse experiences and needs, ensuring that marginalized groups are included in resilience strategies. The synergistic effect of this connection enhances the significance of *Urbanimmunology* by emphasizing the role of diversity in both the social and ecological fabric of urban environments, ensuring that resilience efforts are both comprehensive and inclusive.

4.6.2.2 Complementary Effects

Complementarity arises when the strengths of existing urban theories provide perspectives that do not overlap but instead add depth to the core principles of *Urbanimmunology*, enriching its overall significance.

Regulation Principle and New Urban Economics The *Regulation Principle*, which focuses on maintaining equilibrium within urban systems, aligns with *New Urban Economics*, which examines how market-driven forces shape urban development. This complementary relationship suggests that *Urbanimmunology* can draw on economic perspectives to understand how regulatory mechanisms interact with market dynamics in fostering urban resilience. By integrating economic theories, *Urbanimmunology*'s regulatory approaches can be better understood within the context of market forces and policy frameworks, ensuring that resilience strategies align with both regulatory and economic realities.

Robustilient Regeneration Principle and Landscape Urbanism The *Robustilient Regeneration Principle*, which emphasizes regeneration and sustainability, aligns with *Landscape Urbanism*, which advocates for nature-based solutions. These concepts complement each other by linking regenerative processes in urban systems to ecological resilience through green infrastructure. This combination enhances the significance of *Urbanimmunology* by promoting urban regeneration that is not only resilient in terms of infrastructure but also ecologically regenerative, ensuring long-term sustainability through nature-based interventions.

Coherence Principle and Postcolonial Urban Theory The *Coherence Principle* focuses on systemic integration across urban systems, while *Postcolonial Urban Theory* introduces a critical dimension by emphasizing the importance of local context and decolonial scholarship. These perspectives complement each other by highlighting the need for coherent urban planning that is rooted in specific historical and cultural contexts, particularly in postcolonial settings. This pairing adds depth to *Urbanimmunology* by ensuring that systemic coherence is not only technologically and structurally sound but also culturally and historically relevant.

Avoidance Principle and Urban Political Economy The *Avoidance Principle*, which focuses on mitigating risks, complements *Urban Political Economy* by examining how political and economic factors contribute to urban vulnerability. Understanding the political economy of urban spaces provides insights into the risks cities face, helping *Urbanimmunology* develop more effective risk-avoidance strategies. This complementary effect enhances *Urbanimmunology*'s significance by grounding risk management in the political and economic realities of urban areas, ensuring that resilience strategies are both practical and context-sensitive.

4.6.2.3 Summary

The synergetic and complementary effects of the nine principles of *Urbanimmunology*, when combined with the insights from existing urban theories, result in a robust, multidimensional framework for urban resilience. While the principles provide a broad, foundational understanding of urban systems' resilience, the integration of specific urban theories enriches this understanding by adding layers of complexity, addressing social, political, and ecological factors that influence urban vulnerability. Together, they enhance the scientific significance of *Urbanimmunology*, ensuring its relevance across disciplines and its applicability to real-world urban challenges.

4.6.3 Comprehensiveness of Urbanimmunology

The nine principles of *Urbanimmunology*, when applied to Kantor's comprehensiveness criterion, form a rich, multidimensional framework for urban resilience and vitality that addresses a wide range of phenomena, including ecological, social, economic, and governance-related aspects. By integrating these principles with the strengths of existing urban theories, the comprehensiveness of *Urbanimmunology* is enhanced in both synergetic and complementary ways. Below is a detailed discussion of how these principles and theories interact to further enrich the comprehensiveness of *Urbanimmunology.*

4.6.3.1 Synergetic Effects

Synergy occurs when the nine principles of *Urbanimmunology* and the strengths of existing urban theories complement each other, reinforcing the overall comprehensiveness of *Urbanimmunology* and ensuring that it captures a wide range of factors essential to urban resilience and vitality.

Critical Dependency Principle and Urban Ecology The *Critical Dependency Principle* focuses on the interconnections between urban systems, while *Urban Ecology* emphasizes the relationship between urban areas and natural environments. By recognizing these interdependencies, *Urbanimmunology* ensures a comprehensive approach to resilience that incorporates both ecological and social systems. The synergistic effect of this integration highlights the need for urban resilience strategies that address both human and ecological systems simultaneously, ensuring that environmental sustainability is a central consideration in urban resilience planning.

Avoidance Principle and Neo-Marxist Urban Theory The *Avoidance Principle* in *Urbanimmunology*, which emphasizes preventing potential disruptions before they occur, aligns with *Neo-Marxist Urban Theory*, which critiques the systemic inequalities that contribute to urban vulnerabilities. Neo-Marxism advocates for addressing the root causes of inequality to prevent crises such as housing shortages or economic disparities. The synergistic effect of these perspectives ensures that resilience strategies not only mitigate risks but also address underlying socio-economic issues, making the urban resilience model more inclusive and equitable.

Tolerance Principle and Feminist Urban Theory The *Tolerance Principle* focuses on absorbing and mitigating challenges, which aligns with *Feminist Urban Theory*'s emphasis on addressing social inequalities, particularly those based on gender, race, and class. This principle broadens the scope of resilience to include how urban systems can adapt not only to environmental challenges but also to social ones. The synergistic effect of this integration reinforces the need for urban resilience strategies that consider diverse social experiences and inequalities, ensuring that marginalized groups are not left behind in resilience planning.

Adaptation Principle and Urban Resilience Theory The *Adaptation Principle* encourages urban systems to evolve in response to changing conditions, aligning well with *Urban Resilience Theory*, which focuses on a city's ability to adapt to shocks and stresses over time. By integrating adaptation into resilience thinking, *Urbanimmunology* can address a wide variety of disruptions, from climate change to economic shifts. The synergistic effect of this combination strengthens the comprehensiveness of *Urbanimmunology* by emphasizing continuous learning and adaptation as essential for the long-term survival of urban systems.

Coordination Principle and Actor-Network Theory (ANT) The *Coordination Principle* focuses on integrating different components within urban systems, while *Actor-Network Theory* examines the role of both human and non-human actors in shaping urban resilience. By acknowledging the various actors in the system, including technologies, institutions, and citizens, *Urbanimmunology* ensures that its approach to coordination is both comprehensive and dynamic. The synergistic effect of this integration ensures that resilience strategies are not only comprehensive in terms of urban systems but also in recognizing the full range of actors that contribute to urban resilience, fostering more inclusive and holistic solutions.

4.6.3.2 Complementary Effects

Complementarity occurs when the nine principles of *Urbanimmunology* and the strengths of existing urban theories offer distinct yet complementary perspectives, enhancing the comprehensiveness of *Urbanimmunology* by addressing various aspects of urban resilience and vitality.

Regulation Principle and Urban Political Economy The *Regulation Principle* in *Urbanimmunology* focuses on fine-tuning responses across various domains—social, environmental, and economic—which complements *Urban Political Economy*'s focus on the role of wealth, power, and governance in shaping urban resilience. While regulation seeks system balance, political economy examines how governance structures and economic forces influence resilience. Together, these perspectives ensure that urban resilience strategies are not only balanced but also responsive to political and economic forces, offering a comprehensive understanding of how urban systems can navigate both external pressures and internal disparities.

Diversity Principle and Postcolonial Urban Theory The *Diversity Principle* in *Urbanimmunology* emphasizes incorporating diverse components of urban systems, including social, economic, and environmental factors, while *Postcolonial Urban Theory* highlights the importance of context-specific resilience strategies and critiques the colonial legacies that shape urban development. By integrating both, *Urbanimmunology* ensures that diversity is not only recognized in terms of variety but also within historical, cultural, and political contexts. This integration broadens the scope of *Urbanimmunology* by emphasizing that resilience strategies must be context-specific, respecting both local realities and global histories, ensuring more culturally appropriate and inclusive urban resilience planning.

Robustilient Regeneration Principle and New Urbanism The *Robustilient Regeneration Principle* emphasizes both immediate recovery and long-term urban vitality, while *New Urbanism* focuses on urban design that promotes community cohesion, sustainability, and accessibility. These perspectives complement each other by highlighting the need for resilient urban regeneration that also supports sustainable, people-centered design. This combination enhances the comprehensiveness of *Urbanimmunology* by ensuring that resilience strategies focus not only on recovery but also on creating urban environments that are socially cohesive, environmentally sustainable, and adaptive to future challenges.

Coherence Principle and Right to the City The *Coherence Principle* ensures the harmonious integration of urban systems, while the *Right to the City* advocates for inclusive, equitable access to urban resources and spaces. By aligning these perspectives, *Urbanimmunology* ensures that resilience strategies are not only integrated but also just and inclusive. This pairing ensures that urban resilience strategies address the full spectrum of urban residents' needs, balancing both systemic integration and equitable access for all.

Tolerance Principle and Post Urbanism The *Tolerance Principle*'s focus on absorbing disruptions complements *Post Urbanism*'s emphasis on flexibility, non-hierarchical planning, and adaptive reuse. This combination ensures that resilience strategies are both robust and flexible, capable of adapting to unpredictable future changes while fostering decentralized urban systems. By integrating the flexibility of Post Urbanism with the tolerance for disruption in *Urbanimmunology*, resilience strategies become more adaptable and better suited to evolving urban needs and challenges.

4.6.3.3 Summary

The integration of the nine principles of *Urbanimmunology* with existing urban theories creates a robust, multidimensional framework for urban resilience. The synergetic effects ensure that resilience strategies address a broad spectrum of interconnected factors, from ecological to social and economic dimensions, while the complementary effects enhance the comprehensiveness of *Urbanimmunology* by adding depth and nuance to its approach. This comprehensive framework makes *Urbanimmunology* a more effective and holistic tool for addressing the complex challenges cities face in an increasingly interconnected and rapidly changing world.

4.7 Concluding Remarks: *Urbanimmunology* as a Game-Changing Discipline for Future Cities and Society

Building on the broad and extensive analyses presented in the previous chapters, along with the discussions in this chapter, *Urbanimmunology* has been conceptually established as a groundbreaking discipline. However, it is essential to acknowledge that real-world evidence must still be accumulated to fully validate and refine its principles. Urban studies must evolve beyond the mere aggregation of individual components (e.g., Acuto et al., 2018). In this context, *Urbanimmunology* not only aligns with this argument but also surpasses it by illustrating how the deeply interconnected components of cities and society can be harnessed to create integrated, self-regulating systems capable of adapting and evolving in response to complex and ever-changing conditions.

At its core, *Urbanimmunology* goes beyond mere survival or resilience in the face of disturbances. It emphasizes a deeper vision for urban and social systems—one that encompasses regeneration, evolution, and the ability to seize opportunities in times of disruption. While resilience is a critical component, the true power of *Urbanimmunology* lies in its broader focus on fostering systems that can adapt, grow, and thrive, even in the face of uncertainty. Through principles like Adaptation, Avoidance, and Tolerance, the framework equips both cities and societies with the tools not only to withstand crises but to emerge stronger, more equitable, and more capable of addressing future challenges.

Urbanimmunology's strength lies in its ability to integrate diverse fields of knowledge—biology, ecology, economics, sociology, and urban studies—into a cohesive, multidisciplinary system that understands urban and social systems as deeply interconnected. It is grounded in the understanding that the well-being of cities and societies is not an isolated phenomenon but the result of the dynamic interactions between various components—social networks, ecological systems, political structures, and economic frameworks. This integrated approach enables *Urbanimmunology* to address the full range of urban and societal challenges, from climate change and inequality to migration and technological disruptions, in a way that is both scientifically rigorous and practically applicable.

The framework's vision of thriving sets it apart from traditional theories that focus solely on resilience. *Urbanimmunology* is about creating systems that don't just endure but actively benefit from disruptions, fostering an environment where cities and societies can grow, evolve, and regenerate. By focusing on the principles of Robust Resilient Regeneration, Diversity, and Coherence, *Urbanimmunology* ensures that urban and societal systems can evolve in response to both predictable and unforeseen challenges, ultimately fostering environments that are sustainable, equitable, and adaptive to the future.

Moreover, *Urbanimmunology* is rooted in the belief that the interconnectedness of urban and social systems is key to their long-term sustainability. The system recognizes that urban resilience cannot be separated from social resilience, as the health of one system directly influences the other. *Urbanimmunology* places strong emphasis on the importance of inclusivity and social justice, ensuring that the benefits of urban and societal resilience are equitably distributed. Through its focus on Diversity, Robust Resilient Regeneration, and social justice principles, it advocates for the creation of cities and societies that are not only resilient but also just, inclusive, and capable of fostering well-being for all, especially marginalized and vulnerable groups.

Urbanimmunology's emphasis on adaptability and regeneration extends far beyond the idea of surviving crises. It encourages cities and societies to design systems that are constantly evolving and capable of leveraging disruptions as opportunities for growth. By embracing Adaptation and Tolerance as core principles, *Urbanimmunology* facilitates the creation of urban and social systems that are not just designed to recover from challenges but to actively regenerate and emerge stronger, more resilient, and more capable of responding to future uncertainties. This dynamic approach ensures that cities and societies can thrive, not just in the present but over the long term, making *Urbanimmunology* an essential framework for addressing the challenges of the twenty-first century and beyond.

Equally important is *Urbanimmunology*'s recognition of the role of community-based, bottom-up practices in fostering resilient urban and social systems. It acknowledges that resilience is not solely built through formal institutional structures but is deeply rooted in the collective actions, networks, and agency of the people who live within these systems. By embracing everyday practices such as mutual aid, community building, and grassroots activism, *Urbanimmunology* strengthens the social fabric of cities and societies, ensuring that resilience is both a top-down and bottom-up process. This inclusive approach not only reinforces the resilience of urban and social systems but also ensures that these systems reflect the lived experiences and needs of all individuals.

In essence, *Urbanimmunology* is not just about resilience; it is about transforming how we think about and design cities and societies. It is a framework that views cities and societies as dynamic, interconnected, and adaptive systems—capable of regeneration, growth, and evolution. Its principles of Adaptation, Robust Resilient Regeneration, Diversity, and Coherence provide a vision for urban and social systems that are not merely capable of surviving disruptions but are actively prepared to thrive through them. This forward-thinking, interdisciplinary framework offers a comprehensive, scientifically grounded approach to building cities and societies that are adaptable, sustainable, and equitable, positioning them to flourish in an ever-changing world.

By integrating these principles, *Urbanimmunology* presents a holistic, actionable framework for urban and social development that moves beyond the narrow focus on resilience. It empowers cities and societies to embrace change, regenerate, and evolve, fostering systems that are not only sustainable but also capable of adapting to the complexities and uncertainties of the future. *Urbanimmunology*, therefore, offers a transformative pathway for shaping the cities and societies of tomorrow—places that can grow, thrive, and actively contribute to a more sustainable, just, and equitable world.

References

Acuto, M., Dinardi, C., & Marx, C. (2019). Transcending (in)formal urbanism. *Urban Studies, 56*(3), 475–487.

Acuto, M., Parnell, S., & Seto, K. C. (2018). Building a global urban science. *Nature Sustainability, 1*, 2–4.

Ancien, D. (2011). Global city theory and the new urban politics twenty years on: The case for a geohistorical materialist approach to the (new) urban politics of global cities. *Urban Studies, 48*(12), 2473–2493.

Brenner, N., Marcuse, P., & Mayer, M. (2009). Cities for people, not for profit. *City, 13*(2–3), 176–184. https://doi.org/10.1080/13604810903020548

Chakrabarti, D. (2023). Urban theory of/from the global south: A systematic review of issues, challenges, and pathways of decolonization. *Frontiers in Sustainable Cities, 5*, 1163534. https://doi.org/10.3389/frsc.2023.1163534

Childers, D. L., Pickett, S. T. A., Grove, J. M., et al. (2014). Advancing urban sustainability theory and action: Challenges and opportunities. *Landscape and Urban Planning, 125*, 320–328.

de Munck, B. (2017). Re-assembling actor-network theory and urban history. *Urban History, 44*(1), 111–122.

Derickson, K. D. (2015). Urban geography I: Locating urban theory in the 'urban age'. *Progress in Human Geography, 39*(5), 647–657.

Edensor, T., & Jayne, M. (2012). *Urban theory beyond the west: A world of cities*. Routledge.

Ellis, C. (2002). The new urbanism: Critiques and rebuttals. *Journal of Urban Design, 7*(3), 261–291.

Fainstein, S. S. (2005). Planning theory and the city. *Journal of Planning Education and Research, 25*(2), 121–130.

Fainstein, S. S., & Campbell, S. (2011). *Readings in urban theory* (3rd ed.). Wiley-Blackwell.

Fryling, M. J., & Hayes, L. J. (2010). An interbehavioral analysis of memory. *European Journal of Behavior Analysis, 11*, 53–68.

Harding, A., & Bloklan, T. (2014). *Urban theory: A critical introduction to power, cities and urbanism in the 21st century*. Sage Publications.

Jacobs, J. (1992). *The death and life of great American cities*. Vintage; Reissue edition.

Jaret, C. (1983). Recent Neo-Marxist urban analysis. *Annual Review of Sociology, 9*, 499–525.

Jayne, M., & Ward, K. (2017). *Urban theory: New critical perspectives*. Routledge.

Jazeel, T. (2018). Urban theory with an outside. *Environment and Planning D: Society and Space, 36*(3), 405–419.

Kantor, J. R. (1958). *Interbehavioral psychology*. The Principia Press.

Kelbaugh, D. (2000). Three paradigms: New urbanism, everyday urbanism, post urbanism—An excerpt from the essential COMMON PLACE. *Bulletin of Science, Technology & Society, 20*(4), 285–289.

Kong, L., Mu, X., Hu, G., et al. (2022). The application of resilience theory in urban development: A literature review. *Environmental Science and Pollution Research, 29*, 49651–49671.
Lin, J., & Mele, C. (2013). *The urban sociology reader*. Routledge.
McPhearson, T., Pickett, S. T. A., Grimm, N. B., et al. (2016). Advancing urban ecology toward a science of cities. *Bioscience, 66*(3), 198–212.
Merrifield, A. (2011). The right to the city and beyond: Notes on a Lefebvrian re-conceptualization. *City, 15*(3–4), 473–481.
Mills, E. S., & MacKinnon, J. (1973). Notes on the new urban economics. *The Bell Journal of Economics and Management Science, 4*(2), 593–601.
Mulligan, G. F., Partridge, M. D., & Carruthers, J. I. (2012). Central place theory and its reemergence in regional science. *The Annals of Regional Science, 48*, 405–431.
Nathan, M. (2008). reative Class Theory and Economic Performance in UK Cities. In: Yigitcanlar, T., Velibeyoglu, K., Baum, S. (eds.), Creative Urban Regions: Harnessing Urban Technologies to Support Knowledge City Initiatives, 80-93. IGI Global.
Peake, L., Koleth, E., Tanyildiz, G. S., et al. (2021). *A feminist urban theory for our time: Rethinking social reproduction and the urban*. Wiley.
Ramalho, C. E., & Hobbs, R. J. (2012). Time for a change: Dynamic urban ecology. *Trends in Ecology & Evolution, 27*(3), 179–188.
Robinson, J., & Roy, A. (2015). Debate on global urbanisms and the nature of urban theory. *International Journal of Urban and Regional Research, 40*(1), 181–186.
Roy, A. (2014). Worlding the south: Toward a post-colonial urban theory. In S. Parnell & S. Oldfield (Eds.), *The Routledge handbook on cities of the global south*. Routledge.
Scott, A. J., & Storper, M. (2014). The nature of cities: The scope and limits of urban theory. *International Journal of Urban and Regional Research, 39*(1), 1–15. https://doi.org/10.1111/1468-2427.12134
Short, J. E. (2014). *Urban theory: A critical assessment*. Palgrave.
Storper, M., & Manville, M. (2006). Behaviour, preferences and cities: Urban theory and urban resurgence. *Urban Studies, 43*(8), 1247–1274.
Straussfogel, D. (1997). A systems perspective on world-systems theory. *Journal of Geography, 96*(2), 119–126.
Swanstrom, T. (1993). Beyond economism: Urban political economy and the postmodern challenge. *Journal of Urban Affairs, 15*(1), 55–78.
Thompson, I. H. (2011). Ten tenets and six questions for landscape urbanism. *Landscape Research, 37*(1), 7–26.

Chapter 5
Spatial Interference Hazards and Solutions Informed by *Urbanimmunology*

5.1 Introduction

In today's rapidly urbanizing world, the complexity of human activities and environmental interactions is increasing at an unprecedented pace (Li et al., 2021). Cities, rural areas, and even more remote regions are becoming increasingly interconnected, with diverse systems, activities, and forces coexisting in close proximity. This convergence often leads to unforeseen risks—where different elements of a space unintentionally overlap or interfere with one another, creating hazardous situations that were not anticipated in the original design or planning process. These unanticipated overlaps are increasingly critical to understand, as they can result in significant harm to individuals, property, and the environment.

In response to the need for addressing such risks, the author introduces a new concept: *spatial interference hazards*. This term is proposed to describe potential dangers that emerge from the misalignment, overlap, or conflict between spatial arrangements and functional systems within shared or adjacent physical environments. These hazards manifest in various forms, such as structural failures, environmental disasters, or human conflicts, often going unnoticed until they cause significant damage or disruption. They can include beneath-the-surface risks, like underground utilities or sinkholes, as well as structural and infrastructure failures, such as the collapse of buildings or roads. Landslides, where shifting soil or rock disrupts human settlements and transportation routes, pose additional threats, as do flooding and stormwater challenges that arise from failed water management systems.

Human-wildlife interaction hazards arise when human activities disrupt wildlife habitats, leading to risks such as physical harm, disease transmission, ecological damage, and socio-economic impacts in urban, rural, and protected areas (Narayan & Rana, 2023). Falling objects from buildings, such as debris, construction materials, or loose architectural elements, pose significant skyfall hazards in urban areas,

J. Zhang, *Urbanimmunology*, Urban Sustainability,
https://doi.org/10.1007/978-981-95-4878-1_5

particularly in densely populated environments. Additionally, windstorm-driven traffic disruptions caused by fallen trees or debris on roads and along their sides present further risks. Conflicts arising from multimodal interactions, where vehicles, bicycles, and pedestrians coexist in narrow spaces, can lead to slow traffic, congestion, and safety hazards, particularly in crowded urban areas with activities such as markets, community gates, and other high-traffic zones. The low-altitude economy, which includes low-altitude aircraft such as drones, poses significant dangers to populated areas and infrastructure, with risks such as collisions, noise pollution, and interference with other forms of transportation. Lastly, mass gatherings and tourism overcrowding become a concern when large crowds lead to accidents or public safety issues (Milano et al., 2019; Kanaujiya & Tiwari, 2022). Together, these diverse hazards highlight the critical need to understand and address potential risks that threaten safety and resilience in both urban and rural environments.

The significance of these hazards has grown as urbanization accelerates (United Nations, 2019), climate change intensifies (Solecki et al., 2024), and technological advancements evolve. Both cities and rural regions are becoming more complex and interconnected, making them increasingly vulnerable to unexpected disruptions. These hazards can result in devastating consequences, threatening human lives, property, and the environment. Furthermore, the growing interdependence between urban and nonurban spaces means that risks in one area can easily spill over into others, amplifying their potential impact. As we continue to develop both urban and rural spaces, it becomes clear that the management of these risks is no longer optional. To ensure safety and sustainability, there is a critical need to rethink how we design, plan, and manage spaces.

In this context, the concept of *Urbanimmunology* emerges as a transformative framework for shaping the development of urban and social systems, including hazard management. Drawing inspiration from the human immune system's capacity to defend against harmful agents while maintaining balance and functionality, *Urbanimmunology* presents a novel approach to managing and mitigating spatial interference hazards. Much like the immune system identifies and neutralizes threats in the body, a city or environment guided by *Urbanimmunology* can develop proactive systems that respond effectively to potential risks, ensuring resilience and stability.

Drawing inspiration from human immune intelligence, *Urbanimmunology* establishes nine core principles: *Critical Dependency*, *Avoidance*, *Tolerance*, *Adaptation*, *Regulation*, *Coordination*, *Diversity*, *Robustilient Regeneration*, and *Coherence*, as shown in Chap. 4. These principles offer a holistic approach to managing risks in both urban and nonurban environments. By incorporating immune-inspired strategies into urban planning, infrastructure design, and policymaking, cities and society can become more resilient, adaptable, and better equipped to face the unforeseen challenges of interconnected spaces.

In the remainder of this chapter, various examples of spatial interference hazards will be explored in greater detail. These hazards will be defined, their underlying causes will be examined, their consequences will be analyzed, and the practical challenges they pose will be discussed. Subsequently, the application of

Urbanimmunology principles to mitigate these risks will be examined. Finally, common themes among these hazards will be identified, and overarching strategies for their prevention will be proposed, with the aim of ensuring safer, more resilient environments across both urban and nonurban landscapes.

5.2 Beneath-the-Surface Hazards

Beneath-the-surface hazards refer to large-scale disasters or emergency situations occurring below the Earth's surface, such as in tunnels, mines, subways, basements, or underground infrastructure. These hazards may result from events such as structural collapses, flooding, gas leaks, explosions, fires, or the failure of critical underground systems. The spatial layout and complexity of underground spaces can significantly contribute to the interference that exacerbates these hazards. Factors like obstructed access routes, poor design, or inadequate ventilation complicate evacuation efforts and delay emergency responses.

Due to the inherent vulnerability and confined nature of these environments, beneath-the-surface hazards pose unique and severe risks. The consequences are not only immediate—manifesting in harm to life and property—but also long-term, affecting infrastructure resilience and creating profound social and psychological impacts. Real-world examples of such incidents, from collapsed tunnels (Hong et al., 2023) to hazardous gas leaks (Li et al., 2023), highlight the dangers of *spatial interference* within these spaces. These events reinforce the need for proactive disaster preparedness (Mojtahedi & Oo, 2017), meticulous infrastructure design (Debrock et al., 2023), and advanced emergency response systems (Cui et al., 2019). To address these challenges effectively, it is crucial to take a scientifically informed approach that accounts for *spatial interference* in underground spaces. Coordinated efforts, robust planning, and continued innovation are vital for preventing and mitigating the impacts of beneath-the-surface hazards.

5.2.1 Consequences of Beneath-the-Surface Hazards

5.2.1.1 Loss of Life and Injury

One of the most immediate and tragic consequences of beneath-the-surface hazards is the potential for significant loss of life and injury (Liu et al., 2021). The confined and often intricate nature of underground spaces, such as tunnels or mines, exacerbates the impact of disasters. Limited or obstructed evacuation routes, combined with the difficulty of accessing these spaces, can prevent swift evacuation and increase the risk of injury or fatality (Takahashi et al., 2022). Additionally, factors such as poor ventilation, high levels of smoke, toxic gases, or structural collapse can rapidly transform these environments into deadly traps, with *spatial interference* playing a key role in hindering rescue operations.

5.2.1.2 Economic Disruption

The damage caused by beneath-the-surface hazards can severely disrupt essential services and economic activities (Cao et al., 2022). For example, tunnel collapses and subway accidents can bring transportation systems to a standstill, disrupting not only daily commuting and freight operations but also the broader movement of goods and people. In the case of mining disasters, the consequences can be even more severe—endangering lives and causing long-term damage to local infrastructure and economies. The *spatial interference* caused by such incidents—such as blocked or damaged access points—can further delay recovery efforts and prolong economic instability, particularly in areas that rely heavily on underground infrastructure.

5.2.1.3 Infrastructure Damage

Failures in underground infrastructure can have cascading effects on other systems (Jiang & Tan, 2022). For instance, a subway tunnel collapse or a failure in underground utility systems can disrupt critical services like power, water, or gas, impacting both the underground space and the surface areas above. The difficulty of accessing and repairing these systems due to spatial constraints can extend the downtime, further compounding the challenges of restoring full functionality. This interference with normal operations creates a ripple effect, disrupting urban areas reliant on these services.

5.2.1.4 Environmental Contamination

Certain beneath-the-surface hazards, particularly those involving chemical spills, gas leaks, or mining accidents, can lead to significant environmental contamination (Feng et al., 2021). If toxic gases or hazardous substances are released underground, they can seep into surrounding soil, groundwater, or the air, resulting in long-term ecological damage and health risks for nearby populations. The complex underground layout and *spatial interference* in these environments can make containment efforts more difficult, with limited access points hindering effective mitigation of the spread of contamination.

5.2.1.5 Psychological Destruction

Beyond the physical harm, survivors of beneath-the-surface hazards often face severe psychological impacts (Goh, 2021). Individuals trapped in confined spaces or those exposed to catastrophic events underground may experience trauma, anxiety, or post-traumatic stress disorder. The difficulty in reaching or rescuing survivors due to *spatial interference* can extend their time in these high-stress environments, compounding the emotional toll. The loss of colleagues, friends, or loved ones in such circumstances can also intensify the psychological burden.

5.2.1.6 Disruption of Services and Recovery Challenges

The recovery and restoration of services after a beneath-the-surface catastrophe can take a significant amount of time (Huang et al., 2022). The confined, hazardous nature of underground spaces makes access to damaged infrastructure more difficult, with *spatial interference* often hindering repairs and safety assessments. Emergency response efforts are frequently delayed or complicated by obstructed entry points, structural damage, or lack of proper evacuation routes. As a result, the recovery process may be prolonged, and essential services may remain interrupted for a more extended period.

5.2.2 Practical Issues in Managing Beneath-the-Surface Hazards

5.2.2.1 Rescue and Evacuation Difficulties

One of the most significant practical challenges in managing subterranean disasters is the difficulty of evacuation and rescue (Yu et al., 2022), which is greatly exacerbated by *spatial interference*. Underground spaces often lack sufficient exit routes, or these routes may be obstructed by debris, structural collapse, or complex layouts, making navigation difficult for both individuals and rescuers. The confined and tangled nature of these environments impedes timely and effective evacuation, and specialized equipment (Tian et al., 2022) and highly trained personnel (Li et al., 2024) are essential for search-and-rescue operations. However, these resources are often not readily available in large numbers, especially during simultaneous urban emergencies, increasing the risk of delayed responses and casualties.

5.2.2.2 Structural Failures and Safety Risks

The design and construction of underground spaces present numerous practical challenges related to disaster risks such as collapse, fire, flooding, and gas accumulation (Cao et al., 2022). Maintaining structural integrity is complicated by *spatial interference*, which can hinder regular inspections and timely detection of damage. The confined nature of underground environments restricts access for repairs and reinforcement, increasing the likelihood that minor damage could escalate into severe structural failures. Safety systems, including support beams, reinforced structures, ventilation, and emergency exits, are difficult to implement effectively due to the inherent spatial constraints of underground spaces, making comprehensive risk management more complex.

5.2.2.3 Monitoring Failures and Delayed Warnings

Preventing and responding to beneath-the-surface hazards is particularly difficult due to the spatial constraints that hinder real-time monitoring and early warning systems (Wang et al., 2022). The layout of underground spaces, with their limited accessibility and complex configurations, complicates the installation and maintenance of comprehensive monitoring infrastructure. Sensors that track structural shifts, water levels, air quality, and temperature can detect potential hazards, but outdated or insufficient technology often delays detection, especially when *spatial interference* blocks signal transmission or sensor placement. This delay in issuing warnings can escalate the risk of disaster, as emergency response teams may not have enough time to intervene effectively.

5.2.2.4 Poor Emergency Preparedness and Inefficient Response

Underground environments pose significant challenges to emergency preparedness and response efforts (Bjelland et al., 2021), largely due to spatial interference and confinement. The planning and implementation of accessible escape routes are complicated by the potential for debris and structural damage, which can block or obscure exits. The limited space restricts movement and hampers both evacuation efforts and the operations of emergency responders. Spatial constraints also limit the effectiveness of conventional emergency training, which may not address the specific conditions found underground (Onifade, 2021). Rescue operations are further hindered by the need for specialized equipment capable of functioning in narrow, obstructed spaces (Cai et al., 2025). Additionally, the effectiveness of public communication during emergencies is often reduced due to signal interference and physical barriers, making it harder to disseminate critical information and manage panic in confined underground settings.

5.2.2.5 Communication Breakdown and Obstructed Access

Underground spaces pose unique challenges for communication during a disaster due to *spatial interference* (Sadeghi et al., 2022; Cai et al., 2025). Traditional communication methods, such as mobile phones or radios, often fail in subterranean environments because the physical structure—thick walls, tunnels, and limited coverage—interferes with signal transmission. Maintaining real-time communication between emergency response teams and those trapped below the surface becomes a critical issue (Cai et al., 2025). The lack of direct access and the presence of obstructions further complicate efforts to coordinate rescue operations, making the ability to maintain effective communication more vital than ever.

5.2.2.6 Coordination Failures and Response Delays

Coordinating the response to subterranean disasters involves collaboration among a range of stakeholders, including urban planners, construction engineers, emergency services, and government agencies (Restrepo et al., 2025). The *spatial interference* inherent in underground spaces can hinder the ability of these groups to effectively communicate and access critical information. Limited physical access to affected areas, complex layouts, and structural obstructions make it difficult to coordinate actions in a timely manner. Responding effectively to these challenges is often hindered by insufficient planning, technological limitations, and a lack of coordinated efforts across sectors, all of which are further complicated by the spatial complexities of underground environments.

5.2.3 **Urbanimmunology**-*Informed Solutions to Beneath-the-Surface Hazards*

Focusing on the aforementioned issues related to beneath-the-surface hazards, the nine core principles of *Urbanimmunology* are strategically applied to develop effective solutions. Each principle is carefully matched with specific characteristics of these hazards, ensuring that the proposed approaches are tailored to address the unique challenges posed by the hidden and often overlooked dangers within underground environments. By applying the core principles of *Urbanimmunology* to the practical issues associated with managing beneath-the-surface hazards, cities and society can enhance their preparedness, response, and recovery capabilities. These principles help create adaptive, resilient, and well-coordinated systems that minimize risks and enhance the overall safety and functionality of underground environments. When applied effectively, these principles foster urban ecosystems that can better withstand subterranean disasters and recover with resilience and strength.

5.2.3.1 Strategic and Tactical Approaches to Rescue and Evacuation: A Framework for Preparedness, Prevention, and Adaptation

Rescue and evacuation efforts in underground spaces are inherently complex due to limited access points, obstructions, and the need for specialized equipment, requiring comprehensive planning and adaptable systems. *Urbanimmunology* principles can help address these challenges effectively. The *Critical Dependency Principle* underscores the importance of establishing robust infrastructure, such as multiple exits, reinforced evacuation shafts, and safe zones, integrated with emergency services to ensure resilience during disasters. This principle stresses that cities should prioritize the construction of critical systems that remain operational under adverse conditions. Such systems should be designed for long-term sustainability and easy maintenance, ensuring they can withstand the pressure of catastrophic events. The *Avoidance Principle* suggests limiting development in high-risk subterranean areas, ensuring clear evacuation routes, and prioritizing multiple safe exits to reduce vulnerability. By proactively identifying and mitigating risks before construction, urban planners can prevent unnecessary harm and reduce the burden on emergency responders. Additionally, this principle advocates for strategic land use that balances urban growth with safety considerations, preventing overcrowded or poorly accessible underground spaces. Finally, the *Adaptation Principle* stresses the need for flexible, adaptable evacuation protocols that can be swiftly adjusted based on changing conditions, such as new information or structural damage, ensuring more effective and timely responses. It emphasizes the necessity of real-time decision-making and dynamic communication between all emergency response teams to ensure the safest outcomes. Furthermore, adaptive systems should incorporate continuous training and simulation exercises for responders to improve their readiness for diverse, evolving scenarios.

5.2.3.2 Building Sturdy and Well-Managed Underground Spaces: Urbanimmunology's Path to Safety and Regeneration

Ensuring that underground spaces can withstand disasters like flooding, earthquakes, or fires is crucial, with structural integrity being a top priority to prevent catastrophic failures. The application of *Urbanimmunology* principles offers effective strategies to enhance this resilience. The *Regulation Principle* emphasizes the importance of establishing rigorous, self-imposed standards, including implementing strict safety regulations for underground spaces, such as mandatory building codes, frequent inspections, and retrofitting older infrastructure with modern safety features. Cities should enforce regulations that ensure underground spaces are designed to withstand various potential hazards. The *Tolerance Principle* advocates for finding balanced solutions when faced with competing demands, such as the need for more underground space to accommodate urban growth. It encourages city planners to strike a balance between safety and development by integrating reinforced structures within urban design while considering the functional needs of the space, ensuring sustainability. Finally, the *Robustilient Regeneration Principle* focuses on creating infrastructure that can regenerate quickly after a disaster. For underground spaces, this means designing structures that not only withstand stress but can recover swiftly, minimizing downtime and allowing for the rapid restoration of services. Cities should prioritize the use of resilient building materials and regenerative features to ensure long-term durability.

5.2.3.3 Enhancing Monitoring and Early Warning Systems Through Urbanimmunology: Adapting, Coordinating, and Diversifying for Safety

Real-time monitoring is essential for preventing and responding to beneath-the-surface hazards. Without a comprehensive early warning system, disasters can escalate without proper intervention. The application of *Urbanimmunology* principles can significantly enhance these efforts. The *Adaptation Principle* emphasizes the need for systems to evolve based on new data and technological advancements. In monitoring, cities should integrate flexible, adaptive sensor networks capable of detecting structural shifts, gas accumulation, air quality, or water levels in underground spaces. These systems should evolve alongside emerging technologies, enabling more accurate and timely predictions of potential hazards. The *Coordination Principle* highlights the importance of sharing data between stakeholders, including emergency responders, urban planners, and government agencies. This coordination ensures that warnings are rapidly disseminated and response efforts are synchronized, helping to prevent disaster escalation. Lastly, the *Diversity Principle* advocates for the use of diverse technologies and systems to reduce reliance on any single solution. In monitoring, this could involve implementing various sensor types and communication networks, ensuring that if one system fails, others can step in to provide critical data. By diversifying monitoring tools, cities can create more robust early warning systems capable of addressing different types of subterranean hazards.

5.2.3.4 Strengthening Emergency Preparedness and Response Through Urbanimmunology: Ensuring Critical Systems, Training Standards, and Cohesive Action

Preparedness and response strategies are crucial for minimizing the impact of subterranean disasters, requiring coordinated action, proper training, and clear communication. The application of *Urbanimmunology* principles plays a vital role in strengthening these efforts. The *Critical Dependency Principle* highlights the importance of ensuring essential systems for emergency preparedness, such as training programs, drills, and coordination between emergency services, are consistently in place. These systems should be regularly tested and maintained to guarantee a high level of readiness for any underground disaster scenario. The *Regulation Principle* advocates for establishing strict regulations for preparedness, including mandatory emergency response training for underground workers, regular evacuation drills, and safety checks. These regulations should be enforced to ensure that all individuals in underground spaces are adequately prepared for emergencies. Finally, the *Coherence Principle* emphasizes the need for harmonious functioning between different urban systems. For emergency preparedness, this involves aligning emergency services, underground infrastructure, and communication systems so that all parts of the urban environment work together seamlessly during a crisis, ensuring an effective and coordinated disaster response.

5.2.3.5 Enhancing Access and Communication in Underground Spaces: Urbanimmunology's Path to System Integration, Flexible Solutions, and Resilient Standards

Underground spaces present unique challenges for communication during a disaster. The use of traditional communication methods, such as mobile phones or radios, is often obstructed by the underground environment. As a result, maintaining real-time communication between emergency response teams and those trapped below the surface is a critical issue. The application of *Urbanimmunology* principles can significantly enhance communication in such situations. The *Coherence Principle* promotes the integration of communication systems to ensure seamless coordination between responders, city infrastructure, and emergency networks. Underground communication systems should be connected to above-ground networks to maintain uninterrupted communication during emergencies, incorporating robust infrastructures like emergency radio networks or underground Wi-Fi systems that can operate in crisis situations. The *Adaptation Principle* emphasizes the need for flexibility in communication systems to meet the evolving needs of underground disaster management. For instance, adapting communication tools to work in low-signal environments and investing in technologies like underground data networks or signal amplifiers will ensure that communication lines remain open during emergencies. Lastly, the *Regulation Principle* stresses the importance of creating standards and regulations that ensure underground communication networks are resilient. These regulations should include fail-safe systems designed to function even in the event of a catastrophe (Ogata et al., 1990), guaranteeing continued communication when it is most needed.

5.2.3.6 Enhancing Stakeholder Collaboration in Underground Disaster Response: Urbanimmunology's Approach to Coordination, Diversity, and Evolutionary Recovery

Coordinating the response to subterranean disasters involves various stakeholders, including urban planners, construction engineers, emergency services, and government agencies. Effective communication and collaboration between these parties are essential for timely responses and recovery efforts. The application of

Urbanimmunology principles enhances the coordination and response to such challenges. The *Coordination Principle* is central to managing collaboration among different parties. Creating clear communication channels, shared responsibilities, and joint action plans among all relevant stakeholders is crucial for ensuring a timely and efficient disaster response. Such collaborative efforts enable cities to address all aspects of underground disaster management more effectively. The *Diversity Principle* encourages incorporating a wide range of perspectives and expertise to address subterranean disasters. Stakeholders from fields like engineering, emergency response, urban planning, and public health should work together to ensure that every facet of disaster response is covered. This diversity of thought fosters resilience by offering a variety of solutions to the challenges posed by underground hazards. Lastly, the *Robustilient Regeneration Principle* emphasizes the need for a rapid, adaptive recovery process after a disaster. This applies not only to the physical recovery of underground infrastructure but also to the regeneration of social networks and services impacted by the event. Effective collaboration between all stakeholders ensures swift recovery and the city's ability to adapt to new circumstances following a catastrophe.

5.3 Structural and Infrastructure Failures

Structural and infrastructure failures (Hasan & Foliente, 2015; Mühlhofer et al., 2023) refer to the risks and vulnerabilities that arise from the degradation, deterioration, or inadequate maintenance of essential infrastructure elements such as buildings, bridges, tunnels, utilities, and transportation networks within urban environments. These failures are particularly prevalent in cities with aging infrastructure or in regions that have not consistently conducted inspections and updates. Many cities, especially those built before the widespread adoption of modern construction standards and engineering practices, face heightened risks due to the outdated nature of their infrastructure, which fails to adapt to the evolving demands of urban life.

The primary cause of these failures is aging infrastructure, which may not have been designed to accommodate the intensified needs of modern urbanization or inadequate maintenance and periodic updates (Rahman et al., 2009). Often, infrastructure was originally intended to support fewer people or lower levels of traffic, making older structures especially prone to interference with the flow of daily life. Furthermore, failures can stem from poor construction practices or the use of substandard materials, compounding risks to public safety and the environment. A lack of investment in preventive maintenance and necessary repairs only worsens these issues, allowing infrastructure to deteriorate beyond its intended lifespan, without the necessary upgrades to meet current safety standards (Mazele & Amoah, 2022). Ultimately, such failures can result in significant disruptions—interfering with the normal function and safety of urban spaces and communities.

5.3.1 Consequences of Structural and Infrastructure Failures

The consequences of structural and infrastructure failures are profound, extending well beyond immediate physical damage to include economic disruption, social instability, and long-term impacts on urban development and resilience. These failures can disrupt the functionality of entire urban spaces, causing interference with daily life and undermining the fabric of communities.

Building Failures (Janssens et al., 2012): Old or poorly constructed buildings are more likely to collapse during natural disasters (such as earthquakes, floods, or storms) or due to long-term neglect. In earthquake-prone regions, buildings lacking modern seismic design features pose particularly dire risks, as they may not withstand ground shaking, leading to building collapses that result in significant casualties and widespread damage. Beyond natural disasters, structural weaknesses can lead to catastrophic failures over time, endangering inhabitants and workers and causing spatial disruptions in neighborhoods. The collapse of a building not only damages property but also disrupts the flow of life around it, displacing residents and businesses, blocking roads, and hindering access to essential services.

Bridge and Tunnel Collapses (Deng et al., 2015; Ou et al., 2021): Aging infrastructure like bridges and tunnels in cities with high levels of traffic represents a major threat to both public safety and the local economy. A collapse or failure in such critical infrastructure can cause immediate loss of life, significant traffic disruptions, and widespread economic losses by halting transportation and trade. The resulting interference in transportation networks causes ripple effects across urban mobility, isolating communities, delaying emergency response times, and stalling economic activities. Rebuilding or repairing these structures often strains public resources and hampers the city's ability to recover, affecting the resilience of urban networks and services.

Utility Disruptions (Al-Bayati et al., 2024): Infrastructure failures in utilities such as water pipes, gas lines, or electrical grids can lead to catastrophic service disruptions, particularly during extreme weather events or in densely populated urban areas. A broken water pipe or gas leak can cause flooding or fires, while power grid failures result in widespread outages that cripple the functionality of entire neighborhoods and sectors of the economy. These disruptions can create significant interference with daily life, halting transportation, healthcare, and communication systems, and severely affecting the operation of businesses and critical services. The ripple effects across other sectors of the economy can further strain recovery efforts, disrupting both spatial and economic flows within the city.

Earthquake-Weak Buildings (Stevenson, 1885; Liu et al., 2021): In seismically active zones, the lack of modern earthquake-resistant construction in buildings is particularly perilous. Buildings that cannot withstand seismic events may collapse or suffer severe structural damage during earthquakes, leading to significant loss of life, injury, and widespread destruction of property. The economic and emotional toll on communities, coupled with the disruption of infrastructure, can interfere with recovery efforts for years. A failure to rebuild or retrofit such buildings creates long-lasting disruptions to spatial and social dynamics, further undermining urban resilience and confidence.

These consequences illustrate the far-reaching and often devastating effects of structural and infrastructure failures. Beyond the immediate physical damage, the long-term interference with the safety, stability, and functionality of urban spaces highlights the critical need for proactive maintenance, stringent building codes, and innovative infrastructure solutions to safeguard communities from these risks. The ability to ensure that infrastructure supports the smooth functioning of urban environments is vital for both recovery and resilience.

5.3.2 Practical Issues in Addressing Structural and Infrastructure Failures

5.3.2.1 The Fragility of Aging Infrastructure: The Hidden Risks of Vulnerable Buildings

Older buildings, particularly those not updated to meet modern safety standards, are vulnerable to structural degradation over time (Janssens et al., 2012). These buildings, especially those constructed without adequate foundations or seismic reinforcements, are prone to wear and tear, making them more susceptible to collapse during extreme weather events or natural disasters. A major practical challenge in addressing building failures is identifying and assessing these at-risk structures, which requires regular inspections to detect early signs of deterioration. However, many cities lack the resources and infrastructure to conduct comprehensive assessments. This problem is even more pronounced in informal settlements, where poor construction practices and substandard materials exacerbate vulnerabilities.

The failure of these structures can lead to significant interference in urban spaces, displacing residents, closing off streets, and disrupting local economies. To mitigate these risks, cities should prioritize retrofitting older buildings to meet modern standards, including strengthening foundations, adding seismic reinforcements, and upgrading essential systems. Establishing maintenance programs to ensure buildings remain structurally sound over time is crucial to preventing failures and minimizing interference with the flow of daily life.

5.3.2.2 The Deterioration of Critical Infrastructure: The Rising Risk of Bridge and Tunnel Failures

Bridges, tunnels, and other vital infrastructure elements face gradual deterioration due to continuous stress from traffic, environmental exposure (e.g., salt, moisture, extreme weather), and the passage of time (Deng et al., 2015; Ou et al., 2021). One of the practical challenges in addressing these failures is the difficulty of accurately assessing the condition of such structures and predicting when major repairs or replacements will be necessary. Monitoring these critical infrastructures is essential, but often costly, and cities in densely populated areas face unique challenges in keeping infrastructure maintained.

Bridge and tunnel failures cause immediate interference in urban mobility—blocking transportation routes, delaying emergency services, and isolating communities. These disruptions ripple throughout the urban network, creating long-lasting economic and social consequences. Advances in technology, such as sensors, drones, and imaging techniques, can help improve monitoring, allowing cities to detect issues early and prioritize necessary interventions. Preventive maintenance, including routine inspections and refurbishments, is critical to reducing the risk of collapse and preventing *spatial interference* caused by transportation shutdowns.

5.3.2.3 The Strain of Aging Utility Systems: The Risk of Widespread Disruptions in Urban Infrastructure

Aging utility systems—such as water pipes, gas lines, and electrical grids—pose a significant challenge for cities where infrastructure hasn't been modernized in decades (Al-Bayati et al., 2024). Many of these systems were designed for smaller populations or have reached the end of their useful life. The practical challenge in addressing utility disruptions lies in upgrading or replacing these systems while maintaining service and minimizing the impact on urban functionality.

Utility failures can cause widespread interference, crippling daily life by disrupting water, power, and gas supply. In densely populated areas, a single disruption can have cascading effects, affecting transportation, healthcare, emergency services, and communication. Cities need to invest in infrastructure upgrades that are costly but essential for preventing service interruptions. Implementing real-time monitoring systems with sensors and automated diagnostics is key to detecting early-stage issues before they escalate, helping to avoid large-scale disruptions. Regular maintenance, including the cleaning of pipes and the replacement of aging components, is essential for keeping utility systems operational and minimizing interference with daily activities.

5.3.2.4 The Vulnerability of Earthquake-Prone Buildings: The Threat of Collapse and Urban Disruptions

In earthquake-prone regions, older buildings constructed before the adoption of modern earthquake-resistant standards face a heightened risk of collapse during seismic events (Stevenson, 1885; Liu et al., 2021). Retrofitting these buildings to meet current seismic safety codes presents a significant challenge, both logistically and financially. The lack of strong enforcement of building codes, especially in developing regions, further complicates efforts to strengthen these structures.

The collapse or severe damage to earthquake-weak buildings creates major *spatial interference*—displacing residents, obstructing streets, and impeding recovery efforts. Solutions may include reinforcing foundations, adding steel braces, or installing seismic dampers. A critical challenge lies in determining the most cost-effective way to carry out these retrofits. Public education, government-led funding initiatives, and incentives for retrofitting programs are essential to reducing the risk of such interference. Ensuring widespread adoption of seismic safety measures will help maintain urban stability and prevent large-scale disruptions in earthquake-prone areas.

5.3.3 **Urbanimmunology-*Induced Solutions to Structural and Infrastructure Failures***

Structural and infrastructure failures pose a significant risk to urban systems, particularly in cities with aging infrastructure. These risks can be effectively addressed by applying principles derived from human immune intelligence. Just as the human immune system defends the body against external threats, urban systems can be designed to protect and regenerate from the deteriorating elements within cities. The core principles of *Urbanimmunology* provide a framework for designing cities that not only prevent infrastructure failures but also recover from them, ensuring urban systems remain resilient and adaptable. The urban immune system—comprising critical dependencies, avoidance strategies, tolerance, adaptation, regulation, coordination, diversity, robustilient regeneration, and coherence—ensures that cities are resilient and capable of thriving, even in the face of challenges related to aging infrastructure and system failures.

5.3.3.1 Strengthening Urban Foundations: Urbanimmunology's Approach to Reinforcing Older Buildings

Older buildings, particularly those that have not been updated to meet modern safety standards, are at a higher risk of degradation due to factors like poor foundations, lack of seismic reinforcements, and neglect. Addressing these challenges requires a comprehensive and adaptive approach to ensure the city's physical and functional integrity is maintained. The *Critical Dependency Principle* emphasizes the importance of buildings as vital nodes within urban spaces, as their failure can create significant *spatial interference*—disrupting not only the immediate area but also surrounding infrastructure and mobility. Retrofitting and strengthening these buildings is essential to preventing such disruptions. The *Adaptation Principle* encourages flexibility when dealing with older structures, advocating for modern solutions like seismic retrofitting and foundation strengthening, which help ensure the longevity of these structures while mitigating the risk of spatial dislocation. The *Tolerance Principle* supports incremental improvements to extend the lifespan of existing buildings, minimizing the need for large-scale, disruptive renovations. The

Diversity Principle encourages the use of diverse retrofitting methods, materials, and technologies, ensuring that each building is adapted to its specific needs, which is essential for avoiding localized interference that could affect the broader urban system.

5.3.3.2 Strengthening Critical Infrastructure: Urbanimmunology's Vision for Long-Lasting Bridges and Tunnels

Bridges, tunnels, and other critical infrastructure elements face deterioration from constant use, weather exposure, and aging. The failure of these systems creates significant *spatial interference* by disrupting transportation networks, which are essential for the smooth flow of people, goods, and services. The *Avoidance Principle* highlights the importance of preventing overuse of existing infrastructure, prioritizing upgrades or replacements to avoid exacerbating strain on these systems. The *Regulation Principle* stresses the need for strict maintenance and safety protocols to prevent failures, ensuring minimal disruption in daily life and maintaining the integrity of urban spatial organization. The *Coordination Principle* emphasizes the need for coordinated efforts across various sectors—government, transportation, and engineering—to manage infrastructure efficiently, ensuring timely repairs that prevent prolonged disruptions. Lastly, the *Robustilient Regeneration Principle* calls for infrastructure upgrades that enhance resilience, allowing systems to adapt to future pressures, such as increased traffic or more extreme weather, without disrupting the urban space they serve.

5.3.3.3 Building Future-Ready Utility Systems: A Robust, Dynamic, and Unified Approach to Urban Resilience

Aging utility systems, such as water pipes, gas lines, and electrical grids, are prone to failures. These failures disrupt not only the provision of essential services but also create far-reaching *spatial interference*, affecting everything from transportation to emergency response. The *Diversity Principle* encourages cities to diversify their utility systems by integrating renewable energy sources, alternative materials, and smart technologies, thus reducing the risks associated with any single failure point. The *Robustilient Regeneration Principle* emphasizes investment in modernizing utility systems to ensure they can quickly recover from disruptions and adapt to future challenges like extreme weather events. The *Coherence Principle* stresses the need to integrate utility systems with other urban infrastructure (e.g., transportation and emergency services), ensuring that service disruptions do not cascade and exacerbate *spatial interference* in the city's functioning. The *Adaptation Principle* highlights the importance of flexibility, encouraging the development of modular, adaptable utility systems that can evolve in response to unforeseen disruptions, thus minimizing their spatial impact.

5.3.3.4 Seismic-Resilient Urban Spaces: Advancing Building Safety Through Coordinated, Robust, Adaptive, and Cohesive Urban Resilience

In regions prone to seismic activity, older buildings that were constructed before the adoption of modern earthquake-resistant standards are at a significant risk of collapse. The *Critical Dependency Principle* underscores the importance of protecting these buildings, as their failure can lead to severe *spatial interference*—blocking roads, displacing communities, and disrupting essential urban services. Retrofitting these buildings to meet modern seismic standards helps reduce the potential for such interference. The *Adaptation Principle* encourages the use of modern seismic technologies like base isolators and wall reinforcements to ensure that these structures remain functional after a seismic event, preventing long-term disruption in the surrounding urban space. The *Robustilient Regeneration Principle* focuses on reinforcing buildings to ensure they remain operational after an earthquake, reducing the overall impact on the city's spatial functionality. The *Coordination Principle* emphasizes the need for cooperation between governments, urban planners, engineers, and property owners to ensure that seismic safety is a collective priority, and that retrofitting efforts are aligned with broader urban planning goals. Finally, the *Coherence Principle* stresses the importance of integrating seismic safety measures into city planning, ensuring that retrofitting efforts contribute to a cohesive urban environment that is both safe and functional, even during and after seismic events.

5.4 Landslide Hazards

Landslide hazards refer to the potential dangers posed by soil and slope instability in regions susceptible to landslides, mudslides, or other types of mass movements (Keefer & Larsen, 2007; Pardeshi et al., 2013; Huang et al., 2024). These hazards extend beyond human settlements and can also affect critical infrastructure such as roads, bridges, and railways, even in areas with little or no human presence. While the immediate risk to human life is obvious in populated areas, the broader impact of landslides on infrastructure systems causes significant *spatial interference*—disrupting transportation networks, blocking essential access routes, and hindering recovery operations. Contributing factors to landslide hazards (Glade et al., 2005; Wang et al., 2005) include urban development in hilly or mountainous regions, especially in areas lacking adequate risk mitigation measures, as well as locations prone to natural instability. Factors such as heavy rainfall, seismic activity, and poor land-use practices—such as deforestation, unregulated construction, or inadequate drainage—can trigger landslides. These events often occur unpredictably and without warning, posing significant risks to both human life and infrastructure systems, disrupting spatial connectivity across affected regions.

5.4.1 Consequences of Landslide Hazards

The consequences of landslide hazards are far-reaching and can result in devastating disruptions, particularly when they impact populated areas or vital infrastructure.

5.4.1.1 The Tragic Impact on Lives and Communities

Landslides can have a profound human impact, leading to the tragic loss of life (Petrucci, 2022), with individuals being trapped or crushed by falling debris or collapsing structures. The injuries sustained can be severe, affecting both residents and emergency responders, and making rescue operations more challenging (Wang et al., 2024). Furthermore, landslides often displace entire communities, especially those living in informal settlements built on unstable terrain, forcing people to evacuate or relocate (Mızrak & Turan, 2023). This disruption of human settlement patterns creates long-term social and economic instability for the affected individuals. The relocation and displacement can also strain other regions or neighborhoods, leading to further *spatial interference* as people are forced to migrate and compete for resources.

5.4.1.2 Devastating Infrastructure Destruction

Landslides can cause significant damage to infrastructure, resulting in the destruction of residential and commercial buildings, and leading to considerable economic losses and interruptions to daily life (Chandrasekaran et al., 2013; Gnyawali et al., 2023). Entire neighborhoods may collapse, exacerbating the extent of the impact. Moreover, landslides that block roads, bridges, or railways isolate communities, disrupting access to essential services, supplies, and evacuation routes. These blockages also obstruct emergency response and recovery efforts, causing delays and increasing the complexity of relief operations. Furthermore, critical public utilities such as water, sewage, and electricity systems can be severely damaged, leading to disruptions in basic services and creating public health hazards that complicate the recovery process. This *spatial interference* hampers not only the immediate functionality of the city or region but also its long-term resilience and ability to recover.

5.4.1.3 Destructive Environmental Degradation

Landslides can result in severe environmental consequences, including substantial soil erosion and land degradation, which further destabilize the area and increase the risk of future landslides (Pradhan et al., 2012; Miccadei et al., 2022). The erosion process can negatively affect agricultural productivity, disrupt water quality in rivers and streams, and lead to sedimentation in water sources. Additionally, landslides can destroy natural habitats, displacing local wildlife and plant species. This disturbance to ecosystems can have long-lasting impacts, particularly in fragile mountainous regions, and disrupt the balance of the environment. The environmental degradation caused by landslides amplifies *spatial interference* in the affected areas, making recovery more challenging and affecting both human and ecological systems.

5.4.1.4 Crippling Economic Consequences

Landslides can impose a significant economic burden on affected regions, requiring substantial financial resources for repairing and rebuilding damaged infrastructure, homes, and public utilities. The destruction of key infrastructure such as transportation networks, commercial areas, and agricultural lands disrupts local economies, resulting in income loss, unemployment, and reduced economic activity (Turner, 2018). The recovery process can be protracted, especially in resource-limited areas, where financial constraints can delay restoration efforts. These economic disruptions exacerbate *spatial interference* by preventing the full functionality of affected areas, hindering mobility and trade, and straining local governments' capacity to manage post-disaster recovery. The broader economic ripple effects can extend beyond the immediate region, further affecting neighboring communities.

5.4.2 *Practical Issues in Addressing Landslide Hazards*

5.4.2.1 Critical Failures in Risk Assessment and Monitoring: The Escalating Danger of Inadequate Preparedness

One of the main practical issues in addressing landslide hazards is effective risk assessment and continuous monitoring (Casagli et al., 2023). Identifying landslide-prone areas requires comprehensive geological studies and real-time monitoring of soil stability. However, in many regions, particularly in developing areas, there may be limited data, insufficient technology, and a lack of financial resources to effectively track changes in terrain stability. Without reliable systems for early detection, preventing or mitigating the effects of landslides becomes increasingly difficult. Moreover, *spatial interference* often occurs when there is a failure to update and maintain accurate hazard maps (Li et al., 2022), which complicates risk management efforts. Inaccurate or outdated hazard maps lead to poor land-use decisions that increase vulnerability and prevent targeted interventions.

5.4.2.2 Dangerous Oversight in Land-Use Planning: Urban Development in High-Risk Zones Exacerbates Vulnerability

Urban development in landslide-prone areas represents a significant challenge in addressing these hazards (Li & Samsudin, 2024). Despite awareness of the risks, cities and informal settlements often encroach upon unstable zones due to population pressures and a shortage of developable land. Even where zoning regulations exist, enforcement may be inconsistent or inadequate, with economic interests frequently outweighing safety concerns. This lack of comprehensive land-use planning and enforcement exacerbates the vulnerability of buildings and infrastructure to landslides. Particularly in informal settlements, where construction practices often lack necessary risk mitigation measures, *spatial interference* is worsened as development pushes into hazardous areas, exacerbating both human and infrastructural risks.

5.4.2.3 Critical Failures in Construction Practices: Inadequate Building Resilience in Landslide-Prone Areas

In landslide-prone areas, the integrity of buildings and infrastructure is often compromised by poor construction practices (Lungu et al., 2018). Many structures were built without considering the geological risks posed by unstable soil or inadequate drainage systems. Retrofitting older buildings and structures to resist landslides is expensive and logistically challenging, especially in informal or economically disadvantaged communities. Additionally, the lack of local expertise and limited financial capacity prevent property owners from implementing the necessary safety upgrades. As a result, these buildings are more vulnerable to collapse or damage during a landslide, contributing to *spatial interference* by disrupting both physical structures and the flow of daily life, particularly in areas where people live and work in proximity to landslide-prone zones.

5.4.2.4 Critical Infrastructure Vulnerability: A Growing and Unaddressed Crisis

Landslides can cause severe damage to critical infrastructure, including roads, bridges, power lines, and water systems (Jirásko et al., 2017). Ensuring the resilience of such infrastructure requires significant investment in durable materials, regular maintenance, and careful engineering. However, in many regions, there is insufficient funding and technical expertise to build or maintain infrastructure capable of withstanding landslide events. Furthermore, transportation and communication networks may not be designed to accommodate the challenges posed by landslides, leading to long-term disruptions and the isolation of affected communities. This creates further *spatial interference* as regions are cut off from critical services, impeding recovery efforts and diminishing the resilience of entire communities.

5.4.2.5 Gaps in Emergency Preparedness and Response: A Dire Threat to Communities

Given the unpredictable nature of landslides, emergency preparedness and response are crucial but challenging (Bhatia, 2013; Tsai et al., 2021). Many communities lack the resources, infrastructure, and trained personnel to adequately prepare for or respond to landslide events. Early warning systems may be underdeveloped or entirely absent, making it difficult for residents to evacuate in time. Even when emergency response mechanisms are in place, local authorities often lack the capacity to coordinate relief efforts effectively, especially in remote or rural areas. The absence of evacuation plans and the inadequacy of emergency shelters further exacerbate the crisis, making *spatial interference* even more pronounced as local infrastructure and emergency systems are overwhelmed by the scale of the event.

5.4.2.6 Escalating Environmental Degradation and Its Critical Role in Land Destabilization

Deforestation, soil erosion, and unsustainable land-use practices contribute significantly to landslide risk. However, addressing environmental degradation through conservation efforts presents numerous practical challenges. Large-scale reforestation or soil stabilization projects require long-term investment, strong political will, and effective community involvement (Glade et al., 2005; Sujatha et al., 2023). In many regions, economic pressures encourage unsustainable practices like mining, deforestation, and unregulated construction, further destabilizing the land. These practices not only increase the likelihood of landslides but also hinder recovery and mitigation once they occur. The ongoing environmental degradation amplifies *spatial interference* by reducing the land's capacity to recover, causing cascading impacts that extend beyond the immediate hazard zone, further destabilizing the surrounding region.

5.4.2.7 Unmanageable Economic Strain and Social Displacement

The economic and social implications of landslides present significant obstacles to mitigating the hazard effectively (Alcántara-Ayala, 2025). The financial burden of landslide prevention, monitoring, and mitigation efforts can overwhelm local governments, especially in low-income regions. Reconstruction costs following a landslide event are substantial, placing further strain on local economies. Moreover, displaced communities often face long-term social instability, particularly if they are forced to relocate to areas without sufficient infrastructure or economic opportunities. The economic disruptions caused by landslides, including the destruction of agricultural land or commercial areas, lead to extended recovery periods and hinder local economic recovery. *Spatial interference* in these contexts means that the movement of displaced populations disrupts neighboring regions, further straining resources and delaying recovery efforts.

5.4.2.8 Fragmented Governance and Poorly Coordinated Response

Addressing landslide hazards requires effective coordination between various stakeholders, including government agencies, engineers, local communities, and other organizations (Scolobig, 2016). This can often be a significant practical issue, as fragmented governance structures, poor communication, and political challenges can impede collaboration. Effective land-use planning, infrastructure investment, and disaster preparedness require the involvement of multiple sectors working together. In many cases, the absence of a clear, coordinated strategy or governance framework leads to ineffective hazard management and delayed responses. The lack of coordinated governance also increases *spatial interference* by causing fragmented recovery efforts that delay mitigation and prevent effective action in vulnerable areas.

5.4.3 Urbanimmunology-*Informed Solutions to Landslide Hazards*

By applying all nine core principles of *Urbanimmunology*, we can craft solutions that are sustainable, adaptable, and resilient, ensuring cities can not only survive but thrive in the face of landslide hazards.

5.4.3.1 Strategic Risk Assessment and Monitoring as Pillars of Urban Immunity

Risk assessment and monitoring are crucial for mitigating the impact of landslides and ensuring the safety and resilience of urban areas. The *Critical Dependency Principle* emphasizes the importance of building essential systems for monitoring soil stability and conducting risk assessments, making these systems foundational for long-term urban survival. By establishing comprehensive geological studies, real-time monitoring systems, and early detection technologies, cities can implement timely interventions to protect communities. The *Adaptation Principle* highlights the need for flexibility in these systems to incorporate emerging technologies, such as AI-based tools for predictive modeling or advanced remote sensing methods. This allows cities to stay proactive in managing risks, even as environmental and technological conditions evolve. The *Coherence Principle* stresses the need for risk monitoring systems to integrate smoothly with other urban systems, such as infrastructure, emergency services, and governance, ensuring a unified and efficient approach to disaster management. Lastly, the *Regulation Principle* underscores the importance of strict enforcement of regulatory standards for monitoring landslide-prone areas and updating hazard maps, ensuring consistency and safety across urban regions.

5.4.3.2 Strategic Land Use and Zoning for Balanced Growth and Risk Reduction

Land-use planning and zoning are essential tools for managing landslide risks and ensuring safe urban development. The *Avoidance Principle* calls for halting development in landslide-prone areas by strictly enforcing zoning regulations, preventing new construction in hazardous zones, and thereby minimizing future risks. The *Tolerance Principle* acknowledges that, while avoiding risky areas is ideal, some development may still occur in less hazardous zones, provided safety features like reinforced infrastructure and proper drainage are incorporated. This allows urban growth to continue without compromising safety. The *Regulation Principle* stresses the importance of strong regulatory frameworks, including zoning laws and building codes that ensure resilience standards are met, preventing the creation of new risks in safer areas. The *Coordination Principle* highlights the need for effective collaboration between urban planners, engineers, and government authorities to design zoning laws that balance risk mitigation with urban development needs. Lastly, the *Diversity Principle* encourages flexibility in zoning laws to accommodate varying levels of risk across different regions, enabling urban areas to evolve in a way that meets both safety and development goals.

5.4.3.3 Strengthening Cities: Resilient Construction and Buildings for Risk-Responsive Development

Construction practices and building resilience are critical in reducing the impact of landslides and ensuring long-term safety in urban areas. The *Critical Dependency Principle* stresses that construction practices should prioritize safety and resilience, ensuring buildings can withstand natural hazards and protect urban communities from future disasters. The *Adaptation Principle* encourages flexibility in construction techniques, utilizing locally available materials, traditional knowledge, and modern engineering to build structures that can be upgraded as risks and technologies evolve. The *Tolerance Principle* acknowledges that in older, vulnerable areas, full retrofitting may not be immediately possible; thus, gradual improvements such as adding retaining walls or improving drainage can begin to enhance resilience. The *Regulation Principle* emphasizes the importance of strong regulations, including mandatory building codes and geological surveys, to ensure that new and existing buildings can withstand landslides. The *Robustilient Regeneration Principle* advocates for recovery efforts that focus not just on rebuilding but on regenerating infrastructure to improve long-term resilience through reinforced foundations and flexible designs. Finally, the *Coherence Principle* highlights the need for collaboration among engineers, architects, and urban planners to ensure that construction practices align with broader urban planning and environmental strategies, creating a unified approach to resilience.

5.4.3.4 Evolutionary Infrastructure Resilience and Maintenance for Future-Proof Urban Systems

Infrastructure resilience and maintenance are essential for ensuring cities can withstand and recover from landslide events. The *Critical Dependency Principle* emphasizes the need for resilient infrastructure, as cities rely on systems like roads, bridges, and power supply to maintain daily functions and safeguard the well-being of their populations. The *Adaptation Principle* encourages flexible infrastructure design, allowing systems to evolve in response to changing environmental conditions, population growth, or technological advancements, ensuring long-term resilience. The *Regulation Principle* advocates for rigorous standards that require thorough risk assessments and the use of durable materials in infrastructure projects, ensuring their ability to withstand landslides while also promoting regular maintenance to prevent failures. The *Robustilient Regeneration Principle* highlights the importance of designing infrastructure that not only withstands immediate events but is also capable of recovering quickly, incorporating redundancies like alternative routes and backup systems to minimize downtime. Finally, the *Coherence Principle* stresses the need for infrastructure to integrate seamlessly with other urban systems, such as emergency response and utilities, ensuring coordinated efforts that align with broader risk management strategies.

5.4.3.5 Built for the Worst: Principles for Emergency Preparedness and Response

Effective emergency preparedness and response depend on a range of principles that ensure readiness and adaptability. The *Coordination Principle* emphasizes the importance of collaboration among local governments, emergency responders, community organizations, and residents, facilitating more efficient evacuations, timely rescues, and quicker recovery. The *Tolerance Principle* acknowledges that despite thorough preparation, risk and uncertainty are inevitable, requiring flexible plans that can adjust to unforeseen challenges in resources or logistics during a landslide event. The *Critical Dependency Principle* highlights the necessity of robust emergency systems, such as strengthened early warning systems, clear evacuation plans, and well-trained responders, particularly in landslide-prone areas. The *Adaptation Principle* advocates for the evolution of emergency systems in response to emerging risks and new technologies, incorporating digital tools, crowdsourced data, and innovative recovery techniques. Finally, the *Resilient Regeneration Principle* ensures that post-disaster recovery not only restores communities but also enhances their resilience, preparing them better for future events.

5.4.3.6 Environmental and Land Protection: From Prevention to Regeneration

Environmental conservation is crucial in addressing land degradation and mitigating landslide risks, guided by several key principles. The *Diversity Principle* emphasizes the need for a range of conservation practices, such as reforestation, soil stabilization, and sustainable land management, tailored to local needs to restore ecosystems and reduce landslide threats. The *Critical Dependency Principle* stresses that communities rely on natural systems to prevent erosion, regulate water flow, and stabilize soil, making the protection and enhancement of these natural barriers essential for long-term land stability. The *Avoidance Principle* advocates for limiting practices like deforestation, unsustainable mining, and construction in ecologically sensitive areas to prevent further environmental degradation and reduce the risk of landslides. Finally, the *Robustilient Regeneration Principle* calls for ecosystem restoration efforts that not only restore environmental integrity but also strengthen ecosystems to endure and recover from future challenges, such as landslides, floods, and other environmental pressures.

5.4.3.7 Healing Economies, Reconnecting Communities: A Coordinated Vision for Sustainable Recovery

Economic recovery after a landslide should focus not only on immediate reconstruction but also on building long-term resilience, as highlighted by the *Resilient Regeneration Principle*. This principle emphasizes the importance of diversifying income sources, enhancing economic independence, and creating a more robust local economy that can withstand future disruptions. The *Coordination Principle* underscores the need for collaboration between local authorities, financial institutions, and community organizations to ensure displaced communities receive both immediate aid and sustained economic support, with well-coordinated recovery programs being crucial to minimizing the social and economic impact of landslides. Lastly, the *Tolerance Principle* acknowledges that recovery is often a slow and incremental process, advocating for patience and support for gradual rebuilding and development strategies that lead to more sustainable economic and social rehabilitation in affected areas.

5.4.3.8 Governance in Harmony: Coordinated Action for Resilient Urban Systems

Effective governance in managing landslide hazards relies on several key principles. The *Coordination Principle* stresses the importance of clear roles and responsibilities among various levels of government, private sector actors, and community organizations, with collaborative decision-making and efficient communication being essential for timely and effective hazard management. The *Regulation Principle* highlights the need for strong governance structures to enforce laws and regulations related to land use, construction, environmental conservation, and risk management, ensuring that landslide mitigation efforts are carried out consistently. Finally, the *Coherence Principle* emphasizes that the governance system should function cohesively across all levels, from local governments to national authorities, ensuring that urban systems such as planning, risk assessment, emergency management, and recovery work together in a unified response to landslide hazards.

5.5 Flooding and Stormwater Hazards

Flooding and stormwater hazards represent significant risks in urban areas, often resulting from *spatial interference* between urban development, drainage systems, and the natural environment (Kvočka et al., 2016). These hazards are exacerbated by inadequate drainage infrastructure, excessive impermeable surfaces, and insufficient flood management measures (La Loggia et al., 2020; Yereseme et al., 2022; Alves et al., 2024). Urbanization, combined with poor planning and limited foresight, leaves many cities ill-prepared for extreme weather events, creating a complex web of vulnerabilities that increase the risk of flooding.

A primary contributing factor to flooding is the inefficiency of urban drainage systems. Many cities rely on outdated or inadequate stormwater infrastructure that struggles to cope with intense rainfall or sudden, heavy storms. When stormwater is not effectively managed, it can lead to the inundation of streets, buildings, and entire neighborhoods, causing significant disruption and damage. The lack of modern, adaptive drainage solutions further amplifies the *spatial interference* between urban growth and natural water management systems.

Another key factor is the proliferation of impermeable surfaces. As urban areas expand, natural land surfaces such as soil and grass are increasingly replaced with impermeable materials like concrete, asphalt, and rooftops. These surfaces hinder the natural absorption of water into the ground, causing rainfall to accumulate and overflow, often flooding streets and properties. This *spatial interference* between natural water systems and urban infrastructure makes it difficult for water to be properly absorbed or redirected, raising the risk of urban flooding.

Furthermore, many cities lack adequate flood management infrastructure, including levees, dams, or modern stormwater systems designed to channel excess water away from vulnerable areas. In cities where such infrastructure is either absent or poorly maintained, the risk of damage during heavy storms or rainfall is significantly amplified. Without proactive, coordinated management, these cities remain highly vulnerable to large-scale flooding, particularly when *spatial interference* prevents proper water flow and mitigation.

5.5.1 Consequences of Flooding and Stormwater Hazards

The consequences of flooding and stormwater hazards can be both immediate and long-lasting, affecting not only the physical environment but also the health and safety of communities. These hazards arise from *spatial interference* between urban infrastructure, natural water management systems, and climate variability, leading to widespread impacts.

5.5.1.1 Flash Flooding

In urban areas where drainage systems are inadequate, flash flooding (Liu et al., 2024, 2024, 2024) can occur suddenly following a heavy downpour. Stormwater quickly accumulates in streets, parks, and buildings, overwhelming local infrastructure and exacerbating existing vulnerabilities. The *spatial interference* between urban development and natural water channels prevents water from being absorbed or redirected, causing significant damage to property, roads, and public utilities. Flash flooding can also lead to dangerous situations, such as vehicles being swept away or structures collapsing under the pressure of rising water, particularly where poorly planned development has exacerbated runoff.

5.5.1.2 Waterborne Diseases

Flooding often results in the contamination of drinking water sources as stormwater mixes with sewage and other pollutants. In areas where stormwater management systems are either insufficient or nonexistent, *spatial interference* between urban growth and the environment creates ideal conditions for the spread of waterborne diseases like cholera, typhoid, and dysentery (Shayo et al., 2023). These diseases are most prevalent in densely populated areas with inadequate sanitation infrastructure. The contamination of water supplies can quickly lead to a public health crisis, amplifying the risks in already vulnerable communities.

5.5.1.3 Public Health Crisis

The immediate physical dangers of flooding—such as drowning, injury from debris, and water-related accidents—can be compounded by a long-term public health crisis (Leppold et al., 2022; Suhr & Steinert, 2022). The *spatial interference* between flood-prone areas and housing or infrastructure often results in the destruction of homes, livelihoods, and public services. The resulting stress, anxiety, and trauma can have significant mental health impacts, particularly in communities already facing social and economic challenges. Furthermore, the disruption of healthcare facilities and emergency services during major floods can hinder the delivery of medical care, escalating the crisis. Overcrowded shelters and unsanitary conditions in the aftermath of floods can facilitate the spread of infections and disease, making recovery efforts more difficult.

5.5.2 Practical Issues in Addressing Flooding and Stormwater Hazards

Addressing flooding and stormwater hazards is a complex and multifaceted challenge, requiring a combination of infrastructure improvements, policy interventions, and community engagement. The causes and consequences of these hazards—ranging from poor urban drainage systems to flash flooding and waterborne diseases—demand a comprehensive, spatially informed approach. Several practical issues, however, can hinder effective solutions. These issues stem from financial constraints, urban planning challenges, and the broader impact of climate change, among other factors.

5.5.2.1 Underfunded and Underequipped: Cities Left Defenseless Against Flooding

One of the most significant challenges in addressing flooding and stormwater hazards is securing adequate funding, particularly in cities that face financial constraints, especially in developing regions (Peck et al., 2022). This lack of funding limits their ability to invest in essential infrastructure, which in turn leads to delayed or insufficient flood mitigation measures such as effective drainage systems, flood barriers, or reservoirs. The *spatial interference* between urban development and natural water systems often exacerbates the need for more extensive stormwater management infrastructure, including levees, dams, and retention ponds. Building or upgrading such infrastructure is costly, and without proper investment, cities remain vulnerable to the devastating effects of flooding, which leads to even higher costs, including property damage, lost productivity, and public health crises.

5.5.2.2 Unchecked Growth, Unnatural Disasters: Urbanization's Role in Flooding

Rapid urbanization and overdevelopment significantly contribute to both the causes and consequences of flooding and stormwater hazards (Feng et al., 2021; Cao et al., 2022). As cities expand, natural landscapes are replaced by impermeable surfaces such as concrete, asphalt, and buildings, further disrupting stormwater management. The *spatial interference* between urban expansion and natural water flow creates an imbalance, where the loss of permeable surfaces—like soil, grass, and wetlands—prevents rainwater from being absorbed into the ground. Without adequate green spaces or permeable pavements, excessive runoff overwhelms existing drainage systems. Moreover, unplanned urban sprawl often leads to the development of flood-prone areas, such as floodplains or regions near water bodies, where the existing infrastructure is ill-suited to manage storms or floods. The lack of foresight in integrating drainage systems and flood protection measures into new developments increases vulnerability, leaving urban areas more prone to flooding.

5.5.2.3 Rising Waters, Failing Systems: The New Reality of Climate-Induced Flooding

Climate change significantly exacerbates the frequency and intensity of storms, heavy rainfall, and flooding (Hirabayashi et al., 2013; Callahan, 2025), making existing stormwater systems increasingly inadequate. Unpredictable weather patterns, including more intense rainfall, longer wet seasons, and flash floods, contribute to *spatial interference* by outpacing the capacity of urban systems to respond. Many cities, built in the twentieth century, were not designed to accommodate these severe and more frequent rainfall patterns. Rising sea levels in coastal areas add another layer of risk, particularly in cities lacking proper flood defenses. The combination of rising sea levels and heavy rainfall compounds flooding risks, overwhelming drainage systems and flood protection infrastructure, which were not originally designed to handle such events.

5.5.2.4 Neglected and Overrun: The Infrastructure Crisis Beneath Our Streets

Even in developed cities, the effectiveness of stormwater management infrastructure can be compromised due to aging systems, undercapacity, and poor maintenance (Chang et al., 2013; Vamvakeridou-Lyroudia et al., 2020). Many cities with older infrastructure face the challenge of drainage systems that were designed decades ago, often when urban populations were smaller, and rainfall patterns were more predictable. As urban populations grow and extreme weather events become more frequent, these outdated systems struggle to cope. Additionally, many urban areas have drainage systems that are too small to handle the volume of water during heavy rainfall. These systems may be designed for average rainfall but fail during extreme events, leading to flash flooding and property damage. Even when infrastructure is in place, inadequate maintenance can create issues such as blocked drains, clogged waterways, and damaged flood defenses, preventing stormwater from flowing efficiently and increasing the risk of flooding. Regular inspection and maintenance are essential to ensure these systems function properly during heavy rains.

5.5.2.5 Disjointed Authority, Unified Risk: The Price of Policy and Governance Failure

Effective flood management requires strong governance, coordination between various levels of government, and clear policies prioritizing flood prevention and mitigation (Alexander et al., 2016; Liu et al., 2025). In many regions, *spatial interference* between fragmented governance systems and urban development creates significant obstacles to effective flood management. The lack of integrated policies contributes to ineffective responses to flooding and stormwater hazards. In some countries, flood management responsibilities are divided among multiple governmental agencies, local governments, national agencies, and private sector entities, leading to fragmented and uncoordinated efforts. This lack of a unified approach can result in conflicting priorities or an inability to work together effectively, causing vital steps in flood risk reduction—such as land-use regulations, flood zoning, or constructing flood defenses—to be overlooked. Furthermore, a lack of public awareness and engagement can hinder the implementation of effective flood prevention strategies. Educating residents about flood risks, encouraging community involvement, and promoting sustainable development practices are critical, yet in many cities, public resistance to change or lack of engagement hinders these efforts.

5.5.2.6 Unprotected and Unequal: The Deepening Social and Economic Divide

The social and economic impacts of flooding and stormwater hazards are often most severe in low-income or marginalized communities, which are disproportionately vulnerable to these events (Hossain et al., 2025). These communities often face *spatial interference* between social inequalities, inadequate infrastructure, and environmental vulnerabilities, leaving them exposed to the dangers of flooding. Poor neighborhoods in flood-prone areas may lack the resources or political power to advocate for infrastructure improvements, further increasing their vulnerability. Additionally, these communities are less likely to have insurance to cover flood damage, making recovery even more difficult. Floods lead to significant property damage, loss of livelihoods, and displacement, particularly in areas without adequate flood defenses. Rebuilding after such events can be prohibitively expensive, and many households, particularly in developing regions, cannot afford recovery costs, creating a cycle of vulnerability that traps communities in poverty due to recurrent flooding.

5.5.2.7 United by Risk, Divided by Action: The Crisis in Global Flood Coordination

Addressing flooding and stormwater hazards requires global cooperation and the sharing of knowledge and best practices, as many cities face similar challenges. A significant barrier to effective flood management is the lack of reliable data on flooding patterns, stormwater management, and urban resilience in many regions, which makes it difficult to predict events, plan infrastructure improvements, or develop effective risk reduction strategies. International cooperation allows cities to share data, technology, and expertise to create better solutions (Alfieri et al., 2018). While flooding and stormwater hazards are local, they are often tied to broader global issues, such as climate change and urbanization. Therefore, effective solutions will require collaboration not only within countries but also between nations and international organizations to address the underlying causes of these risks.

5.5.3 Urbanimmunology-*Informed Solutions to Flooding and Stormwater Hazards*

5.5.3.1 Essential Investments: Coordinating Funding and Resources for Flood-Resilient Cities

To address the issue of inadequate funding for flood management, the *Critical Dependency Principle* stresses the importance of treating flood management infrastructure as a vital urban system that should be prioritized. Just as the immune system relies on critical organs to maintain health, cities should treat flood management infrastructure—such as drainage systems and flood barriers—as indispensable to their long-term survival. The *Regulation Principle* supports this by advocating for strong policies that ensure ongoing funding and maintenance for flood prevention measures, such as tax incentives, public–private partnerships, and climate adaptation funds. This can help cities secure the resources necessary to implement large-scale infrastructure projects. Furthermore, the *Coordination Principle* encourages collaboration among various stakeholders—governments, the private sector, and international organizations—to pool resources and share financial responsibilities. This collaborative effort can ensure that funds are used efficiently, reducing *spatial interference* between fragmented initiatives and creating a unified approach toward common flood resilience goals.

5.5.3.2 Designing with Nature: Harmonizing Urban Growth and Flood Resilience

To tackle the challenge of rapid urbanization contributing to flooding, the *Avoidance Principle* is key. This principle advocates for halting further development in flood-prone areas, such as floodplains or areas with insufficient drainage capacity. By restricting construction in vulnerable zones, cities can reduce the risk of overburdening already fragile flood management systems. The *Tolerance Principle* plays a vital role in already developed areas, encouraging cities to adapt to the existing urban landscape by incorporating flood-resistant features into buildings and infrastructure. This may include the use of green roofs, permeable pavements, and increased green spaces, which help mitigate *spatial interference* between built environments and natural water flow. Additionally, the *Coherence Principle* ensures that urban development is harmonized with nature-based flood management strategies. By integrating natural solutions, like wetlands or urban forests, into the urban planning process, cities can improve stormwater retention and reduce the strain on drainage systems.

5.5.3.3 Climate-Resilient Futures: Adapting and Regenerating Cities Through Diverse Solutions

The *Adaptation Principle* is essential in responding to climate change-induced flood risks. As weather patterns become more unpredictable, cities should build flexible, adaptive flood management systems that can accommodate the changing climate. Infrastructure should be designed with future adjustments in mind, enabling cities to adapt to increasing rainfall, stronger storms, and rising sea levels. The *Diversity Principle* supports this by recommending a diverse array of flood management strategies, combining engineered solutions like dams and levees with natural approaches such as coastal wetlands or urban forests. This diversity enhances the city's ability to cope with different types of flood risks. Additionally, the *Robustilient Regeneration Principle* focuses on rebuilding after climate-induced flooding in a way that enhances resilience. Recovery efforts should incorporate flood-resilient designs that not only restore the city but also ensure its long-term capacity to withstand future climate challenges, minimizing *spatial interference* between development and environmental factors.

5.5.3.4 Vital Drainage: Strengthening the Backbone of Urban Resilience

To improve drainage systems, the *Critical Dependency Principle* emphasizes that stormwater management infrastructure is critical to the city's functioning and resilience. Outdated or insufficient drainage systems should be replaced or upgraded to handle the volume of water during extreme weather events. Regular maintenance of these systems is also essential to prevent blockages or inefficiencies. The *Adaptation*

Principle supports the need for flexible infrastructure that can grow with the city's evolving population and changing environmental conditions. Cities should invest in scalable systems that can be expanded or modified as needed. The *Regulation Principle* is also crucial, as it calls for robust urban policies that mandate improvements in drainage design and maintenance standards. Regulations should ensure that drainage systems are able to accommodate both average rainfall and extreme weather events, preventing widespread flooding and minimizing property damage, thereby reducing *spatial interference* between urban development and water flow.

5.5.3.5 From Silos to Synergy: Aligned Policies and Responsive Governance for Flood Resilience

The *Coordination Principle* is fundamental in addressing fragmented policies and improving governance. Effective flood management requires that various government agencies work together, aligning their goals and coordinating their actions to ensure a unified and cohesive approach to flood risk management. This may involve creating integrated flood management plans that include zoning regulations, land-use policies, and flood defenses. The *Regulation Principle* ensures that governments enforce clear policies and guidelines related to flood mitigation, such as land-use planning in flood-prone areas and building codes that require flood-resistant designs. Furthermore, the *Tolerance Principle* encourages cities to be more flexible in their governance by allowing for changes in regulations and strategies as the urban landscape evolves. This flexibility enables cities to quickly respond to emerging flood risks without being constrained by rigid policies, reducing *spatial interference* between governance structures and urban growth.

5.5.3.6 Inclusive Recovery: Strengthening Social and Economic Resilience to Flooding

The *Diversity Principle* is key in addressing the social and economic vulnerabilities that flood-prone communities face. By ensuring that flood management strategies are inclusive and cater to diverse populations—especially vulnerable groups such as low-income communities—cities can better manage the risks of displacement and property loss. Socially diverse and inclusive solutions, such as community-based flood prevention initiatives and affordable housing in safer areas, can reduce the disproportionate impacts of flooding. Additionally, the *Robustilient Regeneration Principle* calls for the rebuilding of flood-affected communities in ways that not only restore infrastructure but also improve the resilience of vulnerable areas. Efforts should focus on providing long-term solutions that ensure these communities can recover and thrive even after repeated flooding events, minimizing *spatial interference* between social vulnerabilities and urban resilience.

5.5.3.7 Uniting Knowledge and Action: Global Coordination for Adaptive Urban Flood Management

The *Coordination Principle* plays a critical role in fostering international cooperation and knowledge sharing. As flooding and stormwater hazards are global challenges, cities can benefit from the experiences, technologies, and best practices of others. Collaborative networks—such as international flood management forums or climate change adaptation initiatives—can help cities access data, research, and technologies that they may not have on their own. The *Adaptation Principle* also supports global knowledge sharing by encouraging cities to adapt flood management strategies that have worked elsewhere while taking into account their unique geographic and climatic conditions. By learning from global experiences, cities can build more resilient flood management systems and be better prepared for future challenges, thus reducing *spatial interference* between different regions' policies and flooding outcomes.

5.6 Human-Wildlife Interaction Hazards

Human-wildlife interaction hazards refer to the risks and challenges that arise when humans and wildlife come into contact, often leading to negative consequences for both parties (Soulsbury & White, 2015; Nyhus, 2016; Pătru-Stupariu et al., 2020; Abrahms et al., 2023; Narayan & Rana, 2023). These interactions can be intentional or unintentional and occur in various settings, including urban, rural, and protected natural environments. The hazards in these contexts are multifaceted, ranging from physical harm and disease transmission to ecological damage and socio-economic impacts. These hazards typically reflect *spatial interference*, where human development or activity disrupts the natural habitats and behaviors of wildlife, increasing the likelihood of conflict.

5.6.1 Types of Hazards Associated with Human-Wildlife Interactions

5.6.1.1 Physical Injury or Death

Large predators, such as bears, wolves, or big cats, pose direct physical risks to humans when encountered unexpectedly (Gulati et al., 2021). Similarly, herbivores like elephants or rhinoceroses may become aggressive if they feel threatened, particularly when protecting their young. In areas where roads and human infrastructure cut through wildlife habitats, animals like deer, moose, and elephants are at risk of being struck by vehicles (Grilo et al., 2025). This creates significant injuries for both animals and people, with human-wildlife "interference" intensifying where transportation corridors cross critical wildlife corridors.

5.6.1.2 Disease Transmission

Wildlife can carry zoonotic diseases—such as rabies, Lyme disease, avian influenza, and SARS-CoV-2—that may be transmitted to humans through direct contact, bites, or the consumption of contaminated animal products (Muehlenbein, 2013; Ahmed et al., 2025). Additionally, certain animals, such as mosquitoes and ticks, act as vectors for diseases like malaria, dengue, and Lyme disease. These risks of disease transmission are exacerbated by human activities that bring people into closer proximity with wildlife, increasing the frequency of *spatial interference* that facilitates cross-species disease spread.

5.6.1.3 Economic and Agricultural Damage

Wildlife can cause significant damage to agriculture, leading to financial losses. Species like elephants, wild boars, and monkeys often destroy crops, which affects farmers' livelihoods (Abed et al., 2025; Wan et al., 2025). Similarly, large predators and scavengers may target livestock, compounding the economic strain on rural communities. Human-wildlife conflicts (HWCs) often arise when wildlife competes with humans for resources like land, water, and food, disrupting agricultural productivity and land management practices. In many cases, the physical presence of human settlements encroaching on wildlife habitats exacerbates these issues, creating ongoing *spatial interference* between wildlife needs and agricultural activities.

5.6.1.4 Ecological Damage

Human activities, particularly habitat destruction and fragmentation, increase the likelihood of human-wildlife conflicts. Fragmented habitats concentrate animals in smaller areas, heightening the chances of negative interactions with humans. Moreover, the introduction of invasive species by humans can disrupt ecosystems, displace native wildlife, and create competition for resources (Keck et al., 2025). Constant human interaction with wildlife can also alter animal behavior, leading some animals to lose their natural fear of humans, while others may become aggressive when their territory is threatened. Such shifts in behavior are often direct results of human-induced *spatial interference*, disrupting natural animal patterns and increasing conflict potential.

5.6.2 Consequences of Human-Wildlife Interaction Hazards

Human-wildlife interactions present a range of serious consequences, affecting human health, ecosystems, economies, and social structures (Schell et al., 2021; Ridwan et al., 2023). These encounters, often intensified by urban expansion and infrastructure development, lead to significant risks to human safety, such as injuries and psychological stress. Wildlife disruption of local ecosystems can result in imbalances, threatening biodiversity. Furthermore, these conflicts have profound economic impacts on agriculture and tourism. This section explores the diverse challenges arising from human-wildlife interactions and the spatial interference between human development and wildlife habitats.

5.6.2.1 Human Health and Safety Risks

Wildlife encounters pose significant health and safety risks to humans, including both physical harm and psychological impacts. Attacks by animals (Gulati et al., 2021; Mayer et al., 2021; Grilo et al., 2025) or vehicle collisions can result in serious injuries or fatalities. For instance, elephant and crocodile attacks in parts of Africa and Asia cause hundreds of human deaths annually. Additionally, people living in areas with frequent human-wildlife conflicts may experience stress, anxiety, and a pervasive sense of insecurity, leading to long-lasting mental health effects. The interference between human infrastructure (roads, settlements) and wildlife habitats increases the frequency and severity of these encounters, heightening the risks to human health and safety.

5.6.2.2 Ecosystem Imbalances

Increased human activity, particularly urbanization and infrastructure development, disrupts local ecosystems, forcing wildlife to adapt to altered landscapes (Abas et al., 2025). Some species may adapt by moving into urban environments (Basak et al., 2023), which can lead to shifts in food webs and ecosystem dynamics. Predators that lose their natural fear of humans may begin scavenging in human settlements, further disturbing the balance between species (O'Bryan et al., 2018). Additionally, frequent human-wildlife interactions may lead to a loss of biodiversity, as animals in conflict with humans face habitat loss, hunting, or retaliatory killings, resulting in the endangerment of species like wolves or big cats. Human encroachment on wildlife habitats creates *spatial interference* that exacerbates these imbalances, displacing animals and increasing the likelihood of negative interactions. This, in turn, raises important questions about how to achieve sustainable human-wildlife coexistence (Thapa et al., 2024).

5.6.2.3 Economic Impact

Human-wildlife conflicts can have a substantial economic impact, particularly in rural areas (Gemeda & Meles, 2018). Crop and livestock damage caused by wildlife, such as elephants or wild boars, can threaten food security and create financial strain for farmers (Conover et al., 2018). Additionally, the costs associated with wildlife management—such as relocating animals or compensating farmers—can be considerable (Fischer & Lindenmayer, 2000; Rondeau & Bulte, 2007; Langridge et al., 2021). Wildlife tourism (Curtin & Kragh, 2014) plays a significant role in the local economy, especially in regions with rich biodiversity. However, poorly managed human-wildlife conflicts can deter tourists, leading to a decline in revenue. The *spatial interference* between human development and wildlife areas creates friction, making it more difficult to manage both agricultural and tourism-based economies.

5.6.2.4 Social Conflict

Social conflict often arises when humans and wildlife compete for shared resources, such as land, water, and food (Yeshey et al., 2024). In these situations, local communities may view wildlife as competitors, while conservationists may prioritize animal welfare over human livelihoods. Additionally, cultural and religious differences can complicate human-wildlife conflict management, as certain animals may be revered in some cultures but viewed as pests or threats in others. These differing perspectives can create tensions that hinder effective conflict resolution. The *spatial interference* of human land use—such as urban expansion or agriculture—often forces wildlife into closer contact with human communities, intensifying competition for resources and escalating social conflict.

5.6.3 Practical Issues in Addressing Human-Wildlife Interaction Hazards

5.6.3.1 Persistent Challenges in Human-Wildlife Conflict and Ineffective Management Practices

Human-wildlife conflict management faces numerous challenges, especially in addressing habitat loss caused by urbanization, agriculture, and infrastructure projects that increasingly encroach on wildlife habitats, forcing animals into closer contact with humans (Ravenelle & Nyhus, 2017; IUCN, 2023; Gross et al., 2025). This *spatial interference* between human expansion and wildlife territories creates higher risks of conflict. While strategies such as wildlife relocation and translocation (Watson & Watson, 2015) are used to mitigate these conflicts, they often fall short. Relocated animals may return to their original habitats due to strong territorial instincts, or they may struggle to adapt to new environments. Additionally, compensation mechanisms, where farmers are reimbursed for damages caused by wildlife (Karanth et al., 2018; Hemminger et al., 2025), are common in many countries but can strain government budgets. Therefore, it is essential to design sustainable and feasible compensation schemes for effective conflict management.

5.6.3.2 Inadequate Strategies for Controlling Zoonotic Diseases in Expanding Human-Wildlife Interfaces

Managing zoonotic diseases, which are increasingly transmitted from wildlife to humans, presents a major challenge, particularly as human activities disrupt wildlife habitats and force animals into closer proximity with human populations. Continuous surveillance of wildlife populations is essential to monitor and prevent disease outbreaks (Watsa, 2020; Cardoso et al., 2022), but this requires significant investment in research and disease management programs. The *spatial interference* of human settlements with wildlife areas also complicates disease control efforts, as urban sprawl can create hotspots for zoonotic transmission (Connolly et al., 2020). Public awareness and education campaigns are critical to inform people about the risks of wildlife-related diseases and to promote safe behaviors when interacting with animals (Decker et al., 2012; Wu et al., 2023). However, ensuring widespread awareness and compliance with health protocols can be difficult, especially in rural or underserved regions where human-wildlife interactions are more frequent due to closer proximity to animal habitats.

5.6.3.3 Insufficient Community Engagement and Livelihood Support in Wildlife Conflict Zones

Effective community engagement is essential for successful conservation and human-wildlife conflict management, as the local communities most impacted by wildlife interactions should be involved in developing and implementing solutions (Mogomotsi et al., 2020; Nkansah-Dwamena, 2023). The *spatial interference* between human settlements and wildlife territories means that communities living near protected areas or wildlife habitats are at greater risk of conflict, making their involvement in conflict mitigation strategies even more critical. Providing alternative livelihoods to these communities can reduce their reliance on harmful practices such as poaching or retaliatory killings (Nkansah-Dwamena, 2023). However, creating sustainable alternatives requires substantial investment in infrastructure, training, and capacity-building to ensure long-term success. The challenge of balancing economic needs with wildlife protection is particularly difficult in regions where human populations are expanding into previously undisturbed natural areas.

5.6.4 **Urbanimmunology-*Informed Solutions to Human-Wildlife Interaction Hazards***

This section explores how the nine core principles of *Urbanimmunology* inform and guide practical solutions to key issues such as human-wildlife conflict and disease management. These principles, which emphasize ecosystem health, community resilience, and the interconnectedness of human and wildlife populations, provide a framework for creating sustainable, integrated solutions. The COVID-19 pandemic has vividly demonstrated how deeply interconnected public health, ecosystems, and human-wildlife interactions are. It has underscored the urgency of adopting solutions that address these challenges holistically, emphasizing the need for systemic changes and long-term preparedness. By applying these principles, we can reduce risks in urban and peri-urban environments, ensuring both ecological and public health are prioritized, while avoiding the same vulnerabilities exposed by the pandemic.

(1) Human-Wildlife Conflict Management

Human-wildlife conflicts represent a significant challenge in urbanized areas, and managing these conflicts requires a holistic approach that accounts for both human and ecological needs. The nine core principles of *Urbanimmunology* lay the foundation for creating sustainable solutions that balance wildlife conservation with human safety and well-being. The COVID-19 pandemic highlighted vulnerabilities in managing both human and environmental health in urban areas, demonstrating how environmental degradation and encroachment on wildlife habitats can contribute to health crises (Zhang & Hayashi, 2022). The lessons learned emphasize the need for enhanced preparedness, cooperation, and flexibility in managing human-wildlife interactions for future resilience. These principles inform strategies that promote coexistence, proactive conflict prevention, community education, and adaptive responses to changing conditions, ultimately contributing to healthier and more resilient urban ecosystems.

5.6.4.1 Critical Dependency Principle: Sustaining Ecosystems to Safeguard Public Health, Livelihoods, and Wildlife Coexistence

The *Critical Dependency Principle* highlights the need for sustainable conflict resolution by recognizing the interdependence between human and wildlife needs. Urban planning plays a crucial role in this by preserving wildlife habitats through green corridors and buffer zones (Senthilkumar et al., 2016; Zellmer & Goto, 2022). The pandemic has underscored how disruptions to ecosystems—whether through habitat loss or wildlife trafficking—can lead to significant public health consequences. The urgency of protecting natural habitats was starkly evident when the COVID-19 pandemic revealed the devastating effects of habitat destruction and wildlife trade. Additionally, compensation systems should be designed with stable funding tied to both conservation goals and community economic stability. For example, when farmers receive reliable payments for crop damage, they are more likely to engage in conservation programs (Karath et al., 2018), which help reduce zoonotic disease risks by keeping wildlife in natural habitats. Moreover, community-based monitoring systems that track wildlife movements (Chouksey et al., 2018) can simultaneously gather data for conflict prediction and disease surveillance, creating a unified early warning network that addresses both human-wildlife conflict and public health concerns, as seen with the global health monitoring systems developed during COVID-19 (Allan et al., 2022).

5.6.4.2 Avoidance Principle: Proactive Strategies to Prevent Conflict and Zoonotic Risk Through Early Intervention and Sustainable Alternatives

The *Avoidance Principle* emphasizes proactive measures to prevent conflicts before they arise. The COVID-19 pandemic revealed the importance of preventing viral transmission before outbreaks occur, showing that early, proactive interventions are critical to safeguarding public health (Zhang, 2020; Zhang & Hayashi, 2022). For instance, wildlife corridors and electric fencing (Feuerbacher et al., 2021) help reduce crop raids while maintaining natural disease barriers by preventing stressed animals from entering urban areas. In addition, introducing alternative livelihoods, such as beekeeping or ecotourism, provides income opportunities that reduce reliance on habitat-destructive practices. These programs can also include disease education as part of their training, contributing to both environmental and public health goals. Furthermore, early warning systems that alert communities to wildlife movements can be expanded to also report sightings of sick wildlife. This would create a shared community network that addresses both conflict and disease risks in a comprehensive, preventative manner, much like the early pandemic monitoring systems used to track viral spread.

5.6.4.3 Tolerance Principle: Encouraging Wildlife Acceptance Through Economic Opportunity and Awareness

The *Tolerance Principle* emphasizes that building community acceptance can transform wildlife from perceived threats into valuable assets. The COVID-19 pandemic reinforced the significance of public education and tolerance in crisis management, as public understanding and acceptance of health measures evolved over time (Hou et al., 2020; Kassen, 2023). Education programs that highlight the beneficial roles of wildlife reduce retaliation and foster greater appreciation for their contributions to regulating diseases. When compensation programs are coupled with training in wildlife-friendly farming techniques, they create economic incentives that encourage coexistence. Similarly, ecotourism initiatives can boost the economic value of healthy wildlife populations, with visitor fees funding both conflict mitigation and disease surveillance programs. This creates a virtuous cycle where tolerance for wildlife leads to both ecological and economic benefits for communities, mirroring how public health education helped communities adapt to pandemic-related measures.

5.6.4.4 Adaptation Principle: Flexible Strategies for Managing Emerging Wildlife Conflicts

The *Adaptation Principle* highlights the importance of flexible strategies in addressing dynamic and evolving challenges. The COVID-19 pandemic showed how quickly new challenges can emerge, requiring adaptive strategies to mitigate risks (Zhang & Hayashi, 2022). For instance, drought-resistant crops not only reduce human-wildlife competition for resources during dry seasons (Kairiza et al., 2024) but also reduce the susceptibility of crops to wildlife-borne pathogens. Additionally, mobile compensation systems that adjust payment amounts based on seasonal conflict patterns can ensure that funding is available for disease prevention measures during high-risk periods. Furthermore, community wildlife monitors, trained in both conflict de-escalation and basic disease recognition, can provide adaptable frontline responses as circumstances evolve. This integrated approach ensures that both human-wildlife conflicts and disease risks are managed effectively, regardless of shifting circumstances, much like the dynamic, evolving responses required during the pandemic.

5.6.4.5 Regulation Principle: Legal Frameworks for Coexistence and Health Security

The *Regulation Principle* stresses the need for effective governance that aligns incentives with clear, enforceable rules to manage human-wildlife interactions. The COVID-19 crisis highlighted the importance of clear regulations in managing wildlife trade, habitat protection, and public health (Borzée et al., 2020; Roe et al., 2020; Aguiar et al., 2025). Without effective governance, issues like wildlife trafficking and habitat destruction can exacerbate risks and lead to public health crises. For instance, zoning laws that establish agricultural buffers around protected areas help reduce conflicts while maintaining natural disease barriers between wildlife and human settlements (Guay et al., 2018). Additionally, licensing systems for alternative livelihoods, such as guided wildlife tours (Spring, 2022), could require participants to complete training in both conflict mediation and zoonotic disease prevention. Co-management agreements with local communities (Stevens et al., 2024), which set clear rules for forest use, not only reduce habitat encroachment but also minimize high-risk human-wildlife contact, helping to prevent disease spillover. These regulatory measures create a cohesive framework that addresses both conflict and disease risks, much like the global regulatory measures introduced to control the spread of COVID-19 (Wong et al., 2021).

5.6.4.6 Coordination Principle: Bridging Sectors to Address Conflict, Conservation, and Public Health

The *Coordination Principle* emphasizes that integrated systems maximize impact by bringing together diverse efforts. The pandemic underscored the value of coordinated, multi-sector responses in mitigating global health risks (El-Jardali et al., 2024). For example, joint task forces that include wildlife rangers, agricultural extension officers, and health workers can collaborate to address conflict incidents, manage habitats, and mitigate disease risks through coordinated site visits. Furthermore, shared data platforms that track wildlife movements and disease outbreaks enable predictive interventions, ensuring timely and effective responses. This integrated approach ensures that conservation, health, and economic objectives align, amplifying their collective impact, much like the coordinated international efforts to manage the COVID-19 pandemic.

5.6.4.7 Diversity Principle: Contextual Solutions for Coexistence and Health Resilience

The *Diversity Principle* stresses that tailored solutions are necessary to address the unique contexts of different communities. The COVID-19 pandemic demonstrated the importance of context-specific responses, as different regions faced unique challenges (Singh et al., 2020). In pastoralist areas, for instance, predator-proof bomas (livestock enclosures) help reduce conflicts with carnivores while preventing disease transmission between wildlife and domestic animals. Similarly, culturally appropriate alternative livelihoods, such as medicinal plant cultivation, provide income and reduce pressure on habitats where high-risk human-wildlife interactions are likely to occur (Nenungwi et al., 2024). Additionally, conflict resolution committees can be trained to recognize and report zoonotic disease risks specific to local wildlife species, ensuring that responses are context-sensitive and effective in managing both conflict and health concerns. This approach underscores the importance of adapting solutions to meet the unique needs and circumstances of each community, much like the varied regional responses to COVID-19.

5.6.4.8 Robustilient Regeneration Principle: Transformative Recovery for Lasting Ecological and Social Balance

The *Robustilient Regeneration Principle* emphasizes the importance of resilience in recovering from shocks and building systems that prevent future conflicts. The COVID-19 pandemic underscored the importance of preparedness and building resilience in systems that can adapt to future crises (Zhang & Hayashi, 2022). For example, reforestation projects that restore wildlife habitats also regenerate natural disease buffers, enhancing ecological resilience (Tsunoda & Enari, 2020). Additionally, wildlife tourism revenue can be allocated to human-wildlife conflict management, reducing tensions and fostering coexistence between communities and wildlife. Similarly, drought-resistant crops help mitigate conflict during resource shortages, ensuring food security. Collectively, these strategies build systems capable of withstanding and recovering from both environmental and health-related shocks, as evidenced by global resilience efforts during the pandemic.

5.6.4.9 Coherence Principle: Bridging Sectors and Policies to Maximize Ecological and Social Gains

The *Coherence Principle* emphasizes that alignment across sectors creates multiplier effects by linking efforts in a coordinated manner. The COVID-19 pandemic highlighted the effectiveness of integrated, comprehensive strategies to tackle complex, multi-sector challenges (Zhang & Hayashi, 2022). Policies that integrate wildlife protection with disease prevention, for example, can create market incentives for holistic land management that benefits both ecosystems and public health. Community conservation agreements can combine conflict resolution and disease surveillance, ensuring that ecosystem protection becomes a shared social value. Furthermore, when international conservation funding acknowledges the livelihood dimensions of conflict reduction, it supports programs that not only safeguard biodiversity but also improve community health. This alignment of goals and actions enhances the impact of conservation efforts across multiple sectors, as demonstrated by global coordination during the pandemic.

(2) Disease Management

In managing zoonotic diseases, it is crucial to apply *Urbanimmunology*'s core principles to create solutions that are both effective and sustainable. The interconnectedness of human health, wildlife, and ecosystems requires an approach that integrates public health measures with wildlife conservation efforts. The COVID-19 pandemic has underscored how quickly zoonotic diseases can spread and disrupt societies, revealing the need for a more integrated approach to disease management. By using the nine core principles, we can reduce the risk of disease transmission, promote ecological health, and ensure that communities are resilient in the face of emerging health threats. Furthermore, the lessons learned from COVID-19 provide invaluable insights that can help prevent future global pandemics by fostering a deeper understanding of the link between human-wildlife interactions, ecosystem health, and the broader public health landscape (Zhang & Hayashi, 2022).

5.6.4.10 Critical Dependency Principle: Linking Public Health to Ecological Stability

The *Critical Dependency Principle* underscores the deep interconnection between the health of human populations and the well-being of wildlife and ecosystems. Just as healthcare systems were overwhelmed during the COVID-19 pandemic, managing human-wildlife interactions requires a balanced approach that prioritizes ecological health. The pandemic revealed how the destruction of habitats and encroachment on wildlife territories increases the risk of diseases spilling over to humans. The global health crisis highlighted the fragile nature of our interconnected systems and the importance of safeguarding ecosystems to prevent future pandemics. Protecting wildlife habitats, conserving biodiversity, and supporting sustainable livelihoods are all essential for minimizing the risks of zoonotic disease spillover and reducing conflicts between humans and wildlife. By nurturing ecosystems, we create a resilient foundation that not only mitigates hazards associated with human-wildlife interactions but also enhances overall public health. This interconnected approach emphasizes that ecological well-being is key to safeguarding both human health and biodiversity, ultimately strengthening our ability to prevent future pandemics.

5.6.4.11 Avoidance Principle: Proactive Strategies to Prevent Zoonotic Disease and Conflict

The *Avoidance Principle* emphasizes that preventing human-wildlife conflicts and the potential spread of zoonotic diseases requires strategic avoidance. The COVID-19 pandemic revealed the catastrophic consequences of failing to proactively manage risks at the human-wildlife-environment interface. In the same way that the world learned to avoid high-risk behaviors that spread the virus, we should also establish buffer zones such as wildlife corridors and protected areas to reduce the opportunities for contact between humans and wildlife that could lead to disease transmission. The pandemic has made it clear that such preventive measures are not just beneficial—they are essential in the fight against emerging infectious diseases. Proactively avoiding risky encounters through thoughtful land-use planning and habitat conservation efforts will be crucial in preventing future zoonotic outbreaks. These lessons emphasize the need for foresight, early detection, and preparedness to prevent potential global health crises before they escalate.

5.6.4.12 Tolerance Principle: Reducing Conflict and Disease Risk Through Cultural Change

The *Tolerance Principle* suggests that human-wildlife conflicts often stem from fear, misunderstanding, and misinformation. Just as the public's understanding of health measures evolved during the COVID-19 pandemic, cultivating tolerance toward wildlife can foster better coexistence and prevent unnecessary human-wildlife interactions that increase the risk of disease transmission. The pandemic showed how misinformation can exacerbate public health crises, making it even more important to educate communities about the ecological benefits of wildlife. For example, bats control insect populations, and certain wildlife help maintain ecosystem balance—these roles reduce human reliance on pesticides and promote a healthier environment that limits disease spread (Tuneu-Corral et al., 2023). Public health campaigns helped people adapt to new practices during the pandemic, and a similar shift in cultural attitudes can reduce fear of wildlife, leading to safer interactions. This cultural shift toward tolerance can be a preventive strategy for managing both wildlife conservation and public health, preventing future zoonotic diseases before they emerge on a global scale.

5.6.4.13 Adaptation Principle: Leveraging Technology and Flexibility for Future Health Security

The *Adaptation Principle* highlights that human-wildlife interactions, including the spread of zoonotic diseases, are dynamic and can change rapidly. The COVID-19 pandemic demonstrated how quickly new threats can emerge and disrupt established systems, emphasizing the importance of adaptive strategies in the face of uncertainty. As we look forward, we need systems in place that can adapt quickly to emerging diseases, just as we did with the rapid development of vaccines during the pandemic. For example, utilizing technology to track wildlife movements, detect early signs of disease in wildlife populations, and predict potential zoonotic spillovers is a powerful tool that can help mitigate the risk of future pandemics. The pandemic has shown that the speed of response can make all the difference, and this principle encourages adaptive, real-time monitoring of wildlife and disease patterns. This preparedness will allow us to better manage potential zoonotic threats and implement immediate control measures before they spread globally.

5.6.4.14 Regulation Principle: Legal Boundaries for a Healthier, Safer Planet

The *Regulation Principle* emphasizes that strong regulations are crucial in minimizing hazardous human-wildlife interactions. The COVID-19 pandemic exposed how a lack of regulation in wildlife trade, habitat destruction, and human encroachment on wildlife territories can contribute to the emergence of diseases that jump from animals to humans. The lessons learned during the pandemic have underscored the need for more rigorous enforcement of wildlife protection laws, zoning regulations, and the sustainable management of ecosystems. Clear regulations on wildlife trade, deforestation, and human-wildlife interaction can drastically reduce the likelihood of zoonotic disease spillovers. Just as COVID-19 highlighted the importance of clear health protocols and global cooperation, regulating and protecting wildlife habitats globally will be key in preventing future pandemics. These efforts will not only help safeguard biodiversity but will also create a more secure and resilient global health system.

5.6.4.15 Coordination Principle: Shared Data, Shared Responsibility, Shared Safety

The *Coordination Principle* emphasizes that effective coordination among stakeholders—such as wildlife managers, health authorities, and local communities—is essential for addressing human-wildlife interaction hazards. Fragmented responses to wildlife management and disease outbreaks often result in missed opportunities for prevention, as witnessed during the early stages of the COVID-19 pandemic. During the pandemic, the importance of coordinated global responses became evident. We should build on these lessons by fostering collaborations between environmental and health organizations to address the risks of zoonotic diseases. Sharing real-time data on wildlife movements, human activity, and health surveillance can allow for predictive models that anticipate and mitigate both human-wildlife conflicts and disease outbreaks. Just as countries collaborated on vaccine distribution, global cooperation in addressing human-wildlife interaction hazards is crucial for preventing future pandemics. A coordinated approach across sectors will ensure more efficient, timely, and impactful responses to emerging health risks.

5.6.4.16 Diversity Principle: Designing Diverse Landscapes to Prevent Wildlife Spillovers

The *Diversity Principle* emphasizes that both ecosystem diversity and local adaptive strategies are essential for reducing human-wildlife conflict risks. While monoculture landscapes near human settlements can discourage wildlife from venturing into these areas in search of food, thereby reducing the risk of conflict and disease transmission (Kremen & Merenlender, 2018), the COVID-19 pandemic highlighted the devastating global consequences of habitat destruction and biodiversity loss. To prevent future pandemics, it is crucial to protect and restore biodiversity in areas far from human settlements by creating diverse landscapes that act as natural buffers against zoonotic diseases, while establishing monoculture landscapes near human settlements to limit wildlife encroachment. By properly promoting biodiversity through targeted conservation efforts, we can reduce the likelihood of wildlife entering human settlements, thereby minimizing potential disease transmission.

5.6.4.17 Robustilient Regeneration Principle: Investing in Nature to Withstand Future Health Crises

The *Robustilient Regeneration Principle* highlights that resilience in addressing human-wildlife hazards involves not only recovering from setbacks but also building robust systems that prevent future conflicts. The COVID-19 pandemic underscored the importance of resilience, as societies worldwide had to adapt quickly to new challenges. One key lesson from the pandemic is that rebuilding ecosystems and restoring wildlife habitats can help prevent the recurrence of zoonotic disease spillovers (Robinson et al., 2022). Reforestation, wildlife corridor creation, and ecological rehabilitation will act as natural barriers to future disease transmission by reducing human-wildlife interactions (Reaser et al., 2021). Furthermore, by building systems that can recover from crises—whether they are environmental, social, or health-related—we ensure that we are better prepared for the next global health challenge (Zhang & Hayashi, 2022). In the same way that nations have learned to adapt their healthcare systems during the pandemic, we should invest in resilient ecosystems to safeguard human health.

5.6.4.18 Coherence Principle: Creating Synergy Across Systems for a Safer Future

The *Coherence Principle* emphasizes that addressing human-wildlife interaction hazards requires a coherent, integrated approach that links conservation, public health, and community well-being. Fragmented policies and actions are ineffective in mitigating the risks associated with these interactions, as was evident in the global response to COVID-19 (Lal et al., 2021). A unified approach, where wildlife protection efforts, disease prevention strategies, and community engagement efforts are seamlessly integrated, is essential for preventing future pandemics. For instance, policies that protect ecosystems and regulate wildlife trade should be aligned with public health guidelines to prevent zoonotic disease spillovers. Additionally, when communities are actively involved in conservation and public health efforts, they contribute to a more resilient, cooperative global health system. Just as the pandemic taught us the value of global solidarity, addressing wildlife-related health risks will require a similarly coordinated and coherent approach.

5.7 Skyfall Hazards

Skyfall hazards refer to the risks posed by objects or materials falling from elevated positions—such as buildings, bridges, construction sites, or other structures—which can result in injuries or fatalities for individuals below (Nardo, 2024). These hazards are particularly prevalent in urban environments, where densely packed buildings and public spaces increase the likelihood of people being affected by falling debris or materials. Skyfall hazards are a significant concern in construction zones, high-rise buildings, and older structures that may lack proper safety measures. The term "skyfall" emphasizes the *spatial interference* created by the falling nature of objects from elevated structures, which can have potentially severe consequences for both workers and the public. These hazards disrupt the space between human activity on the ground and the elevated structures above, creating a dangerous overlap that poses direct threats to both life and property.

5.7.1 Consequences of Skyfall Hazards

5.7.1.1 Injury or Death

The most immediate and severe consequence of skyfall hazards is the risk of injury or death. Even small items, such as tools, glass, or debris, can cause significant harm depending on the height from which they fall. In urban areas, especially crowded streets and pedestrian zones, the risk of falling debris hitting individuals below is heightened. High-rise buildings, construction zones, and older structures with insufficient safety measures are often hotspots for such incidents. The falling objects interfere with the safe movement of people within urban spaces, introducing a deadly risk from above.

5.7.1.2 Damage to Property

Falling objects not only pose risks to human life but can also cause substantial damage to buildings, vehicles, and infrastructure below. This damage can range from broken windows and dented vehicles to more severe structural damage to buildings or public amenities. These events disrupt the spatial integrity of urban spaces, leading to costly repairs and long-term disruptions to daily operations and urban functionality.

5.7.1.3 Liability and Legal Consequences

When a falling object causes injury or damage, the responsible party—whether the property owner, contractor, or other stakeholders—may face legal consequences. Lawsuits for negligence, failure to adhere to safety protocols, or noncompliance with building codes can result in significant financial losses, reputational damage, and regulatory scrutiny. The complexity of liability in skyfall hazard cases, especially in multi-tenant buildings or construction zones, is amplified by the interference these hazards create between the spaces where people live, work, and move.

5.7.1.4 Disruption of Public Life

Skyfall hazards can disrupt the daily life of residents. Pedestrian traffic may be halted, businesses in affected areas may experience operational downtime, and transportation systems may be rerouted to ensure safety. These disruptions are more pronounced in densely populated urban areas, where commercial activity is sensitive to delays and closures. The *spatial interference* caused by falling objects results in a ripple effect across public spaces, affecting the flow of both people and goods.

5.7.1.5 Psychological Impact

Even when no injury occurs, the threat or actual occurrence of falling objects can cause significant psychological distress. People may experience heightened anxiety or trauma when navigating areas prone to such risks. Frequent exposure to the possibility of skyfall hazards can deteriorate residents' sense of security, affecting their mental well-being and altering their perception of urban safety. The psychological impact extends beyond immediate fear and interferes with the overall quality of life in urban spaces.

5.7.2 Practical Issues Related to Skyfall Hazards

5.7.2.1 Neglected Building Safety Poses Deadly Risks

A major practical issue contributing to skyfall hazards is the inadequate maintenance of buildings and structural elements. Aging buildings, especially those with deteriorating facades, windows, or roofs, are more likely to shed debris or materials that endanger pedestrians and vehicles in the spaces below. In many urban areas, inspections are infrequent or insufficient, allowing loose elements to go unnoticed. Structural vulnerabilities in older buildings increase the risk of falling objects, particularly in high-density environments where s*patial interference* amplifies the danger. The lack of proper design updates and retrofitting further exacerbates these risks, making it difficult to manage safety in rapidly growing cities.

5.7.2.2 Unsafe Construction Practices Endanger Lives

Construction zones remain a significant source of falling object hazards, particularly when materials, tools, or scaffolding are left unsecured. Objects frequently fall due to the absence of safety barriers, careless storage, or human error, resulting in dangerous spatial overlap between elevated work areas and pedestrian or vehicular pathways below. In many cases, safety protocols are poorly implemented, with inadequate use of protective measures such as nets, barriers, or warning signs in these high-risk areas. Additionally, insufficient worker training often leads to unsecured materials and accidental tool drops, increasing the risk of injury.

5.7.2.3 Failure to Warn Increases Public Vulnerability

In many urban environments, particularly those with ongoing construction or renovation, the absence of effective warning systems exposes pedestrians and workers to the danger of falling objects. Inadequate signage, missing barriers, and poor communication often result in individuals unknowingly entering areas where spatial interference from overhead work poses serious risks. This lack of clear visual or auditory alerts contributes to confusion and unintentional access to hazardous zones, where the boundary between safe and unsafe areas is not clearly defined.

5.7.2.4 Weak Enforcement Enables Persistent Hazards

Despite the existence of safety regulations, enforcement often remains weak, especially in densely populated urban areas or regions with limited oversight. This inconsistency allows skyfall hazards to persist, as property owners, developers, and contractors frequently ignore or bypass safety protocols. The absence of regular inspections and accountability contributes to unsafe building and construction practices, increasing the risk of falling objects. Without reliable enforcement mechanisms, hazardous conditions continue to interfere with public safety and intensify spatial risks in busy urban environments.

5.7.2.5 Widespread Public Unawareness Fuels Preventable Injuries

Many residents and pedestrians remain unaware of the dangers posed by falling objects, often failing to take necessary precautions in hazardous areas. This lack of awareness increases the risk of spatial interference hazards, particularly in high-traffic zones near construction sites or deteriorating buildings. Without a clear understanding of where these risks exist, individuals may unknowingly expose themselves to danger. The absence of effective public communication leaves people vulnerable to injuries, especially in areas where falling objects are a frequent threat.

5.7.3 Urbanimmunology-*Informed Solutions to Skyfall Hazards*

To address the practical challenges posed by skyfall hazards, the core principles of *Urbanimmunology* are applied to develop solutions that create safer urban environments and more resilient communities by minimizing spatial interference, as illustrated below.

5.7.3.1 Building Resilience: Proactive Safety, Strengthened Regulations, and Transformative Recovery for Urban Structures

Urban buildings play a critical role in sustaining daily life, and their structural safety is fundamental to urban resilience. Applying the *Critical Dependency Principle*, high-risk components such as façades, windows, and rooftops must undergo regular inspections and timely reinforcement to mitigate the risk of falling debris. This proactive safety approach depends on robust core systems—including standardized maintenance protocols, routine inspection mechanisms, and comprehensive regulatory frameworks—supported by coordinated governance across municipal authorities and property managers.

To institutionalize these practices, the *Regulation Principle* emphasizes the need for stricter regulatory standards, mandating periodic safety assessments, risk registration, and enforcement mechanisms, including penalties for noncompliance. In parallel, buildings should be designed or retrofitted for long-term durability, integrating materials and systems that prevent structural degradation.

Going beyond recovery, the *Robustilient Regeneration Principle* calls for a transformative approach following structural failures or extreme weather events. Restoration efforts must not simply aim to return buildings to their previous condition, but rather address underlying vulnerabilities, upgrade wind-resistance standards, and incorporate adaptive features such as real-time monitoring and fall-prevention systems—especially in aging residential areas. Through these integrated strategies, cities can strengthen their capacity to anticipate, absorb, and adapt to future skyfall hazards.

5.7.3.2 Elevating Urban Construction Standards for Public Safety

To mitigate skyfall hazards and enhance safety around construction sites, urban construction practices must be guided by integrated, dynamic governance frameworks. The *Avoidance Principle* recommends that construction activities be strategically located in low-foot-traffic areas wherever possible. In cases where such avoidance is not feasible, robust protective measures—such as reinforced barriers, secure anchoring systems, and enclosed scaffolding—must be in place before operations begin. Construction in high-risk zones should be suspended until safety risks are adequately addressed.

The *Coordination Principle* emphasizes the importance of interdepartmental collaboration among construction firms, regulatory agencies, safety inspectors, and urban planners. This collaboration is vital to standardizing safety protocols, conducting routine inspections, and ensuring unified enforcement of safety standards across all active sites.

Acknowledging that construction in dense urban environments is often unavoidable, the *Tolerance Principle* supports adaptive risk management. This involves deploying intelligent warning systems at worksite boundaries, using modular protective installations, and maintaining real-time construction information sharing to manage dynamic risks effectively.

Finally, the *Diversity Principle* calls for context-sensitive safety strategies that reflect the unique characteristics of surrounding communities. These strategies must account for demographic diversity—including the needs of vulnerable groups such as the elderly and children—and physical environments. By integrating tailored protective solutions such as enhanced barriers, accessible pathways, and clearer multilingual signage, cities can ensure inclusive and equitable safety outcomes across construction zones.

5.7.3.3 Smart Urban Warning Systems: Advancing Real-Time Protection in Public Spaces

To effectively address skyfall hazards in dynamic urban environments, smart warning systems must be integrated into the broader fabric of city management. Guided by the *Adaptation Principle*, these systems should be flexible and responsive, evolving alongside construction timelines, maintenance activities, and changing weather conditions. Real-time alerts—delivered via digital platforms, public signage, or automated alarms—can help reduce spatial conflicts and provide the public with timely risk information.

The *Coherence Principle* stresses that these systems should not operate in isolation but as part of an interconnected urban safety network. This includes seamless integration with meteorological data, emergency services, transportation systems, and municipal governance to enable coordinated responses during high-risk events.

Under the *Regulation Principle*, clear and enforceable policies are essential to mandate the installation, operation, and routine maintenance of standardized warning systems in all high-risk construction and maintenance areas. Regulatory oversight should include penalties for noncompliance to ensure consistent implementation across urban zones.

Finally, the *Robustilient Regeneration Principle* highlights the importance of fail-safe recovery mechanisms. In the event of system failure, immediate response protocols must be activated, accompanied by adaptive learning processes to identify system gaps and continuously improve urban risk management infrastructure. Together, these principles support the development of a real-time, intelligent protection network that enhances public safety and urban resilience.

5.7.3.4 Strengthening Urban Safety Through Smart Enforcement and Integrated Oversight

To effectively reduce skyfall hazards resulting from weak enforcement of safety standards, smart regulatory mechanisms must be embedded within urban governance frameworks. Guided by the *Regulation Principle*, construction and building safety enforcement should be strengthened through routine inspections, transparent reporting, and strict penalties for noncompliance. Empowering regulatory bodies with the authority, tools, and data to monitor adherence ensures that risks related to falling objects are proactively addressed.

The *Coordination Principle* underscores the importance of collaboration among municipal authorities, regulatory agencies, construction firms, and community stakeholders. A coordinated approach promotes real-time information sharing, unified monitoring efforts, and consistent enforcement of safety protocols across different urban areas.

The *Coherence Principle* advocates for integrating safety enforcement into a city-wide, digital governance system, ensuring that all districts operate under standardized regulations and enforcement procedures. A centralized platform can enable full-cycle tracking—from violation detection to corrective action—thereby enhancing transparency and accountability.

Lastly, the *Robustilient Regeneration Principle* calls for the establishment of responsive mechanisms to address enforcement lapses. In the event of regulatory failure or safety incidents, cities must have rapid response systems in place, supported by adaptive learning processes that refine future enforcement strategies and bolster urban resilience. Through these integrated measures, urban areas can build a safer, more accountable built environment.

5.7.3.5 Navigating Urban Risk: Smart Public Awareness for Safer Urban Living

Effective prevention of skyfall hazards hinges on public understanding and engagement with safety risks. The *Critical Dependency Principle* underscores the importance of foundational systems—such as safety regulations, inspection protocols, communication networks, and emergency response mechanisms—coordinated by local authorities and urban planners to elevate public risk awareness. Public education campaigns should inform residents about high-risk areas and promote safe behaviors around construction and maintenance sites.

The *Tolerance Principle* supports the cultivation of a resilience-oriented urban culture, empowering individuals to navigate the complexities of urban development while remaining vigilant and informed. The *Diversity Principle* emphasizes that public safety initiatives must be inclusive and tailored to varied community needs, offering information in multiple languages and formats to ensure accessibility for vulnerable groups such as the elderly, children, and migrant workers.

Through the *Adaptation Principle*, public education efforts should remain flexible and responsive, evolving in real time to reflect emerging risks and changes in the urban environment. Lastly, the *Coordination Principle* highlights the need for cross-sector collaboration—among local governments, schools, construction companies, and public safety organizations—to deliver consistent, trustworthy, and targeted messaging. Together, these principles support the creation of a shared culture of risk awareness and proactive participation, strengthening urban resilience at the community level.

5.8 Windstorm-Induced Traffic Hazards

Windstorm-induced traffic hazards refer to the dangers and disruptions caused by strong winds that result in various objects—such as trees (Nyberg & Johansson, 2013), electric poles, power lines, traffic signals, and other roadside infrastructure—being toppled or damaged. These objects can obstruct roads, impair visibility, and pose significant risks to vehicles, pedestrians, and cyclists. Windstorm-induced hazards interfere with the normal functioning of traffic systems, leading to severe safety issues and property damage. The interference these hazards cause can disrupt the flow of traffic, block critical pathways, and create unsafe conditions for both drivers and pedestrians.

The causes of these hazards are multifaceted and often tied to urban planning, infrastructure design, and maintenance practices. Some common causes include the following.

Improper or Inadequate Tree and Infrastructure Placement Trees planted too close to roadways, electric poles installed in vulnerable locations, or traffic signals placed in areas prone to high winds are more likely to be knocked down or damaged during a storm. These placements can create *spatial interference* in the form of physical obstructions and decreased visibility during windstorms.

Weak or Unstable Structures Trees with shallow root systems, decaying or rotting trees, poorly maintained electric poles, and traffic signals that are not designed to withstand strong winds are more susceptible to being uprooted or toppled. These unstable structures can cause severe *spatial interference* by blocking roadways or disrupting transportation routes.

Lack of Regular Maintenance Trees that are not pruned regularly, power lines that are not inspected for wear, and traffic signals that aren't reinforced to handle extreme weather conditions can all create significant risks. Poor maintenance and infrequent checks on roadside infrastructure can leave these elements more vulnerable to windstorm damage, thereby increasing the likelihood of interference in traffic systems.

Inappropriate Materials or Construction Practices Some infrastructure may not be designed with sufficient resilience to extreme weather events. For example, using materials that deteriorate quickly or fail to provide adequate support can leave structures vulnerable to damage during windstorms. This can lead to greater disruption of traffic flow and additional hazards, as these structures can collapse or be displaced during storms.

When these factors come into play, they create a perfect storm of potential risks, turning everyday roadside elements into hazardous obstacles that interfere with the safe movement of people and vehicles during windstorms. The damage and disruption caused by such hazards can severely affect the functionality of the urban traffic system, turning routine travel into a dangerous challenge.

5.8.1 Consequences of Windstorm-Induced Traffic Hazards

The consequences of windstorm-induced traffic hazards are varied, with significant interference in traffic systems, public safety, infrastructure, and emergency services. These consequences can lead to both immediate and long-term disruptions.

5.8.1.1 Traffic Disruptions

When objects such as trees, electric poles, power lines, and traffic signals fall onto roadways, they can obstruct lanes and create severe traffic disruptions. These blockages can result in road closures, detours, and significant delays for commuters, emergency vehicles, and transportation services. In extreme cases, entire sections of cities may become immobilized, interfering with daily traffic and disrupting essential services like healthcare, logistics, and law enforcement. These disruptions hinder the normal functioning of urban spaces, creating *spatial interference* that extends throughout the city's transport network.

5.8.1.2 Property Damage

Fallen trees and damaged infrastructure can cause substantial property damage. Trees that crash into vehicles can lead to significant harm, resulting in expensive repairs or even totaled vehicles. Similarly, when electric poles or power lines fall, they can damage nearby buildings, cars, and other infrastructure, resulting in costly repairs, insurance claims, and potential long-term service outages. The damage to traffic signals, in particular, can be costly, and the delay in restoring proper traffic control can exacerbate existing traffic disruptions, causing further interference in the transportation system.

5.8.1.3 Injury or Fatality Risks

The most severe consequence is the heightened risk of injury or fatalities. When trees or electric poles fall onto roads or sidewalks, they can strike pedestrians, cyclists, or drivers. In high-traffic areas, falling objects present an immediate and significant danger to everyone in the vicinity. Live power lines that snap pose electrocution risks, further intensifying the hazards. The sudden impact of falling infrastructure can lead to severe injuries or fatalities, especially in densely populated urban areas, where the risk of *spatial interference* in daily movement is elevated.

5.8.1.4 Emergency Response Pressure

The aftermath of a windstorm can overwhelm local emergency response teams. Services may be stretched to their limits as they work to clear roads, rescue trapped individuals, restore power, and repair damaged infrastructure—all of which demand significant time and resources. In the event of widespread damage, responders may struggle to manage multiple priorities at once, leading to delays in addressing accidents, rescues, and essential repairs. This pressure on emergency services not only hampers public safety but also delays critical actions, increasing the risk of further harm.

5.8.1.5 Public Safety Concerns

The presence of unsecured or poorly maintained infrastructure near roadways can create an ongoing safety concern for the public. If residents and commuters perceive areas as high-risk zones, their sense of security may be compromised, especially in locations prone to high winds or where infrastructure is inadequately maintained. The public's trust in the safety and reliability of urban infrastructure can diminish, leading to dissatisfaction with local authorities' preparedness for storm events. Issues such as downed power lines or malfunctioning traffic signals can confuse road users, increasing the likelihood of accidents. These ongoing hazards cause interference in the public's perception of safety, which can have long-term impacts on mobility and well-being.

5.8.2 Practical Issues of Windstorm-Induced Traffic Hazards

5.8.2.1 Unmanaged Roadside Vegetation: A Hidden Threat to Urban Safety

Trees and vegetation, while providing aesthetic and environmental benefits, can create significant challenges during windstorms. Urban trees, especially those with shallow root systems, are particularly vulnerable to high winds, increasing the likelihood of uprooting and the obstruction of roads and sidewalks. This can disrupt traffic flow and impede pedestrian movement, creating hazards for both. Additionally, overhanging branches are prone to breaking off in storms, leading to blockages in roadways and posing direct risks to vehicles and pedestrians. These issues can escalate if tree health is not regularly monitored, and failure to address these risks in advance can result in costly damage and pose serious threats to safety and urban mobility.

5.8.2.2 Vulnerable Power Infrastructure: A Persistent Hazard During Storms

Power lines and electric poles are critical sources of disruption during windstorms. When power poles topple or lines snap, they can block roadways, interrupt essential services, and create severe electrocution hazards. The risk of these failures increases significantly in areas with outdated or weakened infrastructure. Additionally, power lines that are too close to trees or vegetation can lead to dangerous collisions, resulting in downed wires and cascading power outages. This can further exacerbate the risk to public safety, with fallen lines becoming deadly obstacles for pedestrians and vehicles alike. Without proactive maintenance and careful planning, these issues can cause widespread disruptions, leaving communities vulnerable to both immediate hazards and long-term service interruptions.

5.8.2.3 Unstable Traffic Signals and Streetlights: Overlooked Risks in Storm Resilience

Traffic signals and streetlights, if inadequately reinforced, pose significant risks during windstorms. High winds can cause these structures to fall or malfunction, leading to disruptions in traffic flow and creating confusion for both drivers and pedestrians. The instability of traffic signals can result in accidents, while downed streetlights can obscure visibility, further complicating safe navigation through urban streets. In addition, power outages during storms may render traffic signals inoperable, exacerbating traffic congestion and increasing the likelihood of accidents at intersections. Without proper reinforcement and maintenance, these critical components of urban infrastructure can quickly become major obstacles to safe and efficient movement within cities. The lack of clear guidance on how to respond to such outages also leaves pedestrians and drivers vulnerable to making unsafe decisions.

5.8.2.4 Unsecured Signage and Overhead Fixtures: A Dangerous Threat During Windstorms

Billboards, signage, and other overhead structures present a serious risk during windstorms, as they can become airborne or fall, creating dangerous debris. These objects, if not properly secured or maintained, can block roadways, damage vehicles, and pose direct threats to pedestrians. The instability of poorly constructed or outdated signage increases the likelihood of these structures detaching or collapsing, leading to potential injuries and significant disruptions in urban areas. Unsecured outdoor objects can exacerbate the problem, turning into hazardous projectiles that further complicate efforts to maintain safety during storms. The widespread presence of such structures without proper oversight contributes to a growing risk of widespread damage and chaos, particularly in areas with high foot and vehicle traffic.

5.8.2.5 Insecure Outdoor Materials: Everyday Objects Turned Storm Dangers

Loose debris, such as unanchored outdoor furniture, trash bins, and construction materials, can cause major disruptions during windstorms. These items can be blown across streets, blocking traffic, and obstructing pedestrian pathways, creating hazards for both drivers and pedestrians. The risk of debris becomes more pronounced when materials, particularly from buildings, become detached during high winds, adding to the chaos. Items left unsecured are more likely to become dangerous projectiles, further hindering safe movement within urban areas. As debris accumulates, it increases the likelihood of accidents, delays, and significant damage to infrastructure. Without proper management, windstorms can turn everyday outdoor objects into serious threats, amplifying the disruption to both mobility and safety.

5.8.2.6 Vulnerable Walkways: Persistent Risks to Pedestrian Safety in Storm Conditions

Pedestrian safety becomes a major concern during windstorms due to the risk posed by unsecured trees, debris, and unstable infrastructure. These hazards can block walkways, collapse onto pedestrian paths, or pose risks of injury if they become airborne. The presence of large, unpruned trees or improperly secured outdoor furniture increases the likelihood of obstacles obstructing safe passage. In areas with high concentrations of vulnerable infrastructure, such as unstable buildings or power lines, pedestrians are at heightened risk of falling debris or even structural collapse. The absence of protective barriers or overhead structures further exposes pedestrians to these dangers, especially in high-risk zones. The lack of clear, real-time information on hazardous areas can leave pedestrians unaware of threats, complicating their ability to avoid dangerous situations.

5.8.2.7 Policy Shortfalls and Planning Failures in Storm Risk Mitigation

An effective approach to windstorm risk management is often hindered by fragmented planning and a lack of coordination between different levels of government and the community. Without a comprehensive framework, identifying hazards, implementing mitigation measures, and developing response strategies becomes disjointed, increasing the potential for widespread disruption. Inadequate urban planning can leave critical infrastructure vulnerable, creating severe service interruptions when storms hit. Poorly planned evacuation routes and the absence of reliable backup systems exacerbate the difficulty of managing emergencies, leaving communities stranded or in chaos. The absence of effective public education also results in citizens being ill-prepared to handle power outages or secure their property in advance, leading to more property damage and unsafe situations. Furthermore, a lack of collaboration between utility providers, urban planners, transportation agencies, and emergency services can lead to uncoordinated efforts, making it difficult to address windstorm risks comprehensively. Without regular drills and community involvement, response times can be slow, further compounding the disruption caused by windstorms, especially in terms of traffic hazards and public safety.

5.8.3 **Urbanimmunology**-*Informed Solutions to Windstorm-Induced Traffic Hazards*

Windstorms pose direct threats to people and properties. To address these threats, the entire built environment, which houses human populations, should be well prepared to respond in a timely and effective manner. In this regard, the core principles of *Urbanimmunology* can guide responses that are tailored to the unique hazards posed by windstorms, particularly those related to roadside trees, power infrastructure, traffic signals, and debris management. As discussed below, by applying these principles through analogies, we can develop strategies that not only mitigate immediate risks but also support the long-term vitality of urban spaces.

5.8.3.1 Branches of Resilience: Urban Trees as the City's Green Backbone

Roadside trees and vegetation are vital components of urban ecosystems, providing ecological, psychological, and environmental benefits that contribute to a city's overall well-being. However, during windstorms, these green elements can pose safety risks by obstructing roads and creating hazards. In line with the *Critical Dependency Principle*, urban infrastructure must function cohesively to ensure that trees and plants contribute positively to urban life while minimizing the potential risks they pose during extreme weather events.

The *Adaptation Principle* emphasizes the importance of selecting tree species and vegetation that can evolve in response to changing climate patterns. This flexibility ensures that roadside greenery remains resilient to unpredictable weather, thriving in varied environmental conditions while minimizing storm-related hazards.

The *Tolerance Principle* encourages a pragmatic approach to urban planning, allowing some degree of risk by strategically placing smaller trees and vegetation in areas with lower exposure to storm impacts. This approach maintains the ecological benefits of urban greenery while prioritizing safety during extreme weather events.

Finally, the *Coherence Principle* advocates for aligning tree planting and vegetation strategies with broader urban planning efforts. This ensures a holistic approach, integrating green infrastructure with other urban resilience initiatives, such as stormwater management and infrastructure planning, to strengthen the city's overall ability to withstand windstorms and other environmental challenges.

5.8.3.2 Wiring Resilience: Power Infrastructure as the Backbone of Urban Stability

To mitigate the risks posed by power lines in storm-prone areas, cities should adopt the *Avoidance Principle* by replacing overhead power lines with underground systems wherever feasible. Underground lines are less vulnerable to windstorms and falling trees, minimizing disruptions to power supply and enhancing resilience in high-risk zones.

The *Coordination Principle* underscores the need for robust collaboration between utility companies, urban planners, and local authorities to ensure timely maintenance and upgrades of power infrastructure, thereby improving emergency response capabilities. A coordinated strategy ensures that essential services remain operational during windstorms.

The *Regulation Principle* calls for stricter standards on power poles and cables, requiring infrastructure to be designed and maintained to withstand extreme weather conditions. Regular inspections and prompt repairs can prevent failures that could lead to power outages and safety hazards.

Lastly, the *Diversity Principle* advocates for a diversified energy infrastructure, integrating alternative energy sources like solar and wind, and combining underground and above-ground systems. This approach enhances the adaptability and resilience of the power grid, reducing reliance on a single infrastructure type and improving overall reliability during storm events.

5.8.3.3 Signals of Resilience: Ensuring Traffic Control and Streetlight Functionality as the City's Lifeline During Storms

To ensure traffic safety and proper streetlight functionality during windstorms, the *Adaptation Principle* calls for traffic signal systems and streetlights designed with wind-resistant materials, as well as the ability for quick repair or replacement. This flexibility enables cities to recover swiftly from damage, ensuring that both traffic control and street lighting remain operational even during emergencies.

The *Robustilient Regeneration Principle* emphasizes the importance of backup systems, such as manual traffic management or solar-powered signals, and battery-backed streetlights, to restore order and illumination promptly and efficiently in the event of damage. These alternative solutions help prevent accidents, reduce congestion, and ensure the continuity of essential services, including safe public lighting during power outages.

The *Coherence Principle* advocates for integrating traffic signals and streetlights into the broader urban resilience strategy, ensuring that all infrastructure works together seamlessly during windstorms. This interconnected approach strengthens the overall reliability of traffic management and street lighting, enhancing the city's ability to respond swiftly to emergencies, maintain safe roadways, and provide necessary public illumination.

5.8.3.4 Signs of Safety: Securing Urban Structures Against Storm Hazards

The *Regulation Principle* emphasizes the need for implementing and strictly enforcing building codes for billboards, signage structures, and traffic signs, with regular inspections and maintenance to ensure their safety during windstorms. Strengthening regulations in high-risk areas and conducting routine audits can significantly reduce the risk of structural failures that could pose dangers during storms. It is essential to ensure that all signage, including traffic-related signs, is securely fastened and designed to withstand strong winds, especially in areas prone to high gusts.

The *Coordination Principle* fosters collaboration between urban planners, business owners, and local authorities to proactively manage signage. Pre-storm measures, such as securing or removing unstable signs, are critical. This cross-sector communication ensures businesses and municipal departments adhere to safety protocols, improving overall preparedness and resilience within the urban environment.

By embracing the *Diversity Principle*, cities can reduce risks by diversifying the types of signage used—such as digital versus static or smaller versus larger signs. This approach ensures that not all signage relies on a single type, providing greater flexibility and resilience to adapt to different storm conditions. For example, adopting flexible or breakaway designs for billboards and signs will minimize the risk of debris while maintaining clear visibility and communication to pedestrians and drivers. This strategy enhances the city's ability to handle windstorm hazards while ensuring the safety of public spaces.

5.8.3.5 Clearing the Path: Managing Urban Debris to Shield Against Storms

The *Critical Dependency Principle* highlights the importance of ensuring that urban infrastructure is resilient, including securing outdoor furniture, construction materials, and other movable items to prevent them from becoming hazardous debris during windstorms. The *Avoidance Principle* stresses the need for proactive measures such as enforcing building codes that limit the presence of high-risk objects in vulnerable areas, thereby reducing the potential for debris accumulation. The *Adaptation Principle* underscores the necessity for urban systems that can swiftly respond to storm conditions, such as deploying rapid debris removal teams or redesigning city layouts to minimize the impact of windstorm-induced hazards. Lastly, the *Coordination Principle* emphasizes the importance of fostering collaboration between city authorities, businesses, and residents to implement community-driven initiatives that secure outdoor objects and ensure the timely cleanup of debris, thereby creating a coordinated and efficient response to windstorm challenges.

5.8.3.6 Safe Steps: Designing Pedestrian Pathways for Resilience in Storms

The *Critical Dependency Principle* emphasizes the importance of maintaining and strengthening pedestrian walkways and sidewalks as vital infrastructure for public safety during windstorms. This involves regular inspections, tree pruning, securing outdoor furniture, and ensuring that walkways remain clear of debris and obstructions.

The *Adaptation Principle* highlights the need for flexible pedestrian pathways that can adjust to varying levels of windstorm severity. This could include installing adjustable barriers or overhead protective structures along sidewalks to shield pedestrians from falling debris and reduce exposure to the elements.

The *Coordination Principle* stresses the value of collaboration between local authorities, businesses, and the community to ensure the effective maintenance and clearance of pedestrian pathways before, during, and after windstorms. Solutions might include coordinated efforts for real-time updates on safe walking routes and ensuring that clear, accessible signage is available during storms to guide pedestrians.

Lastly, the *Robustilient Regeneration Principle* focuses on the need for rapid recovery and long-term resilience of pedestrian walkways and sidewalks after a windstorm. Solutions include swift debris removal, restoration of damaged pathways, and reinforcement of infrastructure to prevent future disruptions, ensuring pedestrian safety remains consistently upheld.

Together, these principles form a comprehensive strategy to protect pedestrian walkways and sidewalks during windstorms, ensuring both immediate safety and long-term resilience.

5.8.3.7 Blueprint for Resilience: Planning for a Storm-Ready City

The *Critical Dependency Principle* underscores that the foundation of any windstorm risk management system relies on maintaining essential infrastructure—such as power, transportation, and communication systems—ensuring their continuous operation during extreme weather events. Strengthening these systems enhances a city's resilience to both immediate disruptions and long-term challenges.

The *Regulation Principle* emphasizes the importance of a robust regulatory framework that enforces safety, planning, and maintenance standards, ensuring cities are adequately prepared for windstorm hazards. By continuously updating building codes and infrastructure standards to reflect evolving climate realities, cities can reduce vulnerabilities and enhance public safety.

The *Coordination Principle* highlights the need for seamless collaboration between government entities and the private sector, ensuring efficient resource allocation and swift action during windstorms. This coordination enhances response times and accelerates recovery efforts.

The *Robustilient Regeneration Principle* emphasizes the importance of embedding a resilient recovery plan within urban policy. This approach allows cities to swiftly rebound and emerge stronger, minimizing long-term damage and facilitating a transition to an upgraded "new normal." By focusing on proactive strategies, it not only accelerates recovery but also ensures the enduring vitality of urban environments through rapid regeneration.

Finally, the *Coherence Principle* stresses that a unified strategy, aligning urban planning with risk management policies, creates a stable and harmonized urban environment. This ensures all systems function seamlessly during windstorm crises, enhancing the city's overall ability to handle extreme weather events.

5.9 Multimodal Interaction Hazards

Multimodal interaction hazards arise from the coexistence of different transportation modes, such as pedestrians, cyclists, e-bikes, and public transport, within shared urban spaces. These hazards occur when these modes of transport share the same physical spaces (such as roads, sidewalks, intersections, or public plazas) and interfere with one another (Karndacharuk et al., 2014; Beitel et al., 2018; Schiermeyer et al., 2019). The diverse characteristics of these modes—such as differences in speed, maneuverability, and user behavior—create opportunities for conflicts, accidents, and injuries. The causes of these hazards often stem from inadequate infrastructure design, such as poorly marked or overlapping lanes, lack of clear signage, and insufficient space for each mode. Additionally, the unpredictable behavior of users, such as pedestrians crossing in non-designated areas or cyclists weaving between traffic, further exacerbates the risk. Environmental factors, such as poor visibility, weather conditions, and road surface quality, can also contribute to the severity of these hazards. These hazards are particularly prevalent in urban areas where various transportation methods converge in densely populated environments, where the demand for shared space is high and regulation or enforcement of safety measures may be insufficient.

5.9.1 Consequences of Multimodal Interaction Hazards

The consequences of multimodal interaction hazards are multifaceted, affecting individuals, infrastructure, urban functionality, aesthetic appeal, and comfort. These consequences can be broken down into the following categories:

5.9.1.1 Physical Injuries and Accidents

Physical injuries and accidents can arise from various multimodal interactions in urban spaces. Pedestrian-cyclist collisions (Mesimäki & Luoma, 2021) are common, especially when cyclists, particularly those on e-bikes, unexpectedly approach pedestrians in shared areas like sidewalks and bike lanes. Due to the speed disparity between cyclists and pedestrians, these incidents can lead to serious injury, especially when pedestrians fail to anticipate a cyclist's presence. Vehicle-pedestrian collisions (Noh et al., 2022) are also hazardous, particularly at intersections where pedestrians may be distracted or unaware of oncoming cars or buses, resulting in severe injuries or fatalities. Similarly, cyclist-motor vehicle collisions (Ling et al., 2020; Osama et al., 2020) pose significant risks, as cyclists sharing lanes with cars or buses can be endangered by inattentive drivers. The speed and mass difference between motor vehicles and cyclists often lead to severe injuries in these accidents.

5.9.1.2 Disruptions in Traffic Flow

Disruptions in traffic flow can occur when different transportation modes interfere with one another, leading to congestion and delays. For example, delivery trucks blocking bike lanes or buses struggling to navigate crowded intersections due to pedestrians crossing inappropriately can significantly disrupt the movement of people and goods, resulting in delays, frustration, and economic inefficiencies. Additionally, cross-modal delays can arise when one mode interferes with others—such as a pedestrian jaywalking in a bike lane or a cyclist blocking a bus lane—which can create a ripple effect, worsening the overall efficiency of the transportation system.

5.9.1.3 Social and Psychological Consequences

Social and psychological consequences can arise from the interactions between different transportation modes in shared spaces, often leading to frustration and conflict. Users frequently experience frustration when their routes are obstructed by others—such as drivers becoming irritated with cyclists weaving through traffic, or cyclists feeling unsafe when cars disregard bike lanes. Similarly, pedestrians may exacerbate risks by jaywalking or ignoring traffic signals. Additionally, fear and anxiety are common, as pedestrians may feel unsafe due to the perceived danger from speeding cyclists or vehicles, while cyclists may experience anxiety about potential collisions with larger vehicles or pedestrians. This creates a reluctance to use public transportation networks and can increase overall stress in shared spaces.

5.9.1.4 Aesthetic and Comfort Impact

The coexistence of various transportation modes, such as bicycles, e-bikes, pedestrians, and wheelchairs, in shared urban spaces can have a significant aesthetic and comfort impact (Ruiz-Apilánez et al., 2017; Chan & Zhang, 2018; Kleeman et al., 2021). Visual clutter and congestion arise when these modes occupy the same space, undermining the aesthetic quality of urban areas and creating chaotic streetscapes that detract from visual harmony. Poorly delineated bike lanes cutting across pedestrian zones can further disrupt the visual flow, contributing to a cluttered and overwhelming appearance. Additionally, the integration of multiple traffic modes can interfere with the architectural intent of public spaces, compromising spatial coherence and diminishing the overall quality of urban design. The mixing of slow-moving transportation modes can also create a sense of crowding, leading to psychological discomfort, as pedestrians feel uneasy navigating busy spaces filled with cyclists and e-scooters. This reduces the overall comfort of the environment and detracts from the enjoyment of public spaces meant for relaxation or social interaction. Furthermore, poorly integrated multimodal spaces can impede accessibility, particularly for individuals with mobility impairments, such as wheelchair users or those with strollers, compromising the convenience and comfort of the space for all users, especially in high-traffic areas.

5.9.1.5 Environmental Consequences

Multimodal conflicts, such as congestion, can have significant environmental consequences. Inefficiencies caused by these conflicts often result in vehicles idling longer, increasing emissions and contributing to air pollution, which negatively impacts both the environment and public health. Additionally, poorly designed shared spaces can lead to inefficient use of urban areas, with roadways, bike lanes, and pedestrian paths poorly integrated. This results in wasted space that could otherwise be better utilized for public transport, green spaces, or recreational areas, further compromising the sustainability and livability of urban environments.

5.9.2 Practical Issues in Managing Multimodal Interaction Hazards

From a practical standpoint, managing multimodal interaction hazards requires addressing a range of issues that span infrastructure design, spatial planning, policy development, and behavioral management. Below are some of the key challenges faced by urban planners and transportation authorities:

5.9.2.1 Critical Flaws in Infrastructure Design Fueling Mobility Conflicts

Inadequate infrastructure design in urban areas contributes significantly to multimodal hazards and inefficiencies. Many cities have outdated infrastructure that fails to accommodate the growing variety of transportation modes, such as cars, cyclists, and pedestrians. For instance, while roads often have designated lanes for cars, there is frequently a lack of clear or safe spaces for cyclists and pedestrians, resulting in conflicts and unsafe conditions. The absence of separation increases the risk of accidents due to unanticipated interactions between modes. Additionally, intersections are often poorly designed, lacking clear markings for pedestrian crossings or bike lanes, which leads to confusion over right-of-way and higher collision risks. This problem is exacerbated when insufficient space is allocated for cyclists or pedestrians with mobility aids, further complicating movement and creating frustration and delays. Furthermore, poorly designed multimodal infrastructure can degrade both the functionality and aesthetic quality of urban spaces. Disorganized bike lanes or poorly planned intersections contribute to visual clutter, disrupting the intended architectural aesthetics and creating uncomfortable environments for users. This lack of spatial separation not only diminishes urban quality but also discourages people from using these spaces effectively.

5.9.2.2 Hazardous Misalignment of Urban Space Design for Multimodal Use

In many cities, poor space design and positioning for various transportation modes can lead to significant conflicts and safety risks. Misplaced shared spaces, such as bike lanes placed next to sidewalks or alongside high-speed traffic, create hazards by increasing the potential for interference between cyclists, pedestrians, and motor vehicles. Similarly, poorly positioned bus stops or shared pedestrian-cyclist spaces can result in pedestrian congestion or vehicles blocking key pathways, raising the likelihood of accidents and delays. Additionally, the integration of multimodal spaces into the broader urban fabric is often inconsistent, with shared spaces like pedestrian streets or bike lanes failing to align with the natural flow of pedestrian movement or connectivity to key urban areas. This misalignment forces users to navigate inefficient or awkward paths, increasing the risk of conflict. Effective urban planning should ensure that multimodal spaces are positioned to enhance connectivity, accessibility, and safety, while promoting spatial harmony with the surrounding environment. Furthermore, in urban areas with high traffic densities, inefficient land-use planning exacerbates multimodal interaction hazards by squeezing pedestrians and cyclists into cramped, uncomfortable spaces. This creates overcrowded infrastructure where competing demands for space increase the likelihood of interference between different modes of transportation.

5.9.2.3 Systemic Failures in Managing Shared Slow Traffic Spaces

Managing mixed slow traffic in urban spaces presents significant challenges, particularly when slower modes of transport, such as pedestrians, cyclists, and e-scooters, interact with faster-moving vehicles (Frosch et al., 2019; Nikiforiadis et al., 2020). Cyclists, often faster than pedestrians, may find themselves navigating crowded pedestrian areas, leading to conflicts, while pedestrians, unaware of fast-moving cyclists, may unpredictably cross bike lanes, increasing the potential for accidents. These issues are further exacerbated when infrastructure is not designed to accommodate the needs of slower-moving users. Additionally, the lack of enforcement of slow traffic rules contributes to the problem, as cyclists may ride on sidewalks or ignore traffic signals, while pedestrians encroach on bike lanes or fail to use crosswalks. This lack of behavioral regulation fosters an environment where users disregard each other's space, escalating multimodal interaction hazards. Furthermore, overcrowding in shared spaces, particularly during peak hours, compounds these challenges by reducing the effectiveness of multimodal infrastructure and creating psychological discomfort. Pedestrians and cyclists may feel rushed or anxious navigating dense crowds, and e-scooter users often struggle to maneuver through these areas, leading to further interference between modes and a heightened sense of insecurity.

5.9.2.4 Dangerous Misconceptions and Behavioral Gaps in Urban Traffic

Insufficient awareness and education about the rules governing shared spaces contribute significantly to multimodal hazards. Cyclists, drivers, and pedestrians often lack knowledge of the appropriate behaviors expected in these spaces, leading to actions that disrupt the flow of other modes. For instance, cyclists may ride on sidewalks, pedestrians may jaywalk, and drivers may fail to yield to cyclists or pedestrians. While public education campaigns could address these issues, many cities still lack effective programs, leaving users confused and increasing the risk of dangerous interactions. To combat this, comprehensive and sustained awareness campaigns are essential, focusing on traffic etiquette, mutual respect among all transportation users, and the importance of safety in shared spaces. These campaigns should target all user groups—drivers, cyclists, and pedestrians—especially in high multimodal traffic areas, and emphasize the need for organized, clutter-free environments to reduce interference. Additionally, a lack of awareness regarding the aesthetic and environmental impact of shared spaces can contribute to multimodal hazards. Users who neglect the importance of maintaining clean and organized environments can create visual disorder, further complicating interactions between different modes. Educational programs promoting shared responsibility for urban space quality can encourage better user behavior and reduce environmental chaos.

5.9.2.5 Outdated and Unenforced Policies Jeopardizing Public Safety

Policy gaps and enforcement issues significantly contribute to multimodal transportation hazards. Many cities still lack comprehensive and up-to-date policies that address the complexities of multimodal transport. Existing regulations regarding the interactions between cyclists, pedestrians, and vehicles are often vague or outdated, leading to ambiguity and making it difficult to manage the increasing variety of transportation modes. Additionally, some policies fail to account for emerging technologies like e-scooters or shared micro-mobility options, leaving regulatory gaps. Even when policies are in place, inconsistent enforcement undermines their effectiveness. Cyclists may continue to ride on sidewalks or ignore traffic signals, while pedestrians may cross outside designated crosswalks, leading to an increased risk of accidents and conflicts. Furthermore, the absence of comprehensive policy frameworks for the design and use of multimodal spaces often results in chaotic, inefficient, and unsafe environments. Without clear, integrated regulations governing the use of these spaces, urban areas risk exacerbating multimodal interaction hazards, compromising both functionality and safety.

5.9.3 Urbanimmunology-*Informed Solutions to Multimodal Interaction Hazards*

For each of the practical issues identified above regarding multimodal interaction hazards, specific core principles of *Urbanimmunology* are carefully selected to align with their unique characteristics. These principles are chosen for their ability to address the complexities of urban environments and the interconnections between various systems. By applying these tailored principles, we can better understand the risks and challenges posed by multimodal interactions. This approach not only ensures a more effective response to the hazards but also provides a framework for sustainable, long-term solutions that adapt to the dynamic nature of urban ecosystems.

5.9.3.1 Enhancing Urban Transport Infrastructure: Principles for Safe, Efficient, and Resilient Design

The *Critical Dependency Principle* emphasizes the importance of a well-designed infrastructure that allows pedestrians, cyclists, and vehicles to coexist safely, mitigating interference between modes and promoting both transport efficiency and public safety. The *Regulation Principle* ensures clear guidelines and safe accommodations for each transport mode, reducing ambiguity and enhancing safety, while adapting to the evolving needs of urban populations. The *Coordination Principle* underscores the need for collaboration across stakeholders, ensuring that urban planning integrates diverse transport modes effectively, preventing fragmentation and inefficiency. By incorporating the *Diversity Principle*, cities can create adaptable infrastructures that cater to a range of users, from cyclists to emerging transport technologies like autonomous vehicles. Lastly, the *Robustilient Regeneration Principle* advocates for resilient infrastructure that evolves and regenerates in response to increasing demand and unforeseen challenges, maintaining functionality and sustainability in the long term.

5.9.3.2 Strategic Urban Design: Enhancing Multimodal Safety, Flexibility, and Spatial Harmony

The *Avoidance Principle* emphasizes the need to strategically separate incompatible transport modes, such as bike lanes from pedestrian zones and high-speed vehicle traffic, preventing interference and potential conflicts. The *Diversity Principle* promotes the creation of distinct pathways for different transport modes, ensuring safe, efficient navigation while reducing stress and confusion. The *Adaptation Principle* advocates for flexible urban designs that can evolve with new transport technologies, such as e-scooters or micro-mobility options, ensuring that infrastructure remains relevant and safe without compromising the needs of existing users. Lastly, the *Coherence Principle* stresses the importance of integrating transport systems into the surrounding urban environment to create seamless, aesthetically pleasing, and efficient spaces.

5.9.3.3 Designing Harmonious Slow Traffic Environments: Balancing Safety, Flow, and Flexibility

The *Tolerance Principle* emphasizes the importance of creating environments that support the coexistence of slow and fast-moving traffic, using tools like clear signage, designated lanes, and reduced speed zones to minimize interference and promote mutual respect. The *Adaptation Principle* ensures that urban spaces adjust to changing traffic conditions, such as modifying traffic signals or lane designs during peak times, to maintain smooth flow and prevent conflicts. The *Robustilient Regeneration Principle* highlights the need for adaptive recovery strategies, like temporary pedestrian zones or pop-up bike lanes, to alleviate congestion and ensure that urban transport systems remain safe and efficient during disruptions. Finally, the *Coherence Principle* advocates for designs that integrate these strategies seamlessly into the cityscape, reducing confusion and complexity while maintaining clarity for all transport modes.

5.9.3.4 Cultivating Smarter Urban Mobility: Clear Rules, Shared Respect, Evolving Awareness

The *Regulation Principle* stresses the importance of clear rules and guidelines for each mode of transport, reinforced by public education campaigns that ensure all users understand and respect each other's space. This fosters a culture of compliance and accountability. The *Coherence Principle* ensures that educational efforts are consistent and unified, teaching users about safe behaviors and shared responsibilities in multimodal spaces, reducing risky behaviors and preventing interference. The *Adaptation Principle* highlights the need for ongoing updates to educational campaigns as new transportation technologies emerge, addressing specific safety concerns and promoting safe usage across all modes. Lastly, the *Tolerance Principle* advocates for fostering mutual respect and understanding among different transport users, teaching them to be mindful of each other's needs and spaces, ultimately reducing conflict and enhancing safety.

5.9.3.5 Innovating Urban Mobility Governance: Toward Integrated, Flexible, and Enforceable Transport Policies

The *Coherence Principle* emphasizes the need for integrated and aligned policy frameworks that comprehensively address all transport modes, including emerging technologies like micro-mobility, ensuring clear and consistent enforcement. The *Regulation Principle* underscores the importance of strong, consistent enforcement of traffic laws, ensuring compliance with bike lane usage, pedestrian crosswalks, and new mobility devices to create a safer urban environment. The *Critical Dependency Principle* underscores the essential role of clear and updated policies for the long-term health of urban transport systems, emphasizing that ineffective or outdated policies can undermine safety and efficiency. The *Coordination Principle* stresses the need for policy coordination across local, regional, and national levels, preventing fragmented regulations and ensuring a unified response to challenges in multimodal transport. Lastly, the *Adaptation Principle* advocates for flexible policies that can evolve to address the specific challenges posed by emerging transport technologies, ensuring that policies remain relevant and effective.

5.9.3.6 Innovative Urban Design for Functional and Visual Balance: Creating Inclusive, Adaptive, and Clutter-Free Public Spaces

The *Diversity Principle* emphasizes the need for designs that accommodate various transport modes without compromising the aesthetic quality of public spaces, ensuring clear and well-integrated infrastructure that enhances the user experience. The *Robustilient Regeneration Principle* advocates for adaptable urban designs that can evolve with shifting transportation demands, allowing spaces to remain both functional and visually appealing as they regenerate over time. The *Coherence Principle* stresses that multimodal spaces should blend seamlessly with the surrounding urban environment, maintaining the city's visual integrity while ensuring usability. Lastly, the *Avoidance Principle* highlights the importance of careful planning to prevent overcrowding or visual clutter, ensuring that urban areas remain inviting and functional.

5.9.3.7 Designing Smarter Shared Spaces for Urban Mobility: Integrating Policy, Coordination, and Flexibility for Safer Coexistence

The *Critical Dependency Principle* emphasizes the importance of building integrated urban systems that support the safe coexistence of pedestrians, cyclists, and vehicles in constrained spaces. The *Coordination Principle* stresses the need for cooperation between local authorities, urban planners, and business owners to create unified traffic flow systems, ensuring safe movement in high-traffic areas. The *Regulation Principle* calls for clear, enforceable rules—such as designated lanes and speed limits—to reduce conflicts and improve safety in these critical spaces. Lastly, the *Adaptation Principle* highlights the need for flexible, adaptive policies that can adjust to changing traffic volumes and the introduction of new transportation modes, ensuring that urban spaces remain safe and functional as they evolve.

5.10 Low-Altitude Aircraft and Aviation Risks

Low-altitude aircraft and aviation risks refer to the hazards associated with aerial vehicles—such as private aircraft, small commercial planes, helicopters, and drones—that operate at low altitudes in urban environments. These risks are driven by two main factors: the growing volume of air traffic and the increasing use of small aircraft and drones. Cities with airports located near urban centers, especially those with high population densities, are more prone to aviation-related risks due to the *spatial interference* between multiple users of urban airspace.

As air traffic continues to increase, both for commercial flights and private aviation, the likelihood of collisions and accidents also rises (Yang et al., 2024). The expansion of drone usage for purposes such as deliveries, surveillance, and recreational activities further contributes to the complexity of managing low-altitude airspace (Bauranov & Rakas, 2019; Ghasri & Maghrebi, 2021). These aerial vehicles should navigate crowded urban environments with limited airspace, creating a significant risk of interference between different aircraft and the physical infrastructure in cities—such as buildings, power lines, and communication towers.

The *spatial interference* between various types of low-altitude aircraft is heightened by the constraints of urban airspace. These vehicles often share confined airspace, making it difficult to avoid collisions, especially in areas where the density of both aerial and ground-based infrastructure is high. As more aerial vehicles operate at lower altitudes, the potential for interference-related accidents increases, with the complexity of avoiding obstacles adding another layer of risk in these urban environments.

5.10.1 Consequences of Low-Altitude Aircraft and Aviation Risks

The consequences of low-altitude aircraft and aviation risks are wide-ranging and can be divided into three primary categories: physical damage, economic impact, and public safety concerns. As urban environments become increasingly crowded with both human activity and low-altitude aircraft, the risks of aviation-related incidents escalate (Sun et al., 2025). Small commercial planes, private aircraft, helicopters, and drones all share the challenge of operating in confined and congested spaces. In cities with high population densities and numerous tall buildings, even minor accidents can have catastrophic results. Aircraft operating at low altitudes are often in close proximity to residential, commercial, and recreational areas, amplifying the potential for harm. The primary concern in these environments is that crashes can result in widespread damage, with severe consequences not just for the aircraft occupants but also for the surrounding community.

5.10.1.1 Physical Damage

When an aircraft crashes, whether it is a small plane, helicopter, or drone, the impacts are often severe, particularly in urban areas. These accidents can cause direct harm to individuals on the ground, leading to injuries or fatalities (Sun et al., 2024). The infrastructure of the city is also at risk, as crashes can lead to substantial property damage, from the destruction of buildings to the disruption of transportation systems. Helicopters, in particular, pose a serious threat as their crashes can spark fires, collapse structures, or result in mass casualties, especially when they are involved in accidents near high-rise buildings or busy commercial districts (Churchwell et al., 2018).

5.10.1.2 Economic Impact

The economic impact of low-altitude aircraft accidents can be equally devastating. The costs associated with repairing physical damage, compensating for loss of business, and addressing the destruction of infrastructure can be extremely high. Cities also face the burden of emergency response services, which can quickly escalate in cost, especially when the incident requires a large-scale rescue operation or multiple departments. Furthermore, accidents can lead to a temporary suspension of air traffic, disrupting not just local transportation, but also commerce, tourism, and other essential services in the area. These disruptions can take a significant financial toll on both the public and private sectors, with businesses and governments facing potential losses and recovery costs.

5.10.1.3 Public Safety Concerns

With the rise in the use of drones for deliveries, surveillance, and recreational activities, there is a growing likelihood of accidents involving drones colliding with pedestrians, buildings, or other aircraft (Hossein Motlagh et al., 2016; Li et al., 2025). These accidents can cause injuries, fatalities, and damage to critical infrastructure. Moreover, the proximity of aircraft in urban airspace increases the likelihood of collisions and complicates emergency response efforts. First responders may face difficulties reaching the site of an accident due to the congestion of city streets and the complexity of navigating through high-traffic areas. The frequent occurrence of aviation-related incidents in densely populated areas erodes public trust in air safety, leaving the general public increasingly concerned about the risks associated with air travel and aerial operations.

5.10.2 Practical Issues in Addressing Low-Altitude Aircraft and Aviation Risks

5.10.2.1 Aircraft Crashes

Aircraft crashes, particularly involving low-altitude flights such as small aircraft or drones, are more prone to mechanical failures, poor weather, or human error. In densely populated urban areas, these crashes can result in severe consequences, including loss of life, property damage, and disruptions to transportation networks (Tang et al., 2021). A significant challenge for urban centers is managing air traffic at low altitudes, especially in cities with airports near residential or commercial areas. The lack of adequate airspace management, clearly defined flight paths, and no-fly zones increases the likelihood of *spatial interference*, leading to a higher risk of crashes. This *spatial interference* arises when low-altitude aircraft are forced to navigate through constrained airspace, competing with other air traffic or encountering obstacles such as buildings, power lines, and bridges (Petrovsky et al., 2018; Yang & Wei, 2021).

5.10.2.2 Helicopter Accidents

Cities with helipads, often located on high-rise buildings or hospitals, face the risk of helicopter accidents due to mechanical issues, poor visibility, or pilot error. These aircraft, typically used for emergency services or transport, can cause significant damage, including fires, building destruction, or casualties, if they crash in densely populated areas. The risk of helicopter accidents (Churchwell et al., 2018) is heightened in locations where helipads are poorly designed or maintained. As urban areas undergo vertical development, the challenge of coordinating helicopter flight paths to avoid *spatial interference* with crowded airspace or high-rise buildings increases. These risks are further exacerbated by the limited space available for emergency maneuvers, making it harder to address potential conflicts with other aircraft or urban structures.

5.10.2.3 Drone-Related Accidents

The rapid proliferation of drones, used for purposes such as delivery services, surveillance, and recreational activities, introduces significant risks in urban areas (Sun et al., 2024). Drones flying at low altitudes can collide with buildings, infrastructure, or other aircraft, posing hazards to people if they malfunction or are flown in restricted areas. The small size, unpredictable flight paths, and accessibility of drones create new challenges for urban safety, especially in densely populated environments where *spatial interference* between drones, buildings, and other low-altitude aircraft is a growing concern. The absence of clear regulations governing drone use increases the likelihood of accidents, which can result in damage to infrastructure, injuries to pedestrians, or disruption of manned air traffic.

5.10.3 Urbanimmunology-*Informed Solutions to Low-Altitude Aircraft and Aviation Risks*

To address the risks associated with low-altitude aircraft, including private planes, helicopters, drones, and small commercial aircraft in urban areas, we can apply the nine core principles of *Urbanimmunology*, derived from human immune intelligence. Each of the principles can inform the solutions to the challenges posed by these aviation risks, helping cities design safer environments for both air and ground traffic while managing these risks effectively.

5.10.3.1 Critical Dependency Principle: Safeguarding Urban Air Traffic and Aviation Infrastructure for Long-Term Safety

Urban areas, especially those with airports near dense residential or commercial zones, rely heavily on their air traffic management systems. As the volume of low-altitude aircraft traffic continues to grow, particularly with the increasing use of drones and helicopters, the *Critical Dependency Principle* highlights the importance of safeguarding foundational air traffic infrastructure. Regular maintenance, monitoring, and upgrades are essential to prevent catastrophic accidents and ensure the continued safe operation of these systems.

To address these risks, cities should develop and implement a comprehensive airspace management system that prioritizes the safety of both manned and unmanned aircraft (Ali, 2019). This includes conducting regular inspections and maintenance of airport facilities, helipads, and drone landing zones to minimize the risk of mechanical failures or infrastructure issues. Additionally, early warning systems for air traffic control (Deb et al., 2025)—similar to the immune system's ability to detect and respond to pathogens—can be implemented. These systems would monitor low-altitude aircraft in real-time, enabling authorities to identify and address emerging risks such as collisions or airspace congestion before they escalate.

5.10.3.2 Avoidance Principle: Halting Development in High-Risk Aviation Zones

In urban areas, the *Avoidance Principle* involves halting or controlling the development of infrastructure or residential areas in zones that are vulnerable to low-altitude aircraft accidents, such as regions near helipads, airports, or flight paths for small aircraft and drones. This preventive strategy reduces exposure to potential risks by ensuring that critical infrastructure is located in areas less likely to be affected by aviation accidents. By focusing on preventing development in high-risk zones, cities can protect their residents and minimize the likelihood of disaster.

To apply this principle, cities should define and enforce no-fly zones or restricted areas where low-altitude flights are prohibited, and strictly control new development in these regions (Hao et al., 2016; Keysar, 2020). For example, no new construction should be allowed near helicopter landing zones or heavily trafficked air corridors. Additionally, retrofitting existing buildings and infrastructure to mitigate aviation risks—such as adding protective barriers or creating disaster response plans—aligns with the *Avoidance Principle*. This approach ensures that densely populated urban spaces are better equipped to prevent unnecessary risks associated with low-altitude aircraft.

5.10.3.3 Tolerance Principle: Managing Aviation Safety Trade-Offs in Urban Environments

Urban aviation risks involve complex trade-offs, particularly as the demand for air travel and drone usage continues to grow, while safety concerns for residents remain a priority. The *Tolerance Principle* teaches that cities should accept certain risks and accommodate infrastructure development, provided that proper safety measures are implemented to mitigate the worst possible outcomes. This principle emphasizes the need to balance the growing need for airspace with the understanding that current safety systems have limitations.

To apply this principle, cities should acknowledge that while complete safety cannot always be guaranteed, some risks associated with low-altitude flights should be tolerated, as long as advanced technologies like collision avoidance systems, geo-fencing, and reliable air traffic control help minimize these risks. Managing the balance between rapid urban development and maintaining air traffic safety requires flexibility, such as allowing drone flights for delivery services, but with strict operational protocols and monitoring in place. This ensures that while accommodating new technologies, safety remains a priority.

5.10.3.4 Adaptation Principle: Adapting Urban Airspace Management to Changing Aviation Technology

The rapid evolution of aviation technology, particularly drones and small aircraft, necessitates an adaptable approach to urban airspace management. The *Adaptation Principle* encourages cities to remain flexible and evolve their infrastructure, regulations, and airspace systems in response to new risks and technological advancements. This adaptability is essential to keep pace with the growing demands of urban aviation and ensure safety while fostering innovation.

To apply this principle, cities should invest in the development of smart air traffic management systems that can accommodate both manned aircraft and unmanned aerial vehicles (UAVs) (Aposporis, 2024). These systems should leverage real-time data and AI algorithms to predict and avoid potential collisions, ensuring smoother air traffic flow. Additionally, as drone technology and low-altitude aircraft continue to advance, cities should adapt their regulations and infrastructure to accommodate innovations such as autonomous aircraft and vertical take-off and landing (VTOL) vehicles (Liu et al., 2024). This may involve modifying zoning laws to establish new flight paths or landing zones, helping to manage air traffic as it continues to increase.

5.10.3.5 Regulation Principle: Strengthening Aviation Safety Through Comprehensive Regulations

The *Regulation Principle* underscores the importance of implementing robust standards and regulations to prevent aviation-related accidents. Just as the immune system self-regulates to prevent harmful overreactions, urban airspace management should be governed by strict rules to ensure safety without stifling advancements in aviation technology. By establishing clear and effective regulations, cities can balance growth and safety in urban aviation.

To apply this principle, comprehensive regulations should be established for both manned and unmanned aircraft in urban airspace (Xu et al., 2020). These rules should define operational guidelines, including flight altitude limits, no-fly zones near sensitive areas, and emergency landing protocols in case of aircraft malfunctions. Drones, for instance, should be subject to geo-fencing restrictions to prevent them from entering restricted airspace or flying over densely populated areas (Liu et al., 2016). Additionally, cities should develop safety protocols for infrastructure such as helipads, ensuring that they meet rigorous standards to minimize the risks associated with helicopter crashes. These measures help create a safe and regulated environment for urban aviation.

5.10.3.6 Coordination Principle: Fostering Cross-Sector Collaboration for Aviation Safety

The *Coordination Principle* emphasizes the need for effective communication and collaboration among various stakeholders, including government bodies, aviation regulators, drone operators, and urban planners. By fostering strong coordination, cities can prevent the fragmentation of air traffic control efforts and ensure a unified approach to safety. This collaborative approach is crucial for managing urban airspace effectively and integrating new technologies while maintaining safety.

To apply this principle, cities should establish multi-stakeholder task forces or advisory boards that bring together experts from aviation, urban planning, emergency services, and technology sectors (Sun et al., 2025). These groups can collaborate to improve airspace management and develop integrated solutions that address aviation risks while supporting technological advancements. Additionally, public–private partnerships will be vital for developing new urban air mobility systems, such as drone delivery services or small aircraft networks. By coordinating with regulators and operators, cities can ensure that safety standards are met without stifling innovation.

5.10.3.7 Diversity Principle: Diversifying Aviation Solutions to Mitigate Risk

The *Diversity Principle* suggests that cities should diversify their infrastructure solutions to reduce dependency on any single type of aviation technology. This approach strengthens resilience by ensuring that the failure of one system does not cause a complete disruption of urban air traffic. By promoting a range of technologies and systems, cities can ensure greater reliability and adaptability in their airspace management.

To apply this principle, cities can integrate various types of low-altitude aircraft, including helicopters, drones, and small planes, into a cohesive air traffic network (Apospori, 2024). For example, in high-demand areas, cities could explore alternative transportation methods like autonomous aerial taxis alongside traditional flight options. Furthermore, cities should encourage the use of different drone models and technologies, ensuring that the failure of one technology does not lead to a citywide crisis. Similarly, small aircraft should be complemented by robust emergency air traffic management systems capable of addressing failures in any part of the network, thereby maintaining the safety and efficiency of urban airspace.

5.10.3.8 Robustilient Regeneration Principle: Building Resilience in Aviation Systems for Future Challenges

The *Robustilient Regeneration Principle* advocates for ensuring that cities not only have strong systems to withstand immediate threats but also possess the capacity to recover and improve over time. In the context of low-altitude aircraft, this principle calls for cities to develop resilient aviation systems that can adapt to unforeseen challenges. By focusing on both strength and recovery, cities can better handle potential disruptions and continue to thrive in the face of adversity.

To apply this principle, cities should establish effective recovery plans for incidents such as helicopter crashes or drone accidents, focusing on minimizing damage and restoring functionality as quickly as possible. This requires well-coordinated emergency response teams trained to manage aviation accidents in densely populated urban environments. Additionally, cities should continuously upgrade their infrastructure to incorporate the lessons learned from past incidents. This could include designing new technologies or flight paths with built-in resilience and adaptability. Implementing "fail-safes"[1,2] or backup air traffic control systems (Huang et al., 2024) would allow for a rapid recovery in case of technological failures, ensuring that urban airspace remains safe and efficient even during challenging times.

---

[1] https://zala-aero.com/en/news/zala-aero-presented-new-koncepciju-ispolzovanija-bespilotnyh-letatelnyh-apparatov-zdd/ (Accessed on January 25, 2025).

[2] https://www.ehang.com/ehangaav/ (Accessed on January 25, 2025).

5.10.3.9 Coherence Principle: Ensuring Unified and Harmonious Functioning of Urban Aviation Systems

The *Coherence Principle* emphasizes the importance of ensuring that all urban infrastructure systems work together in harmony. In the context of aviation, this means integrating air traffic control, emergency services, and urban infrastructure to function seamlessly. Such integration is crucial for preventing accidents and ensuring the safety of both residents and air traffic.

To apply this principle, cities should implement smart infrastructure systems that integrate real-time data on both aviation and ground traffic (Huang et al., 2024). For example, emergency services should have access to live air traffic data to respond swiftly in the event of a low-altitude aircraft accident. Additionally, airports, helipads, and drone landing zones should be designed and operated in a way that complements the city's existing infrastructure, including transportation, utilities, and emergency response systems. This integrated approach would create a unified urban environment, enhancing both safety and functionality for all city systems.

5.11 Mass Gathering and Tourism Overcrowding Hazards

Mass gathering and tourism overcrowding hazards refer to the safety risks that arise when large numbers of people converge in confined or high-traffic spaces, such as public events or popular tourist destinations (Drury & Stott, 2013; Rahman et al., 2014; Vanumu et al., 2020; Kanaujiya & Tiwari, 2022). These hazards primarily stem from *spatial interference* caused by inadequate infrastructure, poor crowd management, and insufficient emergency preparedness. When crowd density exceeds the capacity of a space, it leads to physical accidents, health risks, and operational disruptions. This is especially true when infrastructure and crowd control measures are not designed to manage large volumes of people, increasing the likelihood of dangerous interference between individuals and the surrounding environment.

The causes of overcrowding hazards are multifaceted and closely linked to the spatial constraints of the environment. Inadequate infrastructure is obviously a major factor, as it leaves venues and tourist destinations without the necessary space, exits, or crowd management systems to accommodate large numbers of people safely. Poor crowd management further exacerbates the situation (Bosta, 2011; Sharma et al., 2018), as the absence of barriers, clear signage, poor warning information, or trained personnel creates chaos and confusion, making it difficult to manage *spatial interference* and keep people from crowding into dangerous areas. Overcrowding is particularly hazardous during peak times or special events when overcapacity occurs at venues or tourist hotspots. The concentration of people in already limited spaces results in *spatial interference* that impedes the flow of movement, contributing to incidents like trampling or difficulty evacuating during emergencies. Emergency preparedness failures (Wang et al., 2021; Sun et al., 2022), such as unmarked evacuation routes or the absence of plans for high-density crowd control, prevent effective responses to crises, allowing *spatial interference* to escalate the danger. Additionally, unregulated visitor flow, whether spontaneous or poorly coordinated during busy seasons, leads to overcrowding and intensifies the risks of interference, particularly when crowd density reaches unsafe levels.

5.11.1 Consequences of Mass Gathering and Tourism Overcrowding Hazards

5.11.1.1 Physical Accidents and Injuries

Overcrowded areas in mass gatherings or tourist sites can lead to dangerous physical accidents, such as stampedes or people being trampled (Feliciani et al., 2023; Kaila, 2024). In tightly packed spaces, panic or sudden movements often trigger chaotic situations, especially when narrow walkways or poorly designed exits amplify risks. Additionally, overcrowding increases the likelihood of slips, trips, and falls, particularly in confined spaces or poorly maintained locations, leading to injuries from collisions or falls.

5.11.1.2 Health and Safety Hazards

Dense crowds create an ideal environment for the rapid spread of infectious diseases (Moosa & Khatatbeh, 2020; Hamidi & Hamidi, 2021; Von Seidlein et al., 2021), particularly in poorly managed tourist sites or during periods of high tourism traffic, like public transportation. In these settings, it becomes easier for illnesses to spread among individuals, potentially leading to larger outbreaks. Moreover, overcrowding can overburden emergency services, causing delays in medical or security responses, and exacerbating the situation during critical incidents in crowded public spaces or major events.

5.11.1.3 Disruption to Normal Operations

Overcrowding can severely hinder emergency evacuations, especially in crowded venues or tourist attractions (Chow & Ng, 2008; Li et al., 2017; Aleksandrov et al., 2018; Mohd Ibrahim et al., 2022). Inefficient crowd control measures or poorly planned evacuation routes can delay or complicate the process, putting lives at risk during emergencies like fires, floods, or terror threats. Additionally, peak tourism seasons or major events that coincide with large crowds can disrupt operations, overwhelming local infrastructure and complicating normal operations at popular destinations.

5.11.1.4 Degradation of the Experience

Overcrowding negatively impacts the overall visitor experience, creating feelings of stress and frustration among attendees (Santana-Jiménez & Hernández, 2011; Pedica et al., 2021; Papadopoulou et al., 2022). With limited space, people are often forced to rush or jostle to move through the crowd, leading to unsafe behaviors and accidents. Additionally, overcrowded conditions, such as long waits and limited rest areas, can cause physical exhaustion, dehydration (Yezli, 2023), or even fainting (Cymes et al., 2021), particularly in settings that lack provisions for comfort and well-being.

5.11.1.5 Long-Term Impact on Destination Sustainability

Overcrowding places immense pressure on local infrastructure, services, and the environment, which can result in long-term damage to the sustainability of tourism destinations and the viability of hosting mass gatherings (Damnjanović, 2021; Butler & Dodds, 2022). This strain may also lead to reputational harm for destinations or event organizers, affecting their ability to attract future visitors. Additionally, repeated overcrowding incidents that result in accidents or health outbreaks can erode public trust in the safety of these venues, potentially decreasing future attendance and participation.

5.11.2 Practical Issues in Addressing Mass Gathering and Tourism Overcrowding Hazards

5.11.2.1 Unsafe and Outdated Infrastructure at Crowded Venues

Venue and Space Design: Many tourist sites and venues were not originally designed to accommodate the large crowds seen in modern mass gatherings, which often leads to congestion and safety risks. Iconic landmarks such as the Eiffel Tower and the narrow streets of historical cities like Venice were built long before mass tourism became prevalent, and as a result, they lack the capacity to safely handle large numbers of visitors. The *spatial interference* caused by the clash between increased foot traffic and infrastructure limitations leads to bottlenecks, restricted movement, and difficulty in emergency evacuations. Additionally, poorly designed spaces, such as narrow walkways or insufficient rest areas, exacerbate crowding and can increase the likelihood of accidents. Some venues may also not have been updated or maintained to accommodate large crowds, leaving them vulnerable to hazards like poorly lit areas, dilapidated staircases, and inadequate signage—all contributing to the overall *spatial interference* within the site.

Emergency Exit and Evacuation Routes: In the event of an emergency, clear and sufficient exits are essential to ensure safety, but many venues suffer from inadequate emergency planning. Often, exits are blocked by crowds or are poorly designed, creating *spatial interference* that forms bottlenecks, hindering the flow of people during critical moments. For historic or culturally significant sites, constraints on modifying the infrastructure, such as preservation laws, further limit the ability to adjust exit routes, which adds to the risks associated with overcrowding. The lack of space or legal restrictions makes it difficult to expand exit capacity, leaving these sites particularly vulnerable during large gatherings.

5.11.2.2 Ineffective Crowd Control and Management Failures

Lack of Training and Resources for Crowd Control: Effective crowd management requires skilled staff who understand crowd behavior and can respond appropriately during emergencies (Ammon & Fried, 1999; Bishop et al., 2020). However, many event organizers or tourist site operators often lack sufficiently trained personnel, resulting in poorly coordinated responses during high-pressure situations. Understaffing, particularly during peak times, exacerbates the challenges of managing the *spatial interference* caused by large numbers of people trying to move through constrained or poorly designed spaces. The lack of coordination between stakeholders (Feliciani et al., 2021)—such as law enforcement, security, emergency medical teams, and venue operators—also complicates efforts to manage crowd dynamics and mitigate risks associated with overcrowding.

Managing Unpredictable Crowd Behavior: Crowd behavior can rapidly escalate into panic and chaos, especially during emergencies (Sharma et al., 2023; Liang et al., 2024). In these situations, *spatial interference*—such as blocked exits, crowded hallways, and lack of mobility—can significantly amplify the risks. Spontaneous surges in crowd size add unpredictability, making it harder to manage the flow of people in already congested areas. Such unexpected crowd movements can overwhelm existing crowd control measures, further complicating safety efforts and increasing the likelihood of accidents or security breaches.

5.11.2.3 Dangerous Overcrowding and Poor Capacity Control

Overcapacity During Peak Times: During peak tourist seasons or special events, many tourist sites experience overcrowding that surpasses their safe capacity, leading to substantial safety risks. This is particularly concerning at historical or cultural venues where crowd control measures may not be flexible enough to handle varying crowd sizes. Overcrowding leads to *spatial interference*, where limited space prevents smooth crowd movement, thereby increasing the risk of accidents or even physical harm. The seasonal nature of tourism further compounds the problem, with unexpected surges of visitors due to unplanned public events, viral trends, or unforeseen factors like social media influencers drawing additional crowds. Managing these fluctuations requires advanced planning, such as reservation systems or timed entry, but these solutions are not always sufficient or feasible to control crowd sizes effectively.

Spatial Constraints and Inefficient Flow Management: Many tourist sites or event venues are spatially constrained, which makes it difficult to manage crowds effectively. Without proper planning for crowd flow, areas can become congested, impeding movement and blocking access to exits in case of an emergency. Ineffective venue layouts that do not account for crowd dynamics—such as narrow aisles, bottlenecks, or blocked pathways—are prime examples of *spatial interference*, which heightens the risk of accidents and reduces the ability to manage crowds safely. This issue is especially prevalent in historic or older venues, where retrofitting the space to accommodate modern crowd management practices can be difficult due to architectural limitations.

5.11.2.4 Severe Health Risks and Emergency Response Failures

Spread of Infectious Diseases: Overcrowding creates environments conducive to the rapid spread of infectious diseases, such as COVID-19 or the flu. In densely packed areas with poor ventilation, viruses spread more easily, and the *spatial interference* caused by restricted movement and limited personal space increases the likelihood of infection transmission. Insufficient sanitation facilities, particularly at outdoor events or in popular tourist locations, further exacerbate the risks. The lack of adequate provisions, such as hand sanitizers, clean water, and emergency medical support, in crowded spaces makes it difficult to contain outbreaks and maintain public health standards.

Delays in Emergency Response: Overcrowded spaces can significantly delay emergency responses. Narrow pathways, blocked exits, and general congestion prevent the swift evacuation of individuals or timely medical interventions. The *spatial interference* caused by large crowds impedes emergency personnel from accessing individuals who need assistance, potentially worsening outcomes during crises. Additionally, the overwhelming number of people requiring help during an emergency can strain emergency services, making it difficult to respond quickly and efficiently.

5.11.2.5 Policy Failures and Weak Regulation in Crowd Safety

Lack of Comprehensive Regulation: A significant challenge in addressing overcrowding risks is the inconsistency in safety regulations (Nunweek & O'Hanlon, 2019; Haghani et al., 2023). Different event organizers, tourism authorities, and local governments often follow varying crowd control standards, which makes it difficult to ensure a uniform level of safety. This lack of regulation contributes to *spatial interference*, as venues may lack sufficient safety measures, and crowd sizes may exceed safe limits, heightening risks at mass gatherings. Even when regulations exist, enforcement is often lacking (Kaila, 2024), especially in informal or smaller tourist areas.

Political and Economic Factors: Economic pressures often encourage event organizers and tourism operators to maximize revenue, sometimes at the expense of safety (Hay, 2019). This can lead to overbooking and overcrowding, especially during peak times, where the focus on profit can lead to disregarding the spatial constraints of the venue. Additionally, political factors (Burns & Novelli, 2007) may prevent the implementation of necessary policies or regulations due to concerns about economic repercussions or public backlash, which delays efforts to address overcrowding.

5.11.2.6 Noncompliance and Ignorance Fuel Overcrowding Risks

Resistance to Regulations: Tourists and event-goers may resist crowd control measures such as timed entry, reservations, or restrictions on access to certain areas. This resistance often contributes to *spatial interference*, where the lack of adherence to safety guidelines leads to overcrowding and unsafe behaviors. Visitors may prioritize convenience or their own desires over the need for safety, which can exacerbate risks in crowded environments.

Lack of Public Awareness: A key contributor to overcrowding hazards is the lack of public awareness about the dangers of crowded environments. Many visitors are unfamiliar with the risks they face and may lack knowledge about crowd etiquette, personal safety, or emergency procedures. Without clear signage or communication regarding crowd density, exits, or emergency protocols, visitors may struggle to navigate crowded spaces safely, which worsens the effects of *spatial interference* and increases the likelihood of accidents.

5.11.3 Urbanimmunology-*Informed Solutions to Mass Gathering and Tourism Overcrowding Hazards*

To address the practical issues of mass gatherings and tourism overcrowding hazards using principles from *Urbanimmunology*, we can apply specific core principles to each issue, informing potential solutions for a safer and more sustainable urban and social environment.

5.11.3.1 Scalable and Responsive Infrastructure for Safer Gatherings: Enhancing Capacity and Real-Time Response in Public Spaces

Cities and tourist sites should focus on building sustainable infrastructure that can support large crowds by adhering to the *Critical Dependency Principle*, which emphasizes the need for scalable infrastructure capable of handling growing population densities. This includes not only expanding venues but also enhancing transportation networks, sanitation facilities, and healthcare services. Long-term urban planning should integrate flexible designs that allow for better adaptation to crowd fluctuations while maintaining safety. According to the *Adaptation Principle*, infrastructure design should incorporate flexibility, such as modular venues that expand during peak times, temporary bleachers, mobile restrooms, and pop-up medical facilities. Investing in adjustable crowd control barriers and resilient, dynamic designs ensures that cities can meet the demands of both tourists and large events. Furthermore, the *Tolerance Principle* encourages cities to accept the inevitability of challenges like unexpected surges in visitors and use real-time data to monitor crowd sizes. This enables dynamic responses, adjusting access, services, or safety measures to maintain a balance between high demand and safe crowd flow.

5.11.3.2 Toward Inclusive and Resilient Crowd Management: Coordinated and Flexible Strategies for Safer, People-Centered Spaces

Effective crowd management and control require the application of various principles to ensure safety and smooth operations at mass gatherings and tourist destinations. The *Coordination Principle* emphasizes the need for effective collaboration among all parties involved, such as local authorities, event organizers, law enforcement, and emergency responders, with a centralized communication system that enables swift decision-making. The *Regulation Principle* advocates for stronger regulatory frameworks that enforce safety measures like maximum crowd density, emergency exits, and crowd flow management, ensuring compliance through regular checks and certifications for event organizers. The *Tolerance Principle* highlights the importance of managing uncertainties, such as sudden surges in attendance, by implementing real-time monitoring systems that track crowd density and enable adaptive responses, like redirecting visitors to less crowded areas. Finally, the *Diversity Principle* stresses the need to recognize the diverse nature of crowds, ensuring inclusive crowd management through multilingual signage, cultural sensitivity training, and tailored measures to accommodate the different behaviors, languages, and needs of various groups, such as local residents, international tourists, children, and the elderly.

5.11.3.3 Forward-Thinking Tourism Flow Management: Smart, Inclusive, and Collaborative Strategies for Reducing Overcrowding

Proactively addressing overcrowding and capacity management involves applying several principles to ensure visitor safety and comfort. The *Avoidance Principle* advocates for systems that control access, such as timed entry tickets, reservation-based access, or visitor quotas, and technology-driven solutions like real-time capacity monitoring apps to prevent overcrowding before it becomes dangerous. The *Diversity Principle* promotes highlighting alternative, less-visited destinations to distribute visitor demand more evenly and reduce pressure on popular sites, ensuring a more enjoyable experience for all. The *Coordination Principle* stresses the importance of collaboration between cities, hotels, tour operators, and transportation services to manage visitor flow, especially during peak seasons, by directing tourists to specific zones or offering incentives to visit nonpeak hours. Finally, the *Regulation Principle* suggests enforcing limits on the number of tourists at specific sites per day or implementing permitting systems for certain activities, ensuring controlled tourism while preserving the safety and integrity of the destination.

5.11.3.4 Building Resilient Cities: Progressive Health and Safety Strategies for Mass Gatherings and Overcrowding Management

Cities should develop a healthcare system that is both robust and resilient to respond effectively to health crises during mass gatherings or overcrowding, in line with the *Robustilient Regeneration Principle*. This could involve deploying pop-up clinics and ensuring clear communication between hospitals and event organizers to provide immediate medical responses. Post-event health screenings and tracking systems can further address health-related concerns. The *Adaptation Principle* emphasizes the need for flexibility in adapting to emerging health risks, such as implementing health checks, mask mandates, or vaccination checkpoints during pandemics or seasonal outbreaks. The *Tolerance Principle* advocates for creating contingency health plans that anticipate sudden health issues, ensuring quick responses through enhanced sanitation, medical support, and flexible measures. Finally, the *Regulation Principle* calls for health safety regulations, such as ensuring sanitation stations, social distancing policies, and capacity limits, to safeguard visitors' health while maintaining crowd control and preventing overcrowding.

5.11.3.5 Optimizing Governance: Evolving Strategies for Effective Regulatory and Policy Solutions in Crowd Management

Regulatory and policy challenges related to tourism and crowd management require a cohesive and coordinated approach to ensure the safety and efficiency of mass gatherings. The *Coherence Principle* emphasizes the need for integrated regulations across sectors such as transportation, public safety, urban planning, and tourism management, ensuring alignment between policies like emergency evacuation routes, transportation schedules, and public health measures for a seamless visitor experience. The *Regulation Principle* stresses the importance of clear and robust regulations that provide a consistent framework for crowd control, emergency preparedness, and sustainable tourism practices, enforced through inspections, permits, and penalties to maintain safety standards. The *Coordination Principle* highlights the need for coordination at local, regional, and national levels, ensuring that policies are consistent and easily adaptable when changes, such as reducing crowd sizes or enhancing health protocols, are needed. Collaboration between various stakeholders, including local governments, event organizers, law enforcement, and tourism agencies, is key to ensuring consistent policy implementation.

5.11.3.6 Enhancing Public Cooperation: Culturally-Aware Approaches for Effective Crowd Management and Engagement

Building public trust and cooperation in crowd management requires effective communication and cultural sensitivity. The *Coordination Principle* emphasizes the importance of clear, open communication through media, social media influencers, and tourism agencies to raise awareness about crowd safety, ensuring that tourists are well-informed about rules, safety measures, and guidelines. Public education campaigns, signage, and announcements can further engage the public, fostering cooperation and compliance. The *Diversity Principle* highlights the need for culturally tailored communication, offering multilingual materials and addressing local customs to ensure that all visitors, regardless of background, understand and follow safety guidelines. The *Tolerance Principle* stresses the importance of accepting different attitudes toward safety measures, particularly when visitors may not be familiar with specific regulations. Providing clear, respectful explanations for protocols and adapting messages to overcome cultural resistance can help increase compliance and foster a more cooperative public environment.

5.12 Commonalities of *Spatial Interference Hazards*

5.12.1 Definition

The first section of this chapter defines *spatial interference hazards* as potential dangers that arise from misalignments, overlaps, or conflicts between spatial arrangements and functional systems within shared or adjacent physical spaces. These hazards typically occur in environments where diverse activities, systems, or forces coexist, such as urban areas, construction sites, industrial zones, or natural landscapes. In other words, when interactions between urban systems, natural processes, and human activities disrupt equilibrium, these hazards can emerge, creating vulnerabilities or intensifying the impacts of disasters.

In the previous sections, *spatial interference hazards* were explored in relation to underground spaces, structures and infrastructure, natural disasters, human-wildlife conflicts, traffic conflicts in public spaces, falling objects from heights, low-altitude aircraft, and overcrowding. While each hazard type has its own unique characteristics, the analyses reveal several common observations. These shared insights are worth summarizing, as they can help inform universal solutions to prevent or mitigate *spatial interference hazards*.

5.12.2 *Causes*

Spatial interference hazards arise from various causes, primarily driven by the complex interactions between different activities, objects, systems, or environmental factors within shared or adjacent spaces. These hazards can occur in any context where multiple elements operate in close proximity. The sources of interference may stem from infrastructural challenges, technological factors, environmental conditions, human actions, and/or governance issues.

5.12.2.1 Infrastructural and Structural Issues

The physical design and maintenance (or lack thereof) of infrastructure contribute significantly to *spatial interference hazards*. Inadequate or improperly maintained infrastructure can directly lead to accidents, particularly when spatial elements are poorly organized or designed without considering safety.

- *Spatial Insufficiency and Inflexibility:* The lack of spatial redundancy or flexibility in infrastructure amplifies risks when a single failure cascades across multiple systems. Physical limitations, such as constrained evacuation routes or insufficient drainage capacities, are examples of design flaws that exacerbate hazards.
- *Structural Failures:* Deteriorating or poorly designed infrastructure, such as bridges, buildings, tunnels, or roads, can collapse or fail, creating interference hazards. For instance, a road cave-in can disrupt traffic and endanger both vehicles and pedestrians. Structural instability in public spaces or event venues can lead to accidents if not promptly addressed.
- *Poor Urban Planning and Zoning:* Inadequate urban planning can create environments where incompatible activities overlap in shared spaces. For example, placing residential areas too close to industrial zones or major traffic routes can expose residents to noise, pollution, and safety risks from accidents or hazardous materials spills. Similarly, pedestrian spaces might not be properly segregated from vehicle lanes, increasing the likelihood of pedestrian-vehicle conflicts.

- *Overburdened Transportation Systems:* In cities with outdated or insufficient transportation infrastructure, the risk of *spatial interference* increases. Overcrowded subways, buses, and roads create conditions where multiple activities (e.g., pedestrian movement, vehicle traffic) are in constant conflict. This can lead to delays, accidents, or unsafe interactions between different modes of transport.

5.12.2.2 Land Use and Spatial Conflicts

Conflicts arise when different land uses or activities share the same space or are located too close to each other. These spatial conflicts can lead to hazards.

- *Tourism Overcrowding:* Tourist hotspots often suffer from overcrowding, especially during peak seasons. When large numbers of tourists gather in small or congested areas (e.g., historic sites, public squares), the risk of accidents—ranging from falls and trampling to traffic disruptions—becomes elevated. Overcrowding also strains local infrastructure, leading to bottlenecks and further safety risks.
- *Large Public Events:* Concerts, sports events, and festivals can generate significant foot traffic and spatial conflicts. If crowd management is not properly handled, overcrowding, pushing, or stampedes can occur, especially in confined spaces like stadiums or open-air venues. Additionally, event-related traffic disruptions can create hazards for both participants and local residents.

5.12.2.3 Technological Factors

Advancements in technology often introduce new hazards, especially when systems and machines interact with each other or with human users in shared spaces.

- *Automation and Robotics:* The use of automated systems, such as self-driving vehicles, drones, or industrial robots, introduces potential hazards when these systems interact with human-operated machinery or vehicles. For example, autonomous drones flying at low altitudes in urban areas can pose risks to pedestrians or other aircraft if their navigation systems are not properly synchronized.
- *Failure of Infrastructure or Systems:* The failure of technological systems can lead to *spatial interference hazards*. For example, a malfunctioning traffic signal system can cause confusion at intersections, increasing the risk of vehicle accidents or pedestrian incidents. Similarly, failures in the infrastructure (e.g., power outages, elevator malfunctions) can lead to hazardous situations in high-traffic spaces.
- *Data or Sensor Failures:* Many systems that manage shared spaces rely on sensors or data systems to avoid interference (e.g., air traffic control systems, vehicle collision-avoidance systems). If these systems malfunction or fail to detect

obstacles, there can be a significant risk of accidents, especially in environments like airports or busy streets.

5.12.2.4 Environmental and Natural Factors

Natural forces and environmental conditions are often unpredictable and can contribute to *spatial interference hazards* by creating disruptions in spaces that were previously safe or predictable.

- *Weather Events:* Severe weather events, such as rainstorms, snowstorms, high winds, and flooding, can rapidly change the conditions of shared spaces. For instance, during a windstorm, flying debris can create hazards for pedestrians or drivers, while heavy rain can flood roads or construction zones, rendering them unsafe or impassable.
- *Flooding and Stormwater Management Challenges:* Improperly designed or overwhelmed stormwater systems can lead to localized flooding, blocking access to streets, subways, or public spaces. In areas with inadequate drainage, water accumulation can lead to accidents, especially when it impedes vehicle or pedestrian movement, creating slip-and-fall risks.
- *Natural Disasters:* Events like landslides, earthquakes, and tornadoes can cause rapid, large-scale *spatial interference hazards*. For example, a landslide may block roads, damage infrastructure, or create hazards in construction zones. Earthquakes might damage buildings, collapse bridges, or disrupt air and rail traffic, leading to dangerous intersections between various spatial elements.
- *Human-Wildlife Interactions:* In urban areas or spaces where human settlements and wildlife habitats overlap, conflicts can arise when animals wander into areas with high-density human activities. These incidents can include wildlife causing traffic accidents (e.g., deer crossing roads) or animals entering urban spaces and creating hazards (e.g., snakes or bears near schools or parks).

5.12.2.5 Human Factors

Human behavior and decision-making are significant contributors to *spatial interference hazards*. These can involve errors, misjudgments, or interactions that unintentionally create conflict or disruption in shared spaces.

- *Negligence or Lack of Awareness:* People might fail to recognize potential hazards in their environment, leading to situations where *spatial interference* occurs. For instance, pedestrians walking into traffic lanes or workers ignoring safety protocols on construction sites can create dangerous situations.
- *Overcrowding:* High-density human activities, such as large public gatherings, festivals, or tourist hotspots, increase the potential for *spatial interference hazards*. When large numbers of people occupy a small area, the risk of physical

collisions, crowd-related accidents (e.g., trampling), and interference with transportation systems becomes higher.
- *Miscommunication or Lack of Coordination:* In settings like construction sites or industrial environments, poor communication between workers or lack of coordination in the use of shared spaces (e.g., pathways, equipment, and work zones) can create hazardous situations. For example, if a worker is not informed about ongoing crane operations in the vicinity, they might inadvertently enter a danger zone, causing collisions or accidents.
- *Conflicts in Multimodal Transportation:* In urban areas, when different modes of transport (e.g., vehicles, bicycles, pedestrians) share the same space, human errors or poor traffic management can result in accidents. Cyclists riding in pedestrian lanes or drivers ignoring traffic signals can lead to dangerous interactions.

5.12.2.6 Regulatory and Governance Failures

Spatial interference hazards can also arise due to inadequate policies, regulations, or enforcement that fail to address the complexities of shared spaces.

- *Lack of Safety Standards or Building Codes:* In some regions, weak enforcement of safety codes can result in hazardous construction practices, leading to risks such as building collapses, fire hazards, or improper zoning. Buildings or public spaces that are not designed with safety in mind heighten the risk of accidents when various activities and systems interact.
- *Inadequate Risk Management Systems:* In environments where multiple activities share the same space, a lack of comprehensive risk management strategies can lead to uncoordinated operations. For example, construction sites without proper oversight or communication between different contractors can create unsafe conditions when machinery or materials conflict with pedestrians or other work zones.
- *Poor Governance:* Poorly managed urban expansion into high-risk zones, inadequate maintenance of critical systems, and fragmented decision-making processes contribute significantly to the prevalence of *spatial interference hazards*. Failures to integrate cross-sectoral perspectives or address the complexities of interdependent systems often exacerbate these risks, making them more difficult to mitigate.

5.12.3 Consequences

The consequences of *spatial interference hazards* can be extensive, including physical harm, financial losses, environmental degradation, and societal disruption. The severity of these consequences depends on the type of hazard, the extent of the interference, the affected area, and the individuals or systems involved. These impacts often intersect, with physical, economic, social, and environmental consequences compounding and amplifying one another.

5.12.3.1 Injuries and Fatalities

Spatial interference hazards can cause severe injuries, fatalities, and disabilities. Collisions between pedestrians, vehicles, or bicycles in urban areas can result in broken bones, head trauma, or internal damage, while poorly maintained infrastructure can lead to falls, sprains, fractures, or concussions. In construction or industrial sites, machinery accidents can cause crushing injuries or fatalities. Structural failures, poor planning, or spatial conflicts in construction zones may lead to building collapses, and natural disasters or traffic mismanagement can trigger multi-vehicle crashes, often resulting in fatalities. High-impact accidents can also cause long-term disabilities, including spinal cord injuries, brain damage, amputations, or permanent limb impairments.

5.12.3.2 Damage to Property and Infrastructure

Spatial interference hazards can cause significant damage to property and infrastructure. Environmental factors such as flooding, landslides, or natural disasters can lead to the destruction of buildings, roads, bridges, and other critical infrastructure, resulting in substantial repair costs and disruption to essential services. In traffic accidents or windstorm-induced crashes, vehicles may sustain heavy damage, imposing repair costs and financial burdens on individuals and businesses. Additionally, public spaces like parks, stadiums, and transportation hubs can experience damage from overcrowding, vandalism, or accidents, leading to costly repairs and temporary closures.

5.12.3.3 Economic Consequences

Spatial interference hazards can have significant economic impacts on individuals, businesses, governments, and entire economies. These hazards can lead to financial losses through increased insurance claims, higher premiums, legal liabilities from accidents, and costly repairs to damaged infrastructure. Events like bridge collapses or prolonged downtime can lead to substantial revenue losses, especially in industries reliant on continuous operations. Additionally, disruptions in transportation, such as road closures or traffic accidents, can delay the flow of goods and services, causing shortages and costly delays. Major events can also impact local businesses, reducing sales and revenue. Furthermore, workplace accidents and transport disruptions can lower productivity, leading to delays and impacting budgets.

5.12.3.4 Environmental Consequences

Spatial interference hazards, particularly those involving natural forces or human activities, can have significant ecological consequences, affecting biodiversity, land use, and ecosystem health. Accidents such as vehicle crashes or infrastructure failures can cause hazardous spills, contaminating soil, water, and air, harming wildlife and plant life. Natural disasters like floods or landslides can lead to soil erosion, deforestation, and loss of biodiversity. Poor stormwater management can damage riparian zones or wetlands, disrupting aquatic ecosystems. Additionally, disasters like flooding or windstorms can devastate agricultural lands, threatening food security, while human-wildlife conflicts in urban areas can displace or kill wildlife, further disrupting ecosystems.

5.12.3.5 Social Consequences

Spatial interference hazards can significantly impact communities and individuals, disrupting daily life and contributing to social inequality. Public health can be severely affected, with injuries leading to long-term physical and mental health issues. Social disruption can occur when hazards displace people or damage community spaces, leading to homelessness or a loss of access to essential services. The closure of public spaces like parks or cultural venues can further strain social cohesion. Vulnerable populations, including low-income groups, the elderly, and people with disabilities, are more likely to suffer from these hazards, facing heightened risks due to limited resources and exposure to environmental dangers.

5.12.4 Practical Issues

Addressing *spatial interference hazards* presents a range of practical challenges that stem from the complexity and variability of the hazards themselves. These issues arise from the interaction of human, technological, environmental, and infrastructural elements within shared spaces, making the identification, prevention, and mitigation of these hazards difficult.

5.12.4.1 Complexity of Identifying and Assessing Hazards

Diverse and Evolving Nature of Hazards: Spatial interference hazards are dynamic and evolve with changes in infrastructure, human activities, and environmental conditions. Urbanization, for example, increases the density of buildings, roads, and public spaces, raising the potential for spatial conflicts (Zhou et al., 2019). Additionally, climate change introduces new environmental hazards, such as more frequent flooding or severe storms, creating interference risks that were previously unanticipated (Barati et al., 2023). This evolving nature of hazards presents a challenge in predicting emerging risks, as risk assessments should continuously adapt to shifting conditions. Traditional hazard assessments may not account for new technologies, changes in land use, or evolving environmental factors, such as altered flooding patterns due to climate change, making it increasingly complex to predict how various activities or objects may interact in the future.

Lack of Comprehensive Data: Accurate data is crucial for assessing and addressing *spatial interference hazards*, but often, data on human activity patterns, environmental conditions, and infrastructural vulnerabilities is incomplete or inaccessible. This lack of comprehensive data hinders the identification of potential risks and the development of targeted mitigation strategies (Wirtz et al., 2014; St. Denis et al., 2020). While real-time monitoring tools, such as sensors and cameras, can provide valuable data, they often fail to cover all potential hazards in a given area. For instance, in urban environments, monitoring pedestrian movement, vehicle traffic, weather conditions, and infrastructure stress simultaneously can be technically challenging and resource-intensive, further complicating the task of gathering comprehensive data.

Inadequate Risk Mapping: Effective spatial risk management depends on accurate spatial mapping (Anelli et al., 2022), but many regions, particularly in developing areas or rapidly growing cities, suffer from outdated or inadequate infrastructure maps. This lack of up-to-date information means that planners and safety officials may not fully understand the interactions between spatial elements or the underlying vulnerabilities of a given space. Additionally, the complex interdependencies between different hazards (Van Wyk de Vries, 2025), such as traffic accidents, environmental disasters, or infrastructure failures, make accurate risk mapping even more challenging. For instance, heavy rainfall may cause flooding, which could disrupt transportation networks, leading to accidents or delays in emergency response, further complicating the prediction and management of risks.

5.12.4.2 Coordination Challenges Among Stakeholders

Lack of Interdisciplinary Collaboration: Addressing *spatial interference hazards* requires collaboration among various stakeholders, such as urban planners, engineers, environmental scientists, transportation officials, emergency responders, and local governments. However, these stakeholders often have different priorities, expertise, and resources (Elkady et al., 2024; Trogrlic et al., 2024), making coordination difficult. Furthermore, in many cities or regions, responsibility for spatial safety is fragmented across different agencies and departments. For example, road safety may fall under the transportation department, while flood management may be overseen by environmental or public works departments. This siloed approach creates barriers to coordinated risk management, leading to inefficient or incomplete responses to *spatial interference hazards*.

Conflicting Interests: Stakeholders involved in managing spatial hazards often face conflicting priorities, which can complicate risk mitigation efforts

(Bahadorestani et al., 2020; Bongo & Sy, 2024). For example, a city may prioritize economic development and urban expansion over environmental protection, leading to increased exposure to risks such as flooding, landslides, or traffic accidents in high-density areas. Similarly, private businesses or developers might focus on cost savings or rapid construction, sometimes at the expense of safety, resulting in structural hazards or inadequate planning for multimodal transport interactions. Balancing development with safety is an ongoing challenge, as cities should integrate effective safety measures into urban design without hindering development or increasing costs. This becomes particularly difficult when immediate development pressures clash with long-term safety goals, making it hard to achieve a balance between economic growth and spatial safety.

5.12.4.3 Technological and Infrastructure Limitations

Outdated or Insufficient Infrastructure: A significant challenge in addressing *spatial interference hazards* is outdated or insufficient infrastructure. Aging systems, such as bridges, roads, utilities, and drainage systems, often fail to meet modern spatial demands, creating hazards that were not anticipated in their original design. Many urban spaces were built before the advent of new technologies, population growth, and increased demand for public spaces and transport systems. For example, roadways or public transport networks designed decades ago may struggle to accommodate the current volume of traffic and pedestrians, leading to spatial conflicts like congestion or accidents. Furthermore, modernizing infrastructure to mitigate these hazards requires substantial investment, which can be difficult to secure. Governments and private companies may lack the resources or political will to fund these upgrades, especially when the benefits of such projects may not be immediately apparent.

Technological Limitations in Hazard Mitigation: Technology plays a crucial role in identifying, monitoring, and mitigating *spatial interference hazards*, but there are significant limitations. For instance, technologies like sensors used for monitoring traffic, weather, and infrastructure integrity can provide inconsistent or inaccurate data, especially in areas with unreliable Internet connections, poor maintenance, or improper calibration. Additionally, while autonomous systems such as drones, autonomous vehicles, and robots can improve safety, they also introduce new risks and challenges (Fisher et al., 2021; Yazdanpanah et al., 2023). Autonomous vehicles (Kenesei et al., 2022; Chougule et al., 2024), for example, may struggle to navigate complex environments or interact safely with pedestrians, creating new types of *spatial interference hazards* not anticipated in traditional transportation systems.

5.12.4.4 Economic and Financial Constraints

High Cost of Mitigation and Prevention: Addressing *spatial interference hazards* requires significant financial investment (Crow et al., 2018; Lee et al., 2022), not only in physical infrastructure but also in research, data collection, and the implementation of safety measures. However, many cities and regions, especially those in developing countries or facing budget constraints, may lack the financial resources to address these hazards comprehensively (Ngoc Su et al., 2021; Liu et al., 2024). Retrofitting existing infrastructure, such as widening roads, adding barriers to prevent vehicle-pedestrian collisions, or improving drainage systems, can be prohibitively expensive, and many governments or organizations may prioritize short-term needs over long-term investments in hazard prevention. Additionally, economic incentives for rapid development, such as real estate expansion or industrial projects, may outweigh the costs of implementing safety measures. Business owners and developers often resist paying for extra safety infrastructure that could delay projects or increase costs, even if the long-term benefits are clear.

Cost-Benefit Imbalance: The financial costs of *spatial interference hazards* are often difficult to quantify, especially when considering long-term damage or broader societal impacts, making it challenging to justify large-scale investments in safety measures. The uncertain long-term benefits of these investments (Fischhoff, 2015), which may reduce the likelihood of accidents or disasters, are not always immediately measurable in terms of short-term financial returns. As a result, gaining political or financial support for extensive safety projects becomes a challenge, as the direct return on investment may not be apparent in the near term.

5.12.4.5 Social and Cultural Barriers

Public Awareness and Engagement: A lack of public awareness or understanding of *spatial interference hazards* can significantly hinder efforts to mitigate them. Without widespread recognition of the risks (Karimiziarani et al., 2022; Reichstein et al., 2025), individuals may fail to take necessary precautions, and communities may not fully support or engage with safety initiatives. Additionally, cultural attitudes toward safety, urban planning, and environmental protection can vary, and in some regions, there may be a tendency to prioritize individual convenience over collective safety.

Social Inequalities and Vulnerable Populations: Social and economic inequalities can amplify the impact of *spatial interference hazards*, leaving vulnerable populations, such as low-income individuals, the elderly, or people with disabilities, more susceptible to their consequences (Hong et al., 2021; Quesada-Román & Campos-Durán, 2022; Smith et al., 2022). These groups may lack the resources to avoid or mitigate these hazards, making them especially vulnerable. Additionally, vulnerable populations are often excluded from safety measures, either due to a lack of resources or because their specific needs are not considered in urban planning processes. For example, poorly designed public spaces that do not account for accessibility needs can expose people with disabilities to additional risks.

5.12.4.6 Regulatory and Policy Challenges

Spatial interference hazards are often exacerbated by inconsistent or weak regulatory frameworks (Parrado, 2018; Bang, 2021). In many cities or countries, outdated or poorly enforced safety regulations fail to adequately address modern spatial risks. Even when regulations are in place, enforcement can be inconsistent, allowing hazards to go unaddressed. For example, traffic laws, zoning codes, or safety protocols for construction sites may be poorly enforced in urban areas, contributing to increased risks. Additionally, as new technologies such as drones, autonomous vehicles, and smart infrastructure emerge, regulations often lag behind (Brady et al., 2025; Sütfeld et al., 2025), creating new *spatial interference hazards* that cannot be adequately managed without proper regulatory frameworks.

5.12.5 Urbanimmunology-*Informed Solutions to Spatial Interference Hazards*

This section summarizes the findings from applying *Urbanimmunology*'s nine core principles, as discussed in previous analyses, by identifying general solutions to spatial interference hazards. This is essential for addressing diverse hazards. While each hazard has unique characteristics, general solutions offer a universal framework that can be adapted to various contexts. This approach promotes efficiency by providing proven strategies for common underlying issues like spatial management and risk reduction. It fosters cross-sectoral collaboration, helping various stakeholders work together. Moreover, categorizing solutions enhances consistency in policy development, ensuring clear guidelines and streamlined implementation. General solutions also offer flexibility, enabling cities and society to respond to evolving challenges over time while supporting long-term sustainability. By establishing adaptable strategies, cities and society can address multiple hazards effectively, balancing short-term needs with future resilience. While each principle serves a specific role in informing solutions and some share common capabilities, the nine principles collectively demonstrate a complementary and synergistic ability to develop integrated, multifaceted solutions to complex spatial interference hazards, ultimately fostering a more resilient and adaptive urban environment.

5.12.5.1 Guiding Spatial Resilience Through the Lens of Critical Dependency Principle: Reinforcing Critical Systems to Prevent Systemic Collapse

The *Critical Dependency Principle* underscores the importance of identifying and prioritizing the systems essential for the safety, functionality, and resilience of urban and social environments—specifically, those that support foundational operations such as transportation, energy, water, and waste management. These core systems act as the "bone marrow" of urban and social life, providing the essential services and resources that allow cities and society to function smoothly. To ensure they remain operational even under stress, these systems should be designed with built-in redundancies and adaptive capacities. This principle stresses that effective solutions to spatial interference hazards should take into account the interconnectedness of these critical systems. A disruption in one can trigger cascading failures across others, creating a domino effect that destabilizes the entire urban and social system, further exacerbating the impact of spatial interferences.

Therefore, solutions informed by this principle should focus on reinforcing these critical dependencies by guaranteeing that each system is resilient to disruptions and capable of quickly recovering. Urban and social systems depend on both physical infrastructure and effective governance structures to maintain operational stability. In this context, governance frameworks, planning institutions, and regulatory bodies should be established, ensuring that essential systems are designed to meet not only current needs but also emerging spatial hazards and evolving challenges. Through cross-sector coordination and robust regulation, cities and society can ensure that these systems function cohesively and can be rapidly restored if disrupted by spatial interference hazards.

Thus, the *Critical Dependency Principle* advocates for building urban and social environments where the fundamental systems and their interdependencies are well understood, carefully managed, and reinforced. This ensures that cities and society remain resilient, adaptable, and capable of sustaining their existence even in the face of spatial disruptions. By prioritizing these critical linkages, cities and society can not only prevent cascading failures but also create a more robust and sustainable urban and social infrastructure capable of enduring and adapting to future spatial challenges.

5.12.5.2 Guiding Spatial Resilience Through the Lens of Avoidance Principle: Preventing Systemic Vulnerabilities Through Physical and Behavioral Risk Mitigation

The *Avoidance Principle* advocates for proactive measures that aim to prevent exposure to high-risk areas. This principle emphasizes identifying regions prone to such hazards before any urban and social development or activities occur, ensuring that critical systems, infrastructure, and communities are not situated in harm's way. By anticipating risks and avoiding them upfront, this approach reduces vulnerability to disruptions and minimizes long-term consequences. This principle employs both physical and behavioral strategies to mitigate risk, with each playing a crucial role in safeguarding urban systems from spatial interferences.

From a physical avoidance perspective, planning and design should prioritize infrastructure that limits exposure to potential threats. For example, cities can incorporate protective designs—such as creating buffer zones, building resilient infrastructure that can withstand external pressures, or relocating sensitive systems away from known hazard-prone areas. This proactive approach ensures that urban areas are less susceptible to disruptions like floods, transportation bottlenecks, or energy blackouts resulting from spatial interferences.

On the behavioral avoidance side, cities and society can adopt data-driven decision-making, adaptive policies, and community behavior modification to avoid high-risk scenarios. By using risk data, predictive models, and public awareness campaigns, urban stakeholders can guide community actions to reduce exposure to hazards. This involves fostering a culture of preparedness and resilience, where individuals and institutions proactively respond to signals of risk, such as adapting land-use planning policies, controlling high-risk activities in vulnerable areas, or providing real-time alerts to mitigate disruptions. The goal is to embed avoidance into everyday practices through smart policies and community engagement, ensuring that actions align with the objective of reducing risk from the outset.

Rather than merely reacting to spatial interference hazards after they occur, the *Avoidance Principle* seeks to design urban and social systems that prevent such threats from taking root in the first place. This integrated approach to risk management emphasizes resilience-building through thoughtful planning, strategic infrastructure investment, and continuous engagement with both the physical environment and community behavior. In doing so, cities and society can minimize long-term damage, preserve resources, and reduce the societal and economic impacts of spatial disruptions.

5.12.5.3 Guiding Spatial Resilience Through the Lens of Tolerance Principle: Absorbing Disruptions and Sustaining Functionality Across Urban and Social Systems

The *Tolerance Principle* advocates for cities and society to focus on withstanding spatial interference hazards rather than solely attempting to prevent them. This principle acknowledges that while total risk elimination may not always be possible, urban and social systems can be designed to absorb shocks and adapt to disruptions, minimizing their negative impact and ensuring ongoing functionality. This approach mirrors the immune system's strategy of withstanding stressors without triggering catastrophic failure, thereby maintaining overall integrity and resilience even in the presence of disruptions.

In terms of physical tolerance, cities and society can integrate resilient infrastructure that is specifically designed to withstand spatial interferences. Examples include flood-resistant buildings, seismic-resistant structures, and adaptable energy grids, all engineered to minimize damage during extreme events such as earthquakes, flooding, or widespread power outages. These infrastructures are designed not only for durability but also for flexibility, allowing urban and social environments to continue functioning despite disruptions. For instance, flood-resistant buildings can be elevated or incorporate water-absorbing materials, while seismic-resistant structures use advanced engineering to absorb shock waves. Additionally, modular and repairable designs enhance recovery by enabling quick restoration, reducing the long-term vulnerability of critical infrastructure to future spatial interferences.

Systemic tolerance extends this concept across urban and social networks. By fostering redundancy and flexibility, cities and society can ensure that essential systems continue to operate even when parts of the network are disrupted. For example, decentralized energy sources, diverse transportation routes, and distributed water supply systems create layers of protection that enable critical services to persist during crises. This ensures that if one part of the system fails due to a spatial interference hazard, other components can take over, preventing cascading failures and minimizing overall disruption.

At a community level, tolerance can be promoted through social safety nets and affordable housing initiatives that enable communities to weather economic, social, or environmental challenges. These strategies allow communities to absorb shocks such as displacement, infrastructure failure, or health crises, while preserving their well-being and stability. Just as tissues in the body adapt to limit damage during stress, communities can develop resilience to ongoing disruptions by reinforcing their social and economic support structures, ensuring that they do not break down under pressure.

Finally, environmental tolerance involves integrating nature-based solutions to mitigate the effects of spatial interferences such as natural disasters, the use of underground spaces, and human-wildlife conflicts. For example, creating floodplains or wetlands to absorb stormwater and reduce flood risks, designing underground infrastructure that minimizes disruption to ecosystems, and implementing wildlife corridors to prevent human-wildlife conflict in urban areas. These strategies enhance the city's and society's ability to manage and coexist with its environment, fostering ecological recovery even during environmental stress.

By embedding tolerance across multiple levels—physical, systemic, community, and environmental—cities and society can enhance their ability to adapt to and absorb spatial interference hazards, ensuring that they remain functional and resilient in the face of ongoing challenges. Through continuous evaluation and incremental improvements, urban and social systems can strengthen their capacity to withstand and recover from disruptions, ultimately ensuring long-term sustainability and stability.

5.12.5.4 Guiding Spatial Resilience Through the Lens of Adaptation Principle: Building Agile Infrastructure, Culture, and Governance for a Changing World

The *Adaptation Principle* emphasizes the importance of enabling urban and social systems and policies to evolve dynamically in response to emerging spatial interference hazards and changing conditions. This principle highlights the need for flexibility and continuous improvement, promoting the development of systems that are not only resilient but also responsive to ongoing challenges. Urban and social environments can be designed to adapt by recognizing patterns, learning from past disruptions, and enhancing their capacity to respond more effectively in the future.

Systemic adaptation plays a critical role in this approach. Cities and society should actively analyze data, reflect on past disruptions, and continuously refine their operations to optimize resilience. This process can be supported by real-time feedback loops, predictive analytics, and scenario modeling, all of which enable cities and society to proactively adjust policies, resource allocations, and operational protocols in anticipation of spatial hazards. These tools enhance decision-making and foster adaptive responses to potential hazards before they escalate.

Structural adaptation is equally crucial for enhancing urban and social resilience. Cities should invest in flexible infrastructure that can be reconfigured or expanded as needed in response to changing conditions. For example, modular systems in sectors such as transportation, energy, or waste management can be scaled or adjusted to meet evolving demands or unforeseen challenges. This scalability ensures that cities and society are not locked into rigid infrastructures but can instead remain agile and adaptive to shifting environmental or social pressures.

At a cultural level, fostering a culture of adaptability within urban communities is vital. This means promoting continuous learning, nurturing diverse skill sets, and encouraging proactive engagement with new technologies and emerging trends. Cultivating an adaptive mindset ensures that individuals, organizations, and entire communities are prepared to embrace change and address the unknown challenges posed by spatial interferences, all while remaining agile in their responses to disruptions.

Moreover, adaptation strategies should be inclusive and equitable, ensuring that all communities, particularly marginalized and vulnerable groups, are not left behind. Spatial hazards often disproportionately affect these populations, and thus, adaptation efforts should prioritize their needs and ensure that they have the resources and support necessary to thrive amid challenges. This requires addressing local socio-economic conditions, investing in capacity-building, and promoting equitable access to adaptive solutions.

Finally, adaptive governance is essential for ensuring that cities and society remain responsive and resilient. Regulatory frameworks should be flexible, continuously evolving in response to new technological advancements, environmental challenges, and emerging risks. By maintaining this flexibility, cities and society can stay ahead of new hazards and ensure equitable and sustainable development while also safeguarding their capacity to adapt to future spatial interference threats.

Through the continuous integration of systemic, structural, cultural, and inclusive adaptation strategies, cities and society can navigate the complexities of spatial interference hazards, ensuring they remain resilient, responsive, and capable of thriving amid dynamic environmental and social changes.

5.12.5.5 Guiding Spatial Resilience Through the Lens of Regulation Principle: Enabling Scalable, Fair, and Responsive Systems Via Multilevel Regulatory Integration

The *Regulation Principle* is essential for fostering urban and social resilience in the face of spatial interference hazards by establishing frameworks that ensure the balance, adaptability, and stability of urban and social systems. Regulation in urban and social contexts ensures that systems remain functional and responsive to evolving risks, such as those posed by climate change, environmental degradation, or infrastructure strain. By fine-tuning responses to challenges, cities and society can optimize resources, promote equity, and maintain stability across sectors, creating a holistic response to spatial disruptions.

The principle emphasizes the need for infrastructure regulation, where cities and society employ adaptive and scalable designs that allow critical infrastructure to adjust in real-time to spatial hazards. This can include, for example, dynamic flood management systems, smart energy grids, and adaptive transportation networks, all of which adjust their operations as needed to mitigate disruptions. These systems are built with resilience in mind, ensuring that cities and society can balance growth and safety, while maintaining functionality in the face of environmental or systemic shocks.

In addition to infrastructure, behavioral and policy regulation plays a crucial role in managing spatial interferences. Urban policies that regulate community behaviors—such as adaptive traffic management systems or zoning laws—ensure that localized challenges, like congestion or resource competition, are addressed proactively. These flexible policies help cities and society avoid or minimize disruptions by adjusting activity levels in response to real-time information, preventing conflicts, and ensuring efficient resource allocation.

Economic and social regulation are equally important for maintaining stability and resilience. Cities and society can use mechanisms like progressive taxation, social safety nets, and inclusive development policies to ensure no part of the urban and social system is overstressed. These measures help reduce systemic inequalities and prevent certain groups, especially marginalized communities, from bearing the brunt of spatial interference hazards. By redistributing resources equitably, cities and society can minimize social vulnerability and enhance overall adaptability.

Finally, environmental regulation mirrors the immune system's ability to self-regulate, preventing harmful overreactions to external threats. Cities and society can, for example, regulate uncontrolled land development, enforce strict building codes, and implement effective risk management systems to operate within sustainable environmental limits. These actions help cities and society not only manage immediate spatial interference hazards but also adapt to long-term climate challenges, ensuring that urban and social environments remain viable and resilient.

Through these multidimensional approaches—ranging from infrastructure and behavioral regulation to economic, social, and environmental regulation—the *Regulation Principle* ensures that urban and social systems are not only able to respond to spatial interference hazards but also evolve in a way that promotes long-term sustainability and equity. This framework allows cities and society to stay ahead of challenges, ensuring that the urban and social environment remains resilient, equitable, and functional in the face of emerging risks.

5.12.5.6 Guiding Spatial Resilience Through the Lens of Coordination Principle: Enhancing Interconnectedness and Synergy Across Urban Systems and Stakeholders

The *Coordination Principle* is crucial for managing the complex, interconnected nature of spatial interference hazards, which often affect multiple sectors and jurisdictions simultaneously. This principle emphasizes the need for a coordinated response across a broad spectrum of stakeholders—including government agencies, private sector organizations, and local communities. In urban and social systems, effective coordination ensures that resources, efforts, and expertise are aligned, optimizing responses to both immediate disruptions and long-term challenges.

At the core of this principle is systemic coordination, which involves integrating communication and operational frameworks across key sectors such as transportation, public health, energy, and emergency services. By ensuring that these sectors can share data and make joint decisions in real-time, cities and society can address spatial interference hazards in a holistic manner. This integrated approach allows for synchronized responses and the efficient allocation of resources, enabling urban and social systems to adapt swiftly to disruptions. The seamless collaboration of various components, systems, and stakeholders fosters an adaptive and agile response, which is essential for addressing evolving spatial hazards.

Infrastructural coordination plays a central role in ensuring resilience during disruptions. Cities and society should design and implement interconnected infrastructure systems that work in harmony to mitigate risks and provide mutual support in times of crisis. For instance, integrating energy grids, transportation networks, and communication systems can enhance the capacity for rapid recovery and maintain essential services even when one component of the system is compromised. This interconnectedness allows for cross-sectoral support that ensures no critical system collapses in isolation, enhancing the overall robustness of urban environments.

Equally important is stakeholder coordination, which ensures that all parties—government agencies, private sector actors, and local communities—are working toward shared goals. Establishing formal collaborative structures such as cross-agency task forces, multi-stakeholder coalitions, and public–private partnerships is vital for fostering a unified approach to tackling spatial interference hazards. Clear communication channels and transparent decision-making processes are essential for ensuring that actions taken by different entities complement one another rather than create conflicts or inefficiencies. Collaborative efforts also improve resource allocation, foster innovation, and ensure that the unique strengths of each sector are leveraged effectively.

Beyond crisis management, the *Coordination Principle* calls for a long-term approach to planning and policy development, ensuring that urban and social systems remain adaptable and resilient as new risks emerge. This includes systematic planning that takes into account the long-term integration of diverse sectors and stakeholders, reinforcing the need for proactive and flexible frameworks. These coordinated strategies ensure that urban and social systems are not only prepared for current hazards but are also equipped to face future uncertainties and emerging challenges.

By embracing the *Coordination Principle*, cities and society can create a cohesive, adaptive, and resilient framework that addresses spatial interference hazards in an integrated and comprehensive manner. This ensures that urban and social systems remain functional, efficient, and responsive, promoting a future where diverse sectors work seamlessly together to safeguard communities and infrastructure.

5.12.5.7 Guiding Spatial Resilience Through the Lens of Diversity Principle: Embracing Multifaceted, Contextual Solutions for Sustainable Adaptation

The *Diversity Principle* underscores the importance of developing tailored, context-specific solutions to address spatial interference hazards, which can vary widely in their causes and impacts across regions and communities. This principle emphasizes the need for multifaceted, adaptable strategies that integrate diverse approaches—technical, ecological, and social—to enhance resilience and long-term sustainability.

One key element of the *Diversity Principle* is functional diversity, which encourages cities and society to incorporate multifunctional infrastructure. For instance, parks that also function as flood mitigation areas or flood barriers that double as recreational spaces promote both resilience and flexibility. This adaptability ensures that urban and social systems can absorb and recover from disruptions while continuing to provide valuable services. Similarly, ecological and demographic diversity strengthens urban and social systems by fostering inclusive communities that support creativity, innovation, and social cohesion. Such communities are better equipped to cope with disruptions such as economic shifts, social changes, or demographic transitions, ensuring that cities and society remain dynamic and responsive in the face of evolving challenges.

Spatial diversity plays a critical role by ensuring that essential services and infrastructure are distributed across multiple locations, reducing the risks associated with overreliance on centralized systems. By spreading critical resources and infrastructure, cities and society can prevent localized disruptions from having widespread impacts, thus enhancing their resilience to spatial hazards. Likewise, systemic diversity ensures that urban and social environments are not overly dependent on a single mode of transportation, energy source, or economic sector. By incorporating redundancy and flexibility, cities and society can absorb shocks more effectively, as diversified systems provide alternative options if one system fails.

Finally, evolutionary diversity encourages cities and society to embrace adaptive planning and dynamic policies. Urban and social systems can continuously evolve by integrating new technologies, responding to new risks, and learning from past experiences. The inclusion of diverse perspectives in decision-making further enhances this evolutionary capacity, helping cities and society navigate future uncertainties while promoting sustained growth and innovation.

In summary, the *Diversity Principle* fosters resilience and adaptability by integrating a variety of approaches, perspectives, and systems. By embracing diversity—whether in infrastructure, demographics, spatial distribution, or planning—cities and society can reduce vulnerability to spatial interference hazards and build a future that is both risk-averse and capable of thriving in the face of change.

5.12.5.8 Guiding Spatial Resilience Through the Lens of Robustilient Regeneration Principle: Evolving Recovery Processes for Sustainable and Adaptive Systems

The *Robustilient Regeneration Principle* emphasizes the need for recovery processes that do not merely restore functionality but actively enhance long-term resilience against spatial interference hazards. These hazards—whether driven by natural disasters, human-made events, or socio-economic shifts—disrupt the interconnected systems of cities and society, challenging both their physical infrastructure and social fabric. Rather than simply returning to a pre-disruption state, robustilient regeneration promotes a dynamic process of adaptation and recovery that strengthens systems to not only withstand but also evolve in response to future spatial challenges.

In addressing spatial interference hazards, recovery efforts should focus on creating redundant and adaptive infrastructure. Cities and society should integrate multiple pathways for recovery, enabling infrastructure to remain flexible and responsive to shifts in urban layouts, accessibility, and spatial configurations. For instance, modular infrastructure systems that can be reconfigured or expanded based on shifting needs ensure faster recovery and greater resilience. This kind of infrastructure enables urban and social systems to adapt more readily to evolving demands, providing the capacity to manage future spatial disruptions more effectively.

However, robustilient regeneration extends beyond the physical dimension. It includes social and economic regeneration, recognizing that spatial interferences often severely impact local communities and economies. When urban spaces are disrupted, social networks and livelihoods are also threatened. Therefore, regeneration efforts should focus on rebuilding social cohesion, supporting local economies, and revitalizing communities. By empowering communities to take an active role in the regeneration process, cities and society can ensure that recovery efforts reflect local needs and priorities. Inclusive governance and local leadership are critical to ensuring that these efforts foster long-term resilience, aligning the needs of all community members with the regeneration process.

The principle of multidimensional regeneration further strengthens this approach by integrating the restoration of physical infrastructure, social networks, and environmental ecosystems. Ecosystem revitalization should be prioritized alongside the rebuilding of urban infrastructure to restore natural functions that mitigate future spatial hazards. This holistic regeneration approach creates interconnected systems, ensuring that urban and social environments are more resilient to both immediate and future disruptions.

Moreover, robustilient regeneration is a continuous, evolving process. It is not merely a reaction to disruption but an ongoing cycle of growth, learning, and adaptation. Cities and society should embrace innovations, sustainable practices, and proactive policies that not only address immediate recovery needs but anticipate future challenges. By integrating real-time feedback loops and predictive modeling, cities and society can refine their regeneration processes over time, continuously improving their capacity to respond to emerging risks.

The empowerment of communities is central to this regenerative process. Communities should be equipped with the knowledge, resources, and networks necessary for recovery, with a focus on building social cohesion and local leadership. By ensuring that communities are active participants in shaping the regeneration process, cities and society can build greater social resilience and ensure that recovery efforts align with the needs of the people they serve.

Thus, the goal of robustilient regeneration is to transform cities and society into more resilient, adaptive, and sustainable environments. By strengthening the connections between people, spaces, and systems, cities and society can not only recover from spatial disruptions but also thrive in the face of future challenges. Through continuous adaptation and learning, cities and society can ensure their long-term vitality, enabling them to grow stronger, more resilient, and better equipped to handle future spatial interferences.

5.12.5.9 Guiding Spatial Resilience Through the Lens of Coherence Principle: Balancing Systemic Interconnection with Redundancy for Robust Response

The *Coherence Principle* emphasizes the importance of alignment and integration among urban and social systems, policies, and strategies to create cohesive, adaptive environments. By ensuring that actions in one area do not inadvertently create vulnerabilities elsewhere, the principle fosters a balanced and unified approach to managing spatial interference hazards. This interconnectedness helps cities and society navigate complexities and adapt to disruptions more effectively.

A core tenet of the *Coherence Principle* is systemic interconnection, which ensures that urban and social systems operate as an integrated whole. This interconnectedness allows each component—whether physical infrastructure, governance, or social networks—to respond to the others, minimizing isolated disruptions and enhancing the overall resilience of the system. Urban and social systems that are functionally and structurally connected can identify emerging risks more efficiently and address potential issues before they escalate, facilitating smoother operations in times of crisis.

Effective communication and feedback mechanisms are essential to the success of this principle. These mechanisms allow for dynamic, real-time adjustments based on the latest data and evolving needs. Transparent communication ensures that various stakeholders—government agencies, private entities, and local communities—collaborate effectively, aligning their actions toward shared goals. The principle also advocates for balance in urban and social systems, meaning that responses to disruptions are proportional and harmonious. Regulatory frameworks should be designed to adapt to changing conditions while ensuring that responses remain consistent, equitable, and effective. By maintaining balance, cities and society can mitigate extreme fluctuations in service availability and ensure that essential functions are maintained despite challenges.

Redundancy is another key aspect of the *Coherence Principle*. By ensuring that critical systems, such as energy grids, transportation networks, and communication infrastructures, are resilient through built-in redundancies, cities and society can continue to function even when individual components fail. This redundancy provides a safety net that minimizes vulnerabilities, ensuring that critical operations remain intact during disruptions, and enabling quicker recovery. Furthermore, learning from past experiences is vital to improving resilience. Urban and social systems should evolve by continuously assessing and refining their processes based on prior challenges, ensuring that they can adapt to both recurring and emerging spatial interference hazards. By embracing a culture of continuous learning, cities and society can develop more effective and innovative solutions over time.

In summary, the *Coherence Principle* guides the creation of resilient, sustainable urban and social environments where infrastructure and communities work in harmony to ensure the long-term well-being of their inhabitants. By fostering systemic interconnection, ensuring effective communication, maintaining balance and redundancy, and prioritizing ongoing learning, cities and society can navigate spatial interference hazards more effectively, ensuring that both urban and social systems remain flexible and adaptive in the face of future challenges.

5.13 Confronting *Spatial Interference Hazards*: *Urbanimmunology*'s Role in Shaping Safer Spaces

As urban and regional environments continue to expand and evolve, the complexity of *spatial interference hazards* presents increasingly intricate challenges. These hazards, ranging from beneath-the-surface risks and structural failures to more visible threats like flooding, human-wildlife conflicts, and mass-gathering dangers, demand more than traditional hazard management approaches. They require a dynamic, integrative strategy that considers the interconnectedness of urban and social systems. *Urbanimmunology* offers a transformative framework for rethinking how we respond to and prevent these spatial hazards. By examining these threats through the core principles of *Urbanimmunology*—Critical Dependency, Avoidance, Tolerance, Adaptation, Regulation, Coordination, Diversity, Robust Resilient Regeneration, and Coherence—this framework enables a deeper understanding of how these hazards interconnect and evolve within urban settings, offering insights that traditional models fail to address comprehensively.

The investigation of *spatial interference hazards*—from structural and infrastructure failures to complex challenges like multimodal interactions and tourism overcrowding—reveals common patterns and unique dynamics for each hazard type. Each hazard presents distinct features that require tailored responses within the *Urbanimmunology* framework. For example, beneath-the-surface risks such as subsidence or underground infrastructure failures call for strategies focused on infrastructure integrity and proactive risk avoidance. In contrast, windstorm-induced traffic disruptions demand adaptive, real-time responses coupled with long-term infrastructure resilience. Human-wildlife interactions highlight the need for coordination between urban planning and ecological preservation, while mass-gathering hazards call for strategies that embrace diversity and adaptation to large-scale human behavior. By analyzing these hazards both individually and collectively, *Urbanimmunology* emphasizes the interconnectedness of urban and social systems and the need for multifaceted, integrated solutions.

This analysis also highlights new and emerging risks as urban and social environments continue to change. With the rise of cyber-physical risks, climate-induced changes, and technological disruptions, the scope of *spatial interference hazards* will likely expand. As new patterns and interdependencies emerge, they provide fresh opportunities to deepen the study of *spatial interference hazards* and further refine *Urbanimmunology*'s framework. This dynamic landscape offers an exciting avenue for enhancing *Urbanimmunology* as a robust scientific discipline, positioning it as an essential tool in addressing increasingly complex urban and social challenges.

By continuing to examine the complexities and nuances of *spatial interference hazards*, *Urbanimmunology* will evolve to better anticipate, mitigate, and manage risks across a broad range of urban environments. As an inherently transdisciplinary field, *Urbanimmunology* integrates insights from urban planning, engineering, environmental science, public health, and more, fostering collaboration and driving innovative, holistic solutions that are both resilient and adaptive. This broad, collaborative approach is crucial for addressing the complexities of modern urbanization and ensuring the long-term safety and sustainability of cities and society.

Looking ahead, future research within *Urbanimmunology* should focus on several key areas. First, enhancing data collection and real-time monitoring technologies will be crucial for predicting and managing *spatial interference hazards* more effectively. Integrating sensor networks, satellite data, and AI-driven analytics will improve hazard forecasting and enable more coordinated, timely responses. Additionally, further research into the integration of emerging technologies—such as autonomous systems, smart infrastructure, and eco-friendly urban designs—will be vital in mitigating risks associated with urban growth. Finally, more attention should be given to the social and psychological dimensions of *spatial interference hazards*, particularly in understanding public engagement, cultural attitudes toward safety, and the needs of vulnerable populations in high-risk areas.

Ultimately, the development of *Urbanimmunology* as a scientific system hinges on its ability to adapt, grow, and integrate new knowledge in response to the evolving landscape of *spatial interference hazards*. As urban centers continue to grow and diversify, *Urbanimmunology* will play a crucial role in ensuring cities and society are not only safe but also resilient and sustainable in the face of future risks. Through continued research, cross-disciplinary collaboration, and the application of innovative solutions, *Urbanimmunology* has the potential to reshape how we design, manage, and protect our urban and social environments for generations to come.

References

Abas, A., Rahman, A. H. A., Md Fauzi, T. A. H. T., & Yusof, A. H. M. (2025). A bibliometric review of global research on the human-wildlife conflicts. *Frontiers in Environmental Science, 12*, 517218. https://doi.org/10.3389/fenvs.2024.1517218

Abed, N., Murgun, R., Deldari, A., et al. (2025). IoT and AI-driven solutions for human-wildlife conflict: Advancing sustainable agriculture and biodiversity conservation. *Smart Agricultural Technology, 10*, 100829. https://doi.org/10.1016/j.atech.2025.100829

Abrahms, B., Carter, N. H., Clark-Wolf, T. J., et al. (2023). Climate change as a global amplifier of human–wildlife conflict. *Nature Climate Change, 13*, 224–234. https://doi.org/10.1038/s41558-023-01608-5

Aguiar, R., Gray, R., Gallo-Cajiao, E., et al. (2025). Preventing zoonotic spillover through regulatory frameworks governing wildlife trade: A scoping review. *PLoS One, 20*(1), e0312012. https://doi.org/10.1371/journal.pone.0312012

Ahmed, A. N., Fornace, K. M., Iwamura, T., et al. (2025). Human animal contact, land use change and zoonotic disease risk: A protocol for systematic review. *Systematic Reviews, 14*, 65. https://doi.org/10.1186/s13643-025-02805-3

Al-Bayati, A. J., Panzer, L., Eiris, R., & Tafazzoli, M. (2024). Third-party damages to underground utilities: The critical need for root cause analysis. In *ASCE – Construction Research Congress 2024*. https://doi.org/10.1061/9780784485279.071

Alcántara-Ayala, I. (2025). Landslides in a changing world. *Landslides, 22*, 2851–2865. https://doi.org/10.1007/s10346-024-02451-1

Aleksandrov, M., Rajabifard, A., Kalantari, M., et al. (2018). People choice modelling for evacuation of tall buildings. *Fire Technology, 54*, 1171–1193. https://doi.org/10.1007/s10694-018-0731-1

Alexander, M., Priest, S., & Mees, H. (2016). A framework for evaluating flood risk governance. *Environmental Science & Policy, 64*, 38–47. https://doi.org/10.1016/j.envsci.2016.06.004

Alfieri, L., Cohen, S., Galantowicz, J., et al. (2018). A global network for operational flood risk reduction. *Environmental Science & Policy, 84*, 149–158. https://doi.org/10.1016/j.envsci.2018.03.014

Ali, B. S. (2019). Traffic management for drones flying in the city. *International Journal of Critical Infrastructure Protection, 26*, 100310. https://doi.org/10.1016/j.ijcip.2019.100310

Allan, M., Lièvre, M., Laurenson-Schafer, H., et al. (2022). The World Health Organization COVID-19 surveillance database. *International Journal for Equity in Health, 21*(Suppl 3), 167. https://doi.org/10.1186/s12939-022-01767-5

Alves, R. A., Rudke, A. P., de Melo Souza, S. T., et al. (2024). Flood vulnerability mapping in an urban area with high levels of impermeable coverage in southern Brazil. *Regional Environmental Change, 24*, 96. https://doi.org/10.1007/s10113-024-02258-9

Ammon, R., & Fried, G. (1999). Crowd management practices. *Journal of Convention & Exhibition Management, 1*(2–3), 119–150. https://doi.org/10.1300/J143v01n02_05

Anelli, D., Tajani, F., & Ranieri, R. (2022). Urban resilience against natural disasters: Mapping the risk with an innovative indicators-based assessment approach. *Journal of Cleaner Production, 371*, 133496. https://doi.org/10.1016/j.jclepro.2022.133496

Apostoris, P. (2024). A review of global and regional frameworks for the integration of an unmanned aircraft system in air traffic management. *Transportation Research Interdisciplinary Perspectives, 24*, 101064. https://doi.org/10.1016/j.trip.2024.101064

Bahadorestani, A., Naderpajouh, N., & Sadiq, R. (2020). Planning for sustainable stakeholder engagement based on the assessment of conflicting interests in projects. *Journal of Cleaner Production, 242*, 118402. https://doi.org/10.1016/j.jclepro.2019.118402

Bang, H. N. (2021). A gap analysis of the legislative, policy, institutional and crises management frameworks for disaster risk management in Cameroon. *Progress in Disaster Science, 11*, 100190. https://doi.org/10.1016/j.pdisas.2021.100190

Barati, A. A., Zhoolideh, M., Azadi, H., et al. (2023). Interactions of land-use cover and climate change at global level: How to mitigate the environmental risks and warming effects. *Ecological Indicators, 146*, 109829. https://doi.org/10.1016/j.ecolind.2022.109829

Basak, S. M., Rostovskaya, E., Birks, J., & Wierzbowska, I. A. (2023). Perceptions and attitudes to understand human-wildlife conflict in an urban landscape—A systematic review. *Ecological Indicators, 151*, 110319. https://doi.org/10.1016/j.ecolind.2023.110319

Bauranov, A., & Rakas, J. (2019). Urban air mobility and manned eVTOLs: Safety implications. In *2019 IEEE/AIAA 38th digital avionics systems conference (DASC), San Diego, CA, USA* (pp. 1–8). https://doi.org/10.1109/DASC43569.2019.9081685

Beitel, D., Stipancic, J., Manaugh, K., & Miranda-Moreno, L. (2018). Assessing safety of shared space using cyclist-pedestrian interactions and automated video conflict analysis. *Transportation Research Part D: Transport and Environment, 65*, 710–724. https://doi.org/10.1016/j.trd.2018.10.001

Bhatia, J. (2013). Landslide awareness, preparedness and response management in India. In C. Margottini, P. Canuti, & K. Sassa (Eds.), *Landslide science and practice*. Springer. https://doi.org/10.1007/978-3-642-31313-4_37

Bishop, P. C., Gibbs, T. M., & Lantz, J. (2020). Crowd management and special event planning. In S. J. Davies & L. J. Fennelly (Eds.), *The professional protection officer* (2nd ed., pp. 283–293). Butterworth-Heinemann. https://doi.org/10.1016/B978-0-12-817748-8.00025-0

Bjelland, H., Njå, O., Heskestad, A. W., & Braut, G. S. (2021). Emergency preparedness for tunnel fires—A systems-oriented approach. *Safety Science, 143*, 105408. https://doi.org/10.1016/j.ssci.2021.105408

Bongo, M. F., & Sy, C. L. (2024). Can diverse and conflicting interests of multiple stakeholders be balanced? *Annals of Operations Research, 339*, 1813–1837. https://doi.org/10.1007/s10479-023-05253-1

Borzée, A., McNeely, J., Magellan, K., et al. (2020). COVID-19 highlights the need for more effective wildlife trade legislation. *Trends in Ecology & Evolution, 35*(12), 1052–1055. https://doi.org/10.1016/j.tree.2020.10.001

Bosta, S. A. (2011). Crowd management based on scientific research to prevent crowd panic and disasters. In R. Peacock, E. Kuligowski, & J. Averill (Eds.), *Pedestrian and evacuation dynamics*. Springer. https://doi.org/10.1007/978-1-4419-9725-8_66

Brady, M., Tranter, K., & Bennett, B. (2025). Automated vehicles, the 'driver dilemma', stopping powers, and paradigms of regulating road traffic. *Computer Law & Security Review, 56*, 106076. https://doi.org/10.1016/j.clsr.2024.106076

Burns, P. M., & Novelli, M. (2007). *Tourism and politics: Global frameworks and local realities*. Elsevier.

Butler, R. W., & Dodds, R. (2022). Overcoming overtourism: A review of failure. *Tourism Review, 77*(1), 35–53. https://doi.org/10.1108/TR-04-2021-0215

Cai, G., Zheng, X., Guo, J., & Gao, W. (2025). Real-time identification of borehole rescue environment situation in underground disaster areas based on multi-source heterogeneous data fusion. *Safety Science, 181*, 106690. https://doi.org/10.1016/j.ssci.2024.106690

Callahan, C. W. (2025). Present and future limits to climate change adaptation. *Nature Sustainability, 8*(4), 1–7. https://doi.org/10.1038/s41893-025-01519-7

Cao, S., Leng, J., Qi, D., et al. (2022). Sustainable underground spaces: Design, environmental control and energy conservation. *Energy and Buildings, 257*, 111779. https://doi.org/10.1016/j.enbuild.2021.111779

Cao, W., Zhou, Y., Güneralp, B., et al. (2022). Increasing global urban exposure to flooding: An analysis of long-term annual dynamics. *Science of the Total Environment, 817*, 153012. https://doi.org/10.1016/j.scitotenv.2022.153012

Cardoso, B., García-Bocanegra, I., Acevedo, P., et al. (2022). Stepping up from wildlife disease surveillance to integrated wildlife monitoring in Europe. *Research in Veterinary Science, 144*, 149–156. https://doi.org/10.1016/j.rvsc.2021.11.003

Casagli, N., Intrieri, E., Tofani, V., et al. (2023). Landslide detection, monitoring and prediction with remote-sensing techniques. *Nature Reviews Earth and Environment, 4*, 51–64. https://doi.org/10.1038/s43017-022-00373-x

Chan, J. K. H., & Zhang, Y. (2018). Sharing space: Urban sharing, sharing a living space, and shared social spaces. *Space and Culture, 24*(1), 157–169. https://doi.org/10.1177/1206331218806160

Chandrasekaran, S. S., Sayed Owaise, R., Ashwin, S., et al. (2013). Investigation on infrastructural damages by rainfall-induced landslides during November 2009 in Nilgiris, India. *Natural Hazards, 65*, 1535–1557. https://doi.org/10.1007/s11069-012-0432-x

Chang, H. K., Tan, Y. C., Lai, J. S., et al. (2013). Improvement of a drainage system for flood management with assessment of the potential effects of climate change. *Hydrological Sciences Journal, 58*(8), 1581–1597. https://doi.org/10.1080/02626667.2013.836276

Chougule, A., Chamola, V., Sam, A., et al. (2024). A comprehensive review on limitations of autonomous driving and its impact on accidents and collisions. *IEEE Open Journal of Vehicular Technology, 5*, 142–161. https://doi.org/10.1109/OJVT.2023.3335180

Chouksey, S., Singh, S., Pandey, R., & Tomer, V. S. (2018). Monitoring the status of human-wildlife conflict and its impact on community based conservation in Bandhavgarh tiger reserve, Madhya Pradesh, India. *Journal of Applied and Natural Science, 10*(2), 710–715. https://doi.org/10.31018/jans.v10i2.1771

Chow, W. K., & Ng, C. M. Y. (2008). Waiting time in emergency evacuation of crowded public transport terminals. *Safety Science, 46*(5), 844–857. https://doi.org/10.1016/j.ssci.2007.01.015

Churchwell, J. S., Zhang, K. S., & Saleh, J. H. (2018). Epidemiology of helicopter accidents: Trends, rates, and covariates. *Reliability Engineering & System Safety, 180*, 373–384. https://doi.org/10.1016/j.ress.2018.08.007

Connolly, C., Keil, R., & Ali, S. H. (2020). Extended urbanisation and the spatialities of infectious disease: Demographic change, infrastructure and governance. *Urban Studies, 58*(2), 245–263. https://doi.org/10.1177/0042098020910873

Conover, M. R., Butikofer, E., & Decker, D. J. (2018). Wildlife damage to crops: Perceptions of agricultural and wildlife leaders in 1957, 1987, and 2017. *Wildlife Society Bulletin, 42*(4), 551–558. https://doi.org/10.1002/wsb.930

Crow, D. A., Albright, E. A., Ely, T., et al. (2018). Do disasters lead to learning? Financial policy change in local government. *Review of Policy Research, 35*(4), 564–589. https://doi.org/10.1111/ropr.12297

Cui, B., Wen, X., & Zhang, D. (2019). The application of intelligent emergency response system for urban underground space disasters based on 3D GIS, BIM and Internet of Things. In *AICS 2019: Proceedings of the 2019 International Conference on Artificial Intelligence and Computer Science* (pp. 745–749). https://doi.org/10.1145/3349341.3349503

Curtin, S., & Kragh, G. (2014). Wildlife tourism: Reconnecting people with nature. *Human Dimensions of Wildlife, 19*(6), 545–554. https://doi.org/10.1080/10871209.2014.921957

Cymes, I., Jalali, R., Glińska-Lewczuk, K., et al. (2021). The association between the biometeorological indicators and emergency interventions due to fainting: A retrospective cohort study. *Science of the Total Environment, 770*, 145376. https://doi.org/10.1016/j.scitotenv.2021.145376

Damnjanović, I. (2021). Overtourism effects: Positive and negative impacts for sustainable development. In W. Leal Filho, A. M. Azul, L. Brandli, et al. (Eds.), *Industry, innovation and infrastructure. Encyclopedia of the UN sustainable development goals*. Springer. https://doi.org/10.1007/978-3-319-95873-6_112

Deb, T., de Laaf, S., La Gatta, V., et al. (2025). A drone early warning system (DEWS) for predicting threatening trajectories. *IEEE Intelligent Systems, 40*(4), 8–18. https://doi.org/10.1109/MIS.2025.3540003

Debrock, S., Van Acker, M., & Admiraal, H. (2023). Design recommendations for sustainable urban underground spaces. *Tunnelling and Underground Space Technology, 140*, 105332. https://doi.org/10.1016/j.tust.2023.105332

Decker, D. J., Siemer, W. F., Evensen, D. T. N., et al. (2012). Public perceptions of wildlife-associated disease: Risk communication matters. *Human-Wildlife Interactions, 6*(1), 112–122. https://www.jstor.org/stable/24874082

Deng, L., Wang, W., & Yu, Y. (2015). State-of-the-art review on the causes and mechanisms of bridge collapse. *Journal of Performance of Constructed Facilities, 30*(2), 04015005. https://doi.org/10.1061/(ASCE)CF.1943-5509.0000731

Drury, J., & Stott, C. (2013). *Crowds in the twenty-first century: Perspectives from contemporary social science*. Routledge.

El-Jardali, F., Fadlallah, R., & Daher, N. (2024). Multi-sectoral collaborations in selected countries of the eastern Mediterranean region: Assessment, enablers and missed opportunities from the COVID-19 pandemic response. *Health Research Policy and Systems, 22*, 14. https://doi.org/10.1186/s12961-023-01098-z

Elkady, S., Mehryar, S., Hernantes, J., & Labaka, L. (2024). Prioritizing stakeholder interactions in disaster management: A TOPSIS-based decision support tool for enhancing community resilience. *Progress in Disaster Science, 22*, 100320. https://doi.org/10.1016/j.pdisas.2024.100320

Feliciani, C., Corbetta, A., Haghani, M., & Nishinari, K. (2023). Trends in crowd accidents based on an analysis of press reports. *Safety Science, 164*, 106174. https://doi.org/10.1016/j.ssci.2023.106174

Feliciani, C., Shimura, K., & Nishinari, K. (2021). What is crowd management? In C. Feliciani, K. Shimura, & K. Nishinari (Eds.), *Introduction to crowd management*. Springer. https://doi.org/10.1007/978-3-030-90012-0_1

Feng, J., Gai, W., & Yan, Y. (2021). Emergency evacuation risk assessment and mitigation strategy for a toxic gas leak in an underground space: The case of a subway station in Guangzhou, China. *Safety Science, 134*, 105039. https://doi.org/10.1016/j.ssci.2020.105039

Feng, B., Zhang, Y., & Bourke, R. (2021). Urbanization impacts on flood risks based on urban growth data and coupled flood models. *Natural Hazards, 106*, 613–627. https://doi.org/10.1007/s11069-020-04480-0

Feuerbacher, A., Lippert, C., Kuenzang, J., & Subedi, K. (2021). Low-cost electric fencing for peaceful coexistence: An analysis of human-wildlife conflict mitigation strategies in smallholder agriculture. *Biological Conservation, 255*, 108919. https://doi.org/10.1016/j.biocon.2020.108919

Fischer, J., & Lindenmayer, D. B. (2000). An assessment of the published results of animal relocations. *Biological Conservation, 96*(1), 1–11. https://doi.org/10.1016/S0006-3207(00)00048-3

Fischhoff, B. (2015). The realities of risk-cost-benefit analysis. *Science, 350*(6260), aaa6516. https://doi.org/10.1126/science.aaa6516

Fisher, M., Mascardi, V., Rozier, K. Y., et al. (2021). Towards a framework for certification of reliable autonomous systems. *Autonomous Agents and Multi-Agent Systems, 35*, 8. https://doi.org/10.1007/s10458-020-09487-2

Frosch, C., Martinelli, D., & Unnikrishnan, A. (2019). Evaluation of shared space to reduce traffic congestion. *Journal of Advanced Transportation, 2019*(1), 6510396. https://doi.org/10.1155/2019/6510396

Gemeda, D. O., & Meles, S. K. (2018). Impacts of human-wildlife conflict in developing countries. *Journal of Applied Sciences and Environmental Management, 22*(8), 1233–1238. https://doi.org/10.4314/jasem.v22i8.14

Ghasri, M., & Maghrebi, M. (2021). Factors affecting unmanned aerial vehicles' safety: A post-occurrence exploratory data analysis of drones' accidents and incidents in Australia. *Safety Science, 139*, 105273. https://doi.org/10.1016/j.ssci.2021.105273

Glade, T., Anderson, M., & Crozier, M. J. (2005). *Landslide hazard and risk*. John Wiley & Sons Ltd. https://doi.org/10.1002/9780470012659

Gnyawali, K., Dahal, K., Talchabhadel, R., & Nirandjan, S. (2023). Framework for rainfall-triggered landslide-prone critical infrastructure zonation. *Science of the Total Environment, 872*, 162242. https://doi.org/10.1016/j.scitotenv.2023.162242

Goh, J. (2021). A literature review of medical support in cave rescue and confined space medicine—Implications in urban underground space development. In *IOP conference series: Earth and environmental science. Vol. 703, 17th World Conference ACUUS 2020 February 3–4 2021, (Helsinki, Finland) virtual conference*. IOP Publishing. https://doi.org/10.1088/1755-1315/703/1/012042

Grilo, C., Neves, T., Bates, J., et al. (2025). Global Roadkill Data: A dataset on terrestrial vertebrate mortality caused by collision with vehicles. *Scientific Data, 12*, 505. https://doi.org/10.1038/s41597-024-04207-x

Gross, E. M., Jayasinghe, N., Dahal, S., et al. (2025). C2C—Conflict to coexistence: A global approach to manage human–wildlife conflict for coexistence. *Conservation Science and Practice, 7*(2), e13292. https://doi.org/10.1111/csp2.13292

Guay, P. J., Van Dongen, W. F. D., McLeod, E. M., et al. (2018). Does zonation and accessibility of wetlands influence human presence and mediate wildlife disturbance? *Journal of Environmental Planning and Management, 62*(8), 1306–1320. https://doi.org/10.1080/09640568.2018.1495066

Gulati, S., Karanth, K. K., Le, N. A., & Noack, F. (2021). Human casualties are the dominant cost of human–wildlife conflict in India. *PNAS, 118*(8), e1921338118. https://doi.org/10.1073/pnas.1921338118

Haghani, M., Coughlan, M., Crabb, B., et al. (2023). A roadmap for the future of crowd safety research and practice: Introducing the Swiss cheese model of crowd safety and the imperative of a vision zero target. *Safety Science, 168*, 106292. https://doi.org/10.1016/j.ssci.2023.106292

Hamidi, S., & Hamidi, I. (2021). Subway ridership, crowding, or population density: Determinants of COVID-19 infection rates in New York City. *American Journal of Preventive Medicine, 60*(5), 614–620. https://doi.org/10.1016/j.amepre.2020.11.016

Hao, L., Guo, C., & Wu, L. (2016). A study on route planning of helicopter in low altitude area. In *2016 IEEE first international conference on Data Science in Cyberspace (DSC), Changsha, China* (pp. 484–488). IEEE. https://doi.org/10.1109/DSC.2016.19

Hasan, S., & Foliente, G. (2015). Modeling infrastructure system interdependencies and socioeconomic impacts of failure in extreme events: Emerging R&D challenges. *Natural Hazards, 78*, 2143–2168. https://doi.org/10.1007/s11069-015-1814-7

Hay, B. (2019). Events management: An international approach. *Journal of Tourism Futures, 5*(1), 104–105. https://doi.org/10.1108/JTF-03-2017-0008

Hemminger, K., Eriksson, L., Nilsson, L., et al. (2025). Farmers' tolerance for crop damage caused by wildlife: The role of compensation. *Wildlife Biology, 2025*(4), e01243. https://doi.org/10.1002/wlb3.01243

Hirabayashi, Y., Mahendran, R., Koirala, S., et al. (2013). Global flood risk under climate change. *Nature Climate Change, 3*, 816–821. https://doi.org/10.1038/nclimate1911

Hong, B., Bonczak, B. J., Gupta, A., et al. (2021). Measuring inequality in community resilience to natural disasters using large-scale mobility data. *Natural Communications, 12*, 1870. https://doi.org/10.1038/s41467-021-22160-w

Hong, Q., Lai, H., & Liu, Y. (2023). Failure analysis and treatments of collapse accidents in loess tunnels. *Engineering Failure Analysis, 145*, 107037. https://doi.org/10.1016/j.engfailanal.2022.107037

Hossain, B., Sarker, M. N. I., & Sohel, M. S. (2025). Flooded lives: Socio-economic implications and adaptation challenges for riverine communities in Bangladesh. *International journal of Environmental Science and Technology, 22*, 4407–4422. https://doi.org/10.1007/s13762-024-05943-8

Hossein Motlagh, N., Taleb, T., & Arouk, O. (2016). Low-altitude unmanned aerial vehicles-based internet of things services: Comprehensive survey and future perspectives. *IEEE Internet of Things Journal, 3*(6), 899–922. https://doi.org/10.1109/JIOT.2016.2612119

Hou, Z., Du, F., Zhou, X., et al. (2020). Cross-country comparison of public awareness, rumors, and behavioral responses to the COVID-19 epidemic: Infodemiology study. *Journal of Medical Internet Research, 22*(8), e21143. https://doi.org/10.2196/21143

Huang, C., Fang, S., Wu, H., Wang, Y., & Yang, Y. (2024). Low-altitude intelligent transportation: System architecture, infrastructure, and key technologies. *Journal of Industrial Information Integration, 42*, 100694. https://doi.org/10.1016/j.jii.2024.100694

Huang, F., Liu, K., Li, Z., et al. (2024). Single landslide risk assessment considering rainfall-induced landslide hazard and the vulnerability of disaster-bearing body. *Geological Journal, 59*(9), 2549–2565. https://doi.org/10.1002/gj.4976

Huang, H., Su, J., & Wang, F.-Y. (2024). The potential of low-altitude airspace: The future of urban air transportation. *IEEE Transactions on Intelligent Vehicles, 9*(8), 5250–5254. https://doi.org/10.1109/TIV.2024.3483889

Huang, H., Zhang, D., & Huang, Z. (2022). Resilience of city underground infrastructure under multi-hazards impact: From structural level to network level. *Resilient Cities and Structures, 1*(2), 76–86. https://doi.org/10.1016/j.rcns.2022.07.003

IUCN (Ed.). (2023). *IUCN SSC guidelines on human-wildlife conflict and coexistence* (1st ed.). IUCN. https://www.hwctf.org/guidelines

Janssens, V., O'Dwyer, D. W., & Chryssanthopoulos, M. K. (2012). Assessing the consequences of building failures. *Structural Engineering International, 22*(1), 99–104. https://doi.org/10.2749/101686612X13216060213473

Jiang, W., & Tan, Y. (2022). Overview on failures of urban underground infrastructures in complex geological conditions due to heavy rainfall in China during 1994–2018. *Sustainable Cities and Society, 76*, 103509. https://doi.org/10.1016/j.scs.2021.103509

Jirásko, D., Vaníček, I., & Vaníček, M. (2017). Interaction of landslide with critical infrastructure. In M. Mikoš, Ž. Arbanas, Y. Yin, & K. Sassa (Eds.), *Advancing culture of living with landslides. WLF 2017*. Springer. https://doi.org/10.1007/978-3-319-53487-9_64

Kaila, H. L. (2024). Worst safety culture at overcrowded religious places kill thousands: Exploring solutions. *IAHRW International Journal of Social Sciences Review, 12*(2), 239–244.

Kairiza, T., Pallegedara, A., & Chigusiwa, L. (2024). The interplay of drought, human–wildlife conflict and food security: Insights from rural Zimbabwe. *The Australian Journal of Agricultural and Resource Economics, 69*(1), 232–250. https://doi.org/10.1111/1467-8489.12592

Kanaujiya, A. K., & Tiwari, V. (2022). Crowd management and strategies for security and surveillance during the large mass gathering events: The Prayagraj Kumbh Mela 2019 experience. *National Academy Science Letters, 45*, 263–273. https://doi.org/10.1007/s40009-022-01114-w

Karanth, K. K., Gupta, S., & Vanamamalai, A. (2018). Compensation payments, procedures and policies towards human-wildlife conflict management: Insights from India. *Biological Conservation, 227*, 383–389. https://doi.org/10.1016/j.biocon.2018.07.006

Karimiziarani, M., Jafarzadegan, K., Abbaszadeh, P., et al. (2022). Hazard risk awareness and disaster management: Extracting the information content of twitter data. *Sustainable Cities and Society, 77*, 103577. https://doi.org/10.1016/j.scs.2021.103577

Karndacharuk, A., Wilson, D. J., & Dunn, R. (2014). A review of the evolution of shared (street) space concepts in urban environments. *Transport Reviews, 34*(2), 190–220. https://doi.org/10.1080/01441647.2014.893038

Kassen, M. (2023). COVID-19 pandemic and digital public awareness platforms: Strategies, solutions, and tools of communication and e-government data management. *Preservation, Digital Technology & Culture, 52*(2), 69–82. https://doi.org/10.1515/pdtc-2023-0012

Keck, F., Peller, T., Alther, R., et al. (2025). The global human impact on biodiversity. *Nature, 641*(8062), 395–400. https://doi.org/10.1038/s41586-025-08752-2

Keefer, D. K., & Larsen, M. C. (2007). Assessing landslide hazards. *Science, 316*, 1136–1138. https://doi.org/10.1126/science.1143308

Kenesei, Z., Ásványi, K., Kökény, L., et al. (2022). Trust and perceived risk: How different manifestations affect the adoption of autonomous vehicles. *Transportation Research Part A: Policy and Practice, 164*, 379–393. https://doi.org/10.1016/j.tra.2022.08.022

Keysar, H. (2020). Who owns the sky? Aerial resistance and the state/corporate no-fly zone. *Visual Studies, 35*(5), 465–477. https://doi.org/10.1080/1472586X.2020.1840094

Kleeman, A., Giles-Corti, B., Gunn, L., Hooper, P., & Foster, S. (2021). Exploring the design, quality and use of communal areas in apartment developments. *Cities & Health, 6*(3), 480–494. https://doi.org/10.1080/23748834.2021.2001169

Kremen, C., & Merenlender, A. M. (2018). Landscapes that work for biodiversity and people. *Science, 362*, eaau6020. https://doi.org/10.1126/science.aau6020

Kvočka, D., Falconer, R. A., & Bray, M. (2016). Flood hazard assessment for extreme flood events. *Natural Hazards, 84*, 1569–1599. https://doi.org/10.1007/s11069-016-2501-z

La Loggia, G., Puleo, V., & Freni, G. (2020). Floodability: A new paradigm for designing urban drainage and achieving sustainable urban growth. *Water Resources Management, 34*, 3411–3424. https://doi.org/10.1007/s11269-020-02620-6

Lal, A., Erondu, N. A., Heymann, D. L., et al. (2021). Fragmented health systems in COVID-19: Rectifying the misalignment between global health security and universal health coverage. *The Lancet, 397*(10268), 61–67. https://doi.org/10.1016/S0140-6736(20)32228-5

Langridge, J., Sordello, R., & Reyjol, Y. (2021). Existing evidence on the outcomes of wildlife translocations in protected areas: A systematic map. *Environmental Evidence, 10*, 29. https://doi.org/10.1186/s13750-021-00236-w

Lee, S., Dodge, J., & Chen, G. (2022). The cost of social vulnerability: An integrative conceptual framework and model for assessing financial risks in natural disaster management. *Natural Hazards, 114*, 691–712. https://doi.org/10.1007/s11069-022-05408-6

Leppold, C., Gibbs, L., Block, K., et al. (2022). Public health implications of multiple disaster exposures. *The Lancet Public Health, 7*(3), e274–e286. https://doi.org/10.1016/S2468-2667(21)00255-3

Li, J., Huang, Z., Wang, H., et al. (2024). Multi-index comprehensive evaluation model for assessing risk to trainees in an emergency rescue training base for building collapse. *Scientific Reports, 14*, 4792. https://doi.org/10.1038/s41598-024-55368-z

Li, J., Liu, S., Huang, W., & Yan, B. (2025). Artificial intelligence in low-altitude flight: Developmental opportunity or ethical challenge? *The Aeronautical Journal, 129*, 1–19. https://doi.org/10.1017/aer.2025.2

Li, Y., Qian, X., Zhang, S., et al. (2023). Assessment of gas explosion risk in underground spaces adjacent to a gas pipeline. *Tunnelling and Underground Space Technology, 131*, 104785. https://doi.org/10.1016/j.tust.2022.104785

Li, H., & Samsudin, N. A. (2024). A systematic review of landslide research in urban planning worldwide. *Natural Hazards, 121*(6), 6391–6411. https://doi.org/10.1007/s11069-024-07064-4
Li, C., Wang, G., He, J., & Wang, Y. (2022). A novel approach to probabilistic seismic landslide hazard mapping using Monte Carlo simulations. *Engineering Geology, 301*, 106616. https://doi.org/10.1016/j.enggeo.2022.106616
Li, L., Xiao, L., Zhang, M., et al. (2017). Agent-based simulation of evacuation considering extreme weather in tourist areas. In *2017 13th international conference on Computational Intelligence and Security (CIS), Hong Kong, China, 2017* (pp. 431–434). IEEE. https://doi.org/10.1109/CIS.2017.00100
Li, X., Zhou, Y., Hejazi, M., et al. (2021). Global urban growth between 1870 and 2100 from integrated high resolution mapped data and urban dynamic modeling. *Communications: Earth & Environment, 2*, 201. https://doi.org/10.1038/s43247-021-00273-w
Liang, H., Lee, S., Sun, J., & Wong, S. C. (2024). Unraveling the causes of the Seoul Halloween crowd-crush disaster. *PLoS One, 19*(7), e0306764. https://doi.org/10.1371/journal.pone.0306764
Ling, R., Rothman, L., Cloutier, M.-C., et al. (2020). Cyclist-motor vehicle collisions before and after implementation of cycle tracks in Toronto. *Canada. Accident Analysis & Prevention, 135*, 105360. https://doi.org/10.1016/j.aap.2019.105360
Liu, C., Fang, D., & Zhao, L. (2021). Reflection on earthquake damage of buildings in 2015 Nepal earthquake and seismic measures for post-earthquake reconstruction. *Structure, 30*, 647–658. https://doi.org/10.1016/j.istruc.2020.12.089
Liu, Z., Felton, T., & Mostafavi, A. (2024). Interpretable machine learning for predicting urban flash flood hotspots using intertwined land and built-environment features. *Computers, Environment and Urban Systems, 110*, 102096. https://doi.org/10.1016/j.compenvurbsys.2024.102096
Liu, S., Li, Y., Wang, B., et al. (2024). Challenges and countermeasures for developing countries in addressing loss and damage caused by climate change. *Advances in Climate Change Research, 15*(2), 353–363. https://doi.org/10.1016/j.accre.2024.02.003
Liu, Y., Lv, R., Guan, X., & Zeng, J. (2016). Path planning for unmanned aerial vehicle under geo-fencing and minimum safe separation constraints. In *2016 12th World Congress on Intelligent Control and Automation (WCICA), Guilin, China* (pp. 28–31). IEEE. https://doi.org/10.1109/WCICA.2016.7578482
Liu, S., Peng, F., Qiao, Y., & Zhang, J. (2021). Evaluating disaster prevention benefits of underground space from the perspective of urban resilience. *International Journal of Disaster Risk Reduction, 58*, 102206. https://doi.org/10.1016/j.ijdrr.2021.102206
Liu, M., Xiao, H., Pan, G., et al. (2024). How to achieve large-scale development in the low-altitude economy. In *2024 IEEE 10th international symposium on Microwave, Antenna, Propagation and EMC Technologies for Wireless Communications (MAPE), Guangzhou, China* (pp. 1–4). IEEE. https://doi.org/10.1109/MAPE62875.2024.10924462
Liu, X., Zhang, X., Ren, R., & Mahmoud, H. N. (2025). Assessment of China's 20 years of infrastructure and policy development for flood disaster management. *International Journal of Disaster Risk Reduction, 117*, 105211. https://doi.org/10.1016/j.ijdrr.2025.105211
Lungu, I., Copilau, J., Stanciu, A., & Bejan, F. (2018). Construction management in zones with landslide risk. *ce/papers, 2*(2–3), 341–346. https://doi.org/10.1002/cepa.694. Special Issue: XVI DECGE 2018 Proceedings of the 16th Danube-European Conference on Geotechnical Engineering.
Mayer, M., Nielsen, J. C., Elmeros, M., & Sunde, P. (2021). Understanding spatio-temporal patterns of deer-vehicle collisions to improve roadkill mitigation. *Journal of Environmental Management, 295*, 113148. https://doi.org/10.1016/j.jenvman.2021.113148
Mazele, O., & Amoah, C. (2022). The causes of poor infrastructure management and maintenance in south African municipalities. *Property Management, 40*(2), 192–206. https://doi.org/10.1108/PM-06-2021-0042
Mesimäki, J., & Luoma, J. (2021). Near accidents and collisions between pedestrians and cyclists. *European Transport Research Review, 13*, 38. https://doi.org/10.1186/s12544-021-00497-z

Miccadei, E., Carabella, C., & Paglia, G. (2022). Landslide hazard and environment risk assessment. *Land, 11*(3), 428. https://doi.org/10.3390/land11030428

Milano, C., Novelli, M., & Cheer, J. M. (2019). Overtourism and tourismphobia: A journey through four decades of tourism development, planning and local concerns. *Tourism Planning & Development, 16*(4), 353–357. https://doi.org/10.1080/21568316.2019.1599604

Mızrak, S., & Turan, M. (2023). Effect of individual characteristics, risk perception, self-efficacy and social support on willingness to relocate due to floods and landslides. *Natural Hazards, 116*, 1615–1637. https://doi.org/10.1007/s11069-022-05731-y

Mogomotsi, P. K., Stone, L. S., Mogomotsi, G. E. J., & Dube, N. (2020). Factors influencing community participation in wildlife conservation. *Human Dimensions of Wildlife, 25*(4), 372–386. https://doi.org/10.1080/10871209.2020.1748769

Mohd Ibrahim, A., Venkat, I., De Wilde, P., et al. (2022). The role of crowd behavior and cooperation strategies during evacuation. *Simulation, 98*(9), 737–751. https://doi.org/10.1177/00375497221075611

Mojtahedi, M., & Oo, B. L. (2017). Critical attributes for proactive engagement of stakeholders in disaster risk management. *International Journal of Disaster Risk Reduction, 21*, 35–43. https://doi.org/10.1016/j.ijdrr.2016.10.017

Moosa, I. A., & Khatatbeh, I. N. (2020). International tourist arrivals as a determinant of the severity of COVID-19: International cross-sectional evidence. *Journal of Policy Research in Tourism, Leisure and Events, 13*(3), 419–434. https://doi.org/10.1080/19407963.2020.1859519

Muehlenbein, M. P. (2013). Human-wildlife contact and emerging infectious diseases. In E. Brondízio & E. Moran (Eds.), *Human-environment interactions. Human-environment interactions* (Vol. 1). Springer. https://doi.org/10.1007/978-94-007-4780-7_4

Mühlhofer, E., Koks, E. E., Kropf, C. M., et al. (2023). A generalized natural hazard risk modelling framework for infrastructure failure cascades. *Reliability Engineering & System Safety, 234*, 109194. https://doi.org/10.1016/j.ress.2023.109194

Narayan, E., & Rana, N. (2023). Human-wildlife interaction: Past, present, and future. *BMC Zoology, 8*, 5. https://doi.org/10.1186/s40850-023-00168-7

Nardo, R. (2024). Rooftop fall hazards: The Most common scenarios. https://www.fallproof.com/blog/rooftop-fall-hazards-the-most-common-scenarios/

Nenungwi, L., Mokotjomela, T. M., Vukeya, L. R., et al. (2024). A human–wildlife conflict: Potential impacts of fatal harvesting approaches on medicinal plants in Free State Province, South Africa. *South African Geographical Journal, 107*(1), 20–38. https://doi.org/10.1080/03736245.2024.2315955

Ngoc Su, D., Luc Tra, D., Thi Huynh, H. M., et al. (2021). Enhancing resilience in the Covid-19 crisis: Lessons from human resource management practices in Vietnam. *Current Issues in Tourism, 24*(22), 3189–3205. https://doi.org/10.1080/13683500.2020.1863930

Nikiforiadis, A., Basbas, S., & Garyfalou, M. I. (2020). A methodology for the assessment of pedestrians-cyclists shared space level of service. *Journal of Cleaner Production, 254*, 120172. https://doi.org/10.1016/j.jclepro.2020.120172

Nkansah-Dwamena, E. (2023). Lessons learned from community engagement and participation in fostering coexistence and minimizing human-wildlife conflict in Ghana. *Trees, Forests and People, 14*, 100430. https://doi.org/10.1016/j.tfp.2023.100430

Noh, B., Park, H., & Yeo, H. (2022). Analyzing vehicle–pedestrian interactions: Combining data cube structure and predictive collision risk estimation model. *Accident Analysis & Prevention, 165*, 106539. https://doi.org/10.1016/j.aap.2021.106539

Nunweek, J., & O'Hanlon, B. (2019). Overcrowding: Regulatory problems and proposed responses. *Parity, 32*(5), 12–14. https://search.informit.org/doi/abs/10.3316/informit.686500264882449

Nyberg, R., & Johansson, M. (2013). Indicators of road network vulnerability to storm-felled trees. *Natural Hazards, 69*, 185–199. https://doi.org/10.1007/s11069-013-0693-z

Nyhus, P. J. (2016). Human–wildlife conflict and coexistence. *Annual Review of Environment and Resources, 41*, 143–171. https://doi.org/10.1146/annurev-environ-110615-085634

O'Bryan, C. J., Braczkowski, A. R., Beyer, H. L., et al. (2018). The contribution of predators and scavengers to human Well-being. *Nature Ecology & Evolution, 2*, 229–236. https://doi.org/10.1038/s41559-017-0421-2

Ogata, Y., Isei, T., & Kuriyagawa, M. (1990). Safety measures for underground space utilization. *Tunnelling and Underground Space Technology, 5*(3), 245–256. https://doi.org/10.1016/0886-7798(90)90012-9

Onifade, M. (2021). Towards an emergency preparedness for self-rescue from underground coal mines. *Process Safety and Environmental Protection, 149*, 946–957. https://doi.org/10.1016/j.psep.2021.03.049

Osama, A., Albitar, M., Sayed, T., & Bigazzi, A. (2020). Determining if walkability and bikeability indices reflect pedestrian and cyclist safety. *Transportation Research Record, 2674*(9), 767–775. https://doi.org/10.1177/0361198120931844

Ou, G., Jiao, Y., Zhang, G., et al. (2021). Collapse risk assessment of deep-buried tunnel during construction and its application. *Tunnelling and Underground Space Technology, 115*, 104019. https://doi.org/10.1016/j.tust.2021.104019

Papadopoulou, N. M., Ribeiro, M. A., & Prayag, G. (2022). Psychological determinants of tourist satisfaction and destination loyalty: The influence of perceived overcrowding and overtourism. *Journal of Travel Research, 62*(3), 644–662. https://doi.org/10.1177/00472875221089049

Pardeshi, S. D., Autade, S. E., & Pardeshi, S. S. (2013). Landslide hazard assessment: Recent trends and techniques. *Springer Plus, 2*, 523. https://doi.org/10.1186/2193-1801-2-523

Parrado, S. (2018). The culture of risk regulation: Responses to environmental disasters. *Regulation and Governance, 14*(3), 599–615. https://doi.org/10.1111/rego.12214

Pătru-Stupariu, I., Nita, A., Mustăţea, M., et al. (2020). Using social network methodological approach to better understand human–wildlife interactions. *Land Use Policy, 99*, 105009. https://doi.org/10.1016/j.landusepol.2020.105009

Peck, A. J., Adams, S. L., Armstrong, A., et al. (2022). A new framework for flood adaptation: Introducing the flood adaptation hierarchy. *Ecology and Society, 27*(4), 5. https://doi.org/10.5751/ES-13544-270405

Pedica, C., Diamanti, M., & Vilhjálmsson, H. H. (2021). Assessing the disturbance from overcrowding in outdoor nature experiences. In *Extended Abstracts of the 2021 CHI Conference on Human Factors in Computing Systems (CHI EA '21)* (pp. 1–8). Association for Computing Machinery, New York, NY, USA, Article 58. https://doi.org/10.1145/3411763.3443439.

Petrovsky, A. M., Doole, J., Ellerbroek, J. M., et al. (2018). Challenges with obstacle data for manned and unmanned aviation. *Sensing and Spatial Information Sciences, XLII-4/W1*, 143–149. https://doi.org/10.5194/isprs-archives-XLII-4-W10-143-2018

Petrucci, O. (2022). Landslide fatality occurrence: A systematic review of research published between January 2010 and March 2022. *Sustainability, 14*(15), 9346. https://doi.org/10.3390/su14159346

Pradhan, B., Chaudhari, A., Adinarayana, J., et al. (2012). Soil erosion assessment and its correlation with landslide events using remote sensing data and GIS: A case study at Penang Island, Malaysia. *Environmental Monitoring and Assessment, 184*, 715–727. https://doi.org/10.1007/s10661-011-1996-8

Quesada-Román, A., & Campos-Durán, D. (2022). Natural disaster risk inequalities in Central America. *Papers in Applied Geography, 9*(1), 36–48. https://doi.org/10.1080/23754931.2022.2081814

Rahman, N. S., Abdul, F., & Rosli, H. Z. (2014). An innovation approach for improving passenger vessels safety level: Overload problem. *International Journal of Business Tourism and Applied Sciences, 2*(2), 61–74.

Rahman, H. A., Beznosov, K., & Marti, J. R. (2009). Identification of sources of failures and their propagation in critical infrastructures from 12 years of public failure reports. *International Journal of Critical Infrastructures, 5*(3), 220–244. https://doi.org/10.1504/IJCIS.2009.024872

Ravenelle, J., & Nyhus, P. J. (2017). Global patterns and trends in human–wildlife conflict compensation. *Conservation Biology, 31*(6), 1247–1256. https://doi.org/10.1111/cobi.12948

Reaser, J. K., Witt, A., Tabor, G. M., et al. (2021). Ecological countermeasures for preventing zoonotic disease outbreaks: When ecological restoration is a human health imperative. *Restoration Ecology, 29*(4), e13357. https://doi.org/10.1111/rec.13357

Reichstein, M., Benson, V., Blunk, J., et al. (2025). Early warning of complex climate risk with integrated artificial intelligence. *Nature Communications, 16*, 2564. https://doi.org/10.1038/s41467-025-57640-w

Restrepo, C. E., Zimmerman, R., Laefer, D. F., et al. (2025). Stakeholder engagement to strengthen underground infrastructure resilience. *Environment Systems and Decision, 45*, 10. https://doi.org/10.1007/s10669-025-10003-8

Ridwan, Q., Wani, Z. A., Anjum, N., et al. (2023). Human-wildlife conflict: A bibliometric analysis during 1991–2023. *Regional Sustainability, 4*(3), 309–321. https://doi.org/10.1016/j.regsus.2023.08.008

Robinson, J. M., Aronson, J., Daniels, C. B., et al. (2022). Ecosystem restoration is integral to humanity's recovery from COVID-19. *The Lancet Planetary Health, 6*(9), e769–e773. https://doi.org/10.1016/S2542-5196(22)00171-1

Roe, D., Dickman, A., Kock, R., et al. (2020). Beyond banning wildlife trade: COVID-19, conservation and development. *World Development, 136*, 105121. https://doi.org/10.1016/j.worlddev.2020.105121

Rondeau, D., & Bulte, E. (2007). Wildlife damage and agriculture: A dynamic analysis of compensation schemes. *American Journal of Agricultural Economics, 89*(2), 490–507. https://doi.org/10.1111/j.1467-8276.2007.00995.x

Ruiz-Apilánez, B., Karimi, K., García-Camacha, I., et al. (2017). Shared space streets: Design, user perception and performance. *Urban Design International, 22*, 267–284. https://doi.org/10.1057/s41289-016-0036-2

Sadeghi, S., Soltanmohammadlou, N., & Nasirzadeh, F. (2022). Applications of wireless sensor networks to improve occupational safety and health in underground mines. *Journal of Safety Research, 83*, 8–25. https://doi.org/10.1016/j.jsr.2022.07.016

Santana-Jiménez, Y., & Hernández, J. M. (2011). Estimating the effect of overcrowding on tourist attraction: The case of Canary Islands. *Tourism Management, 32*(2), 415–425. https://doi.org/10.1016/j.tourman.2010.03.013

Schell, C. J., Stanton, L. A., Young, J. K., et al. (2021). The evolutionary consequences of human–wildlife conflict in cities. *Evolutionary Applications, 14*(1), 178–197. https://doi.org/10.1111/eva.13131

Schiermeyer, C., Pascucci, F., Rinke, N., Berkhahn, V., & Friedrich, B. (2019). Modeling and solving of multiple conflict situations in shared spaces. In S. Hamdar (Ed.), *Traffic and granular flow '17. TGF 2017*. Springer. https://doi.org/10.1007/978-3-030-11440-4_49

Scolobig, A. (2016). Stakeholder perspectives on barriers to landslide risk governance. *Natural Hazards, 81*(Suppl 1), 27–43. https://doi.org/10.1007/s11069-015-1787-6

Senthilkumar, K., Mathialagan, P., Manivannan, C., et al. (2016). A study on the tolerance level of farmers toward human-wildlife conflict in the forest buffer zones of Tamil Nadu. *Veterinary World, 9*(7), 747–752. https://doi.org/10.14202/vetworld.2016.747-752

Sharma, D., Bhondekar, A. P., Shukla, A. K., et al. (2018). A review on technological advancements in crowd management. *Journal of Ambient Intelligence and Humanized Computing, 9*, 485–495. https://doi.org/10.1007/s12652-016-0432-x

Sharma, A., McCloskey, B., Hui, D. S., et al. (2023). Global mass gathering events and deaths due to crowd surge, stampedes, crush and physical injuries—Lessons from the Seoul Halloween and other disasters. *Travel Medicine and Infectious Disease, 52*, 102524. https://doi.org/10.1016/j.tmaid.2022.102524

Shayo, G. M., Elimbinzi, E., Shao, G. N., et al. (2023). Severity of waterborne diseases in developing countries and the effectiveness of ceramic filters for improving water quality. *Bulletin of the National Research Centre, 47*, 113. https://doi.org/10.1186/s42269-023-01088-9

Singh, N. S., Abrahim, O., Altare, C., et al. (2020). COVID-19 in humanitarian settings: Documenting and sharing context-specific programmatic experiences. *Conflict and Health, 14*, 79. https://doi.org/10.1186/s13031-020-00321-w

Smith, G. S., Anjum, E., Francis, C., et al. (2022). Climate change, environmental disasters, and health inequities: The underlying role of structural inequalities. *Current Environmental Health Reports, 9*, 80–89. https://doi.org/10.1007/s40572-022-00336-w

Solecki, W., Roberts, D., & Seto, K. C. (2024). Strategies to improve the impact of the IPCC special report on climate change and cities. *Nature Climate Change, 14*, 685–691. https://doi.org/10.1038/s41558-024-02060-9

Soulsbury, C. D., & White, P. C. L. (2015). Human–wildlife interactions in urban areas: A review of conflicts, benefits and opportunities. *Wildlife Research, 42*(7), 541–553. https://doi.org/10.1071/WR14229

Spring, J. (2022). Nature-based tourism and guided wildlife tours: Designing wildlife tour experiences that optimise sustainable learning opportunities. *Journal of Ecotourism, 22*(1), 187–207. https://doi.org/10.1080/14724049.2022.2098963

St. Denis, L. A., Mietkiewicz, N. P., Short, K. C., et al. (2020). All-hazards dataset mined from the US National Incident Management System 1999–2014. *Scientific Data, 7*, 64. https://doi.org/10.1038/s41597-020-0403-0

Stevens, M., Rawat, S., & Satterfield, T. (2024). Care, conflict, and coexistence: Human–wildlife relations in community forests. *People and Nature, 7*(1), 231–246. https://doi.org/10.1002/pan3.10760

Stevenson, D. (1885). Earthquake-proof buildings. *Nature, 32*, 316. https://doi.org/10.1038/032316b0

Suhr, F., & Steinert, J. I. (2022). Epidemiology of floods in sub-Saharan Africa: A systematic review of health outcomes. *BMC Public Health, 22*, 268. https://doi.org/10.1186/s12889-022-12584-4

Sujatha, E. R., Sudarsan, J. S., & Nithiyanantham, S. (2023). A review on sustainable reinforcing techniques to stabilize slopes against landslides. *International Journal of Environmental Science and Technology, 20*, 13873–13882. https://doi.org/10.1007/s13762-023-04832-w

Sun, X., Hu, Y., Qin, Y., & Zhang, Y. (2024). Risk assessment of unmanned aerial vehicle accidents based on data-driven Bayesian networks. *Reliability Engineering & System Safety, 248*, 110185. https://doi.org/10.1016/j.ress.2024.110185

Sun, H., Li, Y., & Zhang, J. (2022). Collaboration-based reliable optimal casualty evacuation network design for large-scale emergency preparedness. *Socio-Economic Planning Sciences, 81*, 101192. https://doi.org/10.1016/j.seps.2021.101192

Sun, X., Wang, S., Zhang, X., & Wandelt, S. (2025). LAERACE: Taking the policy fast-track towards low-altitude economy. *Journal of the Air Transport Research Society, 4*, 100058. https://doi.org/10.1016/j.jatrs.2025.100058

Sütfeld, L. R., Bronson, J., & Kirchmair, L. (2025). Automated vehicle regulation needs to speak to code, not to humans: Keeping safety and ethics in the public domain. *Philosophy & Technology, 38*, 15. https://doi.org/10.1007/s13347-025-00846-z

Takahashi, A., Yasufuku, K., & Abe, H. (2022). Evaluation of underground space for tsunami evacuation safety with route obstacles by agent-based simulation. *GEOMATE Journal, 22*(92), 83–91. https://geomatejournal.com/geomate/article/view/1657

Tang, H., Zhang, Y., Mohmoodian, V., & Charkhgard, H. (2021). Automated flight planning of high-density urban air mobility. *Transportation Research Part C: Emerging Technologies, 131*, 103324. https://doi.org/10.1016/j.trc.2021.103324

Thapa, A., Mukherjee, T., Pradhan, A., et al. (2024). Understanding the prospects of human-wildlife coexistence: A conceptual framework. *Biodiversity and Conservation, 33*, 3583–3615. https://doi.org/10.1007/s10531-024-02922-w

Tian, H., Zou, Z., Hao, S., et al. (2022). Key technologies and equipment for quickly and safely building life support and rescue channels in mine disasters. *Coal Geology & Exploration, 50*(11), 1–13. https://doi.org/10.12363/issn.1001-1986.22.05.0415

Trogrlic, R. S., Reiter, K., Ciurean, R. L., et al. (2024). Challenges in assessing and managing multi-hazard risks: A European stakeholders perspective. *Environmental Science & Policy, 157*, 103774. https://doi.org/10.1016/j.envsci.2024.103774

Tsai, Y.-J., Syu, F.-T., Shieh, C.-L., et al. (2021). Framework of emergency response system for potential large-scale landslide in Taiwan. *Water, 13*(5), 712. https://doi.org/10.3390/w13050712

Tsunoda, H., & Enari, H. (2020). A strategy for wildlife management in depopulating rural areas of Japan. *Conservation Biology, 34*(4), 819–828. https://doi.org/10.1111/cobi.13470

Tuneu-Corral, C., Puig-Montserrat, X., Riba-Bertolín, D., et al. (2023). Pest suppression by bats and management strategies to favour it: A global review. *Biological Reviews, 98*(5), 1564–1582. https://doi.org/10.1111/brv.12967

Turner, A. K. (2018). Social and environmental impacts of landslides. *Innovative Infrastructure Solution, 3*, 70. https://doi.org/10.1007/s41062-018-0175-y

United Nations. (2019). World urbanization prospects: The 2018 revision.

Vamvakeridou-Lyroudia, L. S., Chen, A. S., Khoury, M., et al. (2020). Assessing and visualising hazard impacts to enhance the resilience of critical infrastructures to urban flooding. *Science of the Total Environment, 707*, 136078. https://doi.org/10.1016/j.scitotenv.2019.136078

Van Wyk de Vries, M. (2025). All hazards are multihazards, few of them are natural. *npj Natural Hazards, 2*, 18. https://doi.org/10.1038/s44304-025-00071-w

Vanumu, L. D., Laxmikant, K., & Rao, K. (2020). Human stampedes at mass gatherings: An overview. *Collective Dynamics, 5*, 502–504. https://doi.org/10.17815/CD.2020.83

Von Seidlein, L., Alabaster, G., Deen, J., & Knudsen, J. (2021). Crowding has consequences: Prevention and management of COVID-19 in informal urban settlements. *Building and Environment, 188*, 107472. https://doi.org/10.1016/j.buildenv.2020.107472

Wan, L., Liu, G., & Su, X. (2025). Evaluating the socioeconomic and psychological implications of human-wildlife conflict within protected areas in China. *Ecological Frontiers, 45*(3), 693–700. https://doi.org/10.1016/j.ecofro.2024.12.012

Wang, Y., Kyriakidis, M., & Dang, V. N. (2021). Incorporating human factors in emergency evacuation—An overview of behavioral factors and models. *International Journal of Disaster Risk Reduction, 60*, 102254. https://doi.org/10.1016/j.ijdrr.2021.102254

Wang, H., Liu, G., Xu, W., & Wang, G. (2005). GIS-based landslide hazard assessment: An overview. *Progress in Physical Geography, 29*(4), 548–567. https://doi.org/10.1191/0309133305pp462ra

Wang, A. Y., Ma, Z., & Zhang, Y. T. (2022). Research on safety early warning standard of large-scale underground utility tunnel in ground fissure active period. *Frontiers in Earth Science, 10*, 828477. https://doi.org/10.3389/feart.2022.828477

Wang, M., Wang, W., Dai, C., et al. (2024). Risk analysis and evaluation of emergency rescue in landslide disaster. *Natural Hazards, 120*, 14809–14835. https://doi.org/10.1007/s11069-024-06811-x

Watsa, M. (2020). Rigorous wildlife disease surveillance. *Science, 369*, 145–147. https://doi.org/10.1126/science.abc0017

Watson, D. M., & Watson, M. J. (2015). Wildlife restoration: Mainstreaming translocations to keep common species common. *Biological Conservation, 191*, 830–838. https://doi.org/10.1016/j.biocon.2015.08.035

Wirtz, A., Kron, W., Löw, P., et al. (2014). The need for data: Natural disasters and the challenges of database management. *Natural Hazards, 70*, 135–157. https://doi.org/10.1007/s11069-012-0312-4

Wong, M. C., Huang, J., Wong, S. H., & Yuen-Chun, T. J. (2021). The potential effectiveness of the WHO international health regulations capacity requirements on control of the COVID-19 pandemic: A cross-sectional study of 114 countries. *Journal of the Royal Society of Medicine, 114*(3), 121–131. https://doi.org/10.1177/0141076821992453

Wu, Y., Luo, L., Wang, Y., et al. (2023). Strengthened public awareness of one health to prevent zoonosis spillover to humans. *Science of the Total Environment, 879*, 163200. https://doi.org/10.1016/j.scitotenv.2023.163200

Xu, C., Liao, X., Tan, J., et al. (2020). Recent research progress of unmanned aerial vehicle regulation policies and technologies in urban low altitude. *IEEE Access, 8*, 74175–74194. https://doi.org/10.1109/ACCESS.2020.2987622.k

Yang, J., Wang, Y., Hang, X., & Delahaye, D. (2024). A review on airspace design and risk assessment for urban air mobility. *IEEE Access, 12*, 157599–157611. https://doi.org/10.1109/ACCESS.2024.3481148

Yang, X., & Wei, P. (2021). Autonomous free flight operations in urban air mobility with computational guidance and collision avoidance. *IEEE Transactions on Intelligent Transportation Systems, 22*(9), 5962–5975. https://doi.org/10.1109/TITS.2020.3048360

Yazdanpanah, V., Gerding, E. H., Stein, S., et al. (2023). Reasoning about responsibility in autonomous systems: Challenges and opportunities. *AI & Society, 38*, 1453–1464. https://doi.org/10.1007/s00146-022-01607-8

Yereseme, A. K., Surendra, H. J., & Kuntoji, G. (2022). Sustainable integrated urban flood management strategies for planning of smart cities: A review. *Sustainable Water Resources Management, 8*, 85. https://doi.org/10.1007/s40899-022-00666-5

Yeshey, K. R. J., Ford, R. M., & Nitschke, C. R. (2024). Social and ecological dimensions are needed to understand human-wildlife conflict in subsistence farming context. *People and Nature, 6*, 2602–2617. https://doi.org/10.1002/pan3.10740

Yezli, S. (2023). Risk factors for heat-related illnesses during the Hajj mass gathering: An expert review. *Reviews on Environmental Health, 38*(1), 33–43. https://doi.org/10.1515/reveh-2021-0097

Yu, R., Mao, Q., & Lv, J. (2022). An extended model for crowd evacuation considering rescue behavior. *Physica A: Statistical Mechanics and Its Applications, 605*, 127989. https://doi.org/10.1016/j.physa.2022.127989

Zellmer, A. J., & Goto, B. S. (2022). Urban wildlife corridors: Building bridges for wildlife and people. *Frontiers in Sustainable Cities, 4*, 954089. https://doi.org/10.3389/frsc.2022.954089

Zhang, J. (2020). Transport policymaking that accounts for COVID-19 and future public health threats: A PASS approach. *Transport Policy, 99*, 405–418. https://doi.org/10.1016/j.tranpol.2020.09.009

Zhang, J., & Hayashi, Y. (2022). *Transportation amid pandemics: Lessons learned from COVID-19*. Elsevier. https://doi.org/10.1016/C2020-0-04079-X

Zhou, D., Lin, Z., & Lim, S. H. (2019). Spatial characteristics and risk factor identification for land use spatial conflicts in a rapid urbanization region in China. *Environmental Monitoring and Assessment, 191*, 677. https://doi.org/10.1007/s10661-019-7809-1

Chapter 6
The HEARTY City: Drawing Parallels with the Human Immune System for Urban Futures

6.1 Reimagining Urban Transformation: The HEARTY City and the Immune System Paradigm

As Zhang (2025) wisely points out in his vision for the HEARTY City, urban planning has often been limited to narrow, fragmented viewpoints. These approaches, like scattered puzzle pieces without a unifying picture, often result in strategies that conflict with or overlook the broader systems they aim to serve. The six letters in the HEARTY City model represent key pillars that form the foundation of this reimagined approach: 'H' stands for Healthy City, 'E' for Eco-City (or Energy-efficient City), 'A' for AI-driven City, 'R' for Resilient City, 'T' for Tourism City, and 'Y' for Yearning City.

To move beyond the limits of traditional urban planning, we need new frameworks—ones that offer a deeper understanding of the complex forces within urban systems. Here, the analogy to the human immune system (Janeway et al., 2001; Delves et al., 2017; Murphy et al., 2022; Abbas et al., 2023) offers a compelling perspective for rethinking urban planning. Just as the immune system adapts, responds, and maintains balance amid constant threats, we can imagine cities as living, breathing entities—dynamic and interconnected, always shifting and evolving. This view allows us to see cities not as static, isolated places, but as vibrant, responsive systems—systems that need resilience, adaptability, and integration, qualities often overlooked in traditional models.

In the following pages, we will explore how the HEARTY City can be compared to the human immune system, grounding this new urban model in a theoretical framework that is both meaningful and relevant to our time. As urbanization grows at an unprecedented rate, cities are facing complex challenges—rapid population growth, environmental decline, and rising social inequalities (e.g., UN, 2020). The siloed approaches of traditional urban planning often fail when dealing with these

J. Zhang, *Urbanimmunology*, Urban Sustainability,
https://doi.org/10.1007/978-981-95-4878-1_6

interconnected crises, resulting in fragmented solutions that are not only ineffective but sometimes even harmful.

By drawing a parallel between the HEARTY City and the human immune system, we emphasize the need for a holistic, adaptive, and self-regulating model of urban development. Just as the immune system constantly scans for threats and responds precisely, cities should continually assess and adjust their strategies in response to the changing environmental, economic, and social forces. This analogy invites us to view the urban landscape not as a collection of infrastructure and resources, but as an organism capable of responding to both external and internal challenges, adapting, and protecting its core.

Achieving the vision of the HEARTY City requires the harmonious integration of various urban planning components—health, sustainability, resilience, and technology—into a unified and cohesive framework. This ambitious goal will require collaborative governance (Dai & Azhar, 2024), the use of advanced technologies like AI and IoT (Ahad et al., 2020), and a steadfast commitment to long-term sustainability. To realize such a transformative model, cross-sector partnerships (Vogel et al., 2022), evidence-based decision-making (HakemZadeh et al., 2024), and a focus on creating urban spaces that nurture human well-being while preserving the ecological balance that supports us all will be essential.

6.2 HEARTY City

In order to better compare the HEARTY City to the human immune system, the following six pillars are outlined (Zhang, 2025). First, the *Healthy City* concept focuses on creating environments that support the overall well-being of all residents, including their physical, mental, and social health. It prioritizes inclusive design, ensuring fair access to resources, particularly for vulnerable groups, while fostering community connections and networks that nurture emotional and psychological well-being.

Second, an *Eco-City* (also known as an *Energy-efficient City*) is a concept in urban planning that focuses on environmental sustainability, optimizing the use of resources, and reducing ecological impact. The objective of an *Eco-City* is to create an urban space that is self-sufficient and promotes the health and prosperity of its inhabitants, while minimizing damage to the natural surroundings.

Third, an *AI-driven City* refers to an urban area where artificial intelligence technologies are integrated throughout the city's systems to improve efficiency, sustainability, and the quality of life for its residents. Often referred to as smart cities, these urban areas harness vast amounts of data—typically collected from sensors and IoT devices embedded throughout the environment—to optimize urban management and adapt dynamically to real-time changes in conditions.

Fourth, a *Resilient City* is an urban environment designed and managed to endure and recover from various types of shocks and stresses—whether natural, social, economic, or technological—while maintaining essential functions and ensuring the welfare of its citizens. In this context, resilience involves not only the city's ability to bounce back from disruptions but also its capacity to anticipate, prepare for, and reduce the impact of potential future risks, particularly in the face of growing urbanization and climate change.

Fifth, a *Tourism City* is a place where the tourism industry plays a central role, shaping the local economy, infrastructure, culture, and identity to attract and serve visitors. These cities offer a wide array of services, attractions, and experiences catering to both leisure and business travelers. Tourism cities can range from iconic global destinations like Paris, New York, and Venice to emerging travel hotspots in developing regions that highlight their unique cultural, historical, or natural attractions.

Sixth, a *Yearning City* refers to cities driven by a strong ambition to transform, evolve, or reinvent themselves in response to challenges such as social, economic, or environmental pressures. These cities are often in a state of constant change, always striving for improvement, expansion, and modernization. The term "yearning" reflects a city's collective desire to advance, whether in terms of sustainability, inclusivity, or economic progress, often as a response to past inequalities, environmental crises, or rapid urban growth.

The HEARTY City framework, as a holistic and unified concept, recognizes the intricate, interconnected nature of its foundational elements, where both benefits and challenges naturally arise in the pursuit of shared urban objectives. These benefits and challenges are not fixed; rather, they are dynamic, meaning that the impact one element has on another can vary depending on various contexts.

This dynamic quality of benefits and challenges highlights the necessity for thoughtful attention to how each element is prioritized and executed. It emphasizes the importance of a more nuanced, context-aware approach to urban planning—one that acknowledges not only the synergies between elements but also the possible tensions that may arise when multiple urban objectives are pursued concurrently. Achieving a truly cohesive and sustainable urban future requires balancing competing demands, ensuring that the advantages of one function do not unintentionally overshadow or conflict with the challenges posed by another. In this way, the HEARTY City model encourages a comprehensive, flexible approach to urban development, effectively addressing both immediate needs and long-term sustainability goals.

6.3 Using Human Immune Intelligence to Analogize the HEARTY City

The HEARTY City concept adopts an integrated and holistic approach to urban planning, aiming to reflect the dynamic and interconnected nature of urban systems. Much like the human body, which is a complex, interdependent system where various functions work together to sustain life, the HEARTY City seeks to establish a balanced and harmonious urban environment. In this environment, different urban functions interact to ensure the well-being and prosperity of its residents.

By comparing the HEARTY City to the human immune system, we gain a deeper insight into how urban functions collaborate to maintain equilibrium and respond to external pressures—just as the immune system protects the body from harmful influences.

The essence of this analogy extends beyond comparing the individual pillars of the HEARTY City to the immune system. It emphasizes the interdependence and connectivity of these pillars. Just as dysfunction in one part of the human body can trigger cascading effects throughout the system, the pillars of the HEARTY City should function in harmony. When one pillar falters, it can disrupt the entire urban ecosystem, much like how the failure of a single element in the immune system can leave the body vulnerable to disease.

By framing each pillar in this way, we highlight how their interconnectedness enables the city to function as an adaptive, self-regulating organism—capable of fostering well-being, resilience, and innovation.

6.3.1 Healthy City: The Immune System's Defense Mechanism

The Healthy City is dedicated to supporting the well-being of all its residents, with a strong focus on physical, mental, and social health. It fosters inclusive, fair access to resources, much like the immune system defends the body from pathogens. The Healthy City protects and nurtures its population's health through preventive healthcare, accessible medical resources, and a design that promotes social connections and overall well-being.

For example, the Healthy City ensures that vulnerable groups have access to essential services such as healthcare, mental health support, clean air, and safe housing—much like the immune system defends the body from infections and promotes healing. The defense is not only physical but also extends to emotional and psychological well-being, paralleling the immune system's ability to respond to mental health stressors. Just as immune cells target invaders and regenerate tissues after an assault, the Healthy City works to restore and strengthen its population.

The emphasis on inclusivity ensures that no group is left exposed to risks, mirroring the immune system's surveillance and targeted defense. A weakened immune system leaves the body vulnerable to a wide range of threats, and similarly, a compromised Healthy City weakens the broader urban system, making it susceptible to various challenges. When the health foundation falters, other aspects of the city—such as economic productivity or resilience to shocks—are likely to suffer.

For instance, poor health among a city's population reduces the ability of citizens to contribute meaningfully to society, both socially and economically. This mirrors how a compromised immune system diminishes the body's ability to function optimally. A population in poor health can result in lower workforce participation, increased absenteeism, and reduced productivity, all of which disrupt the city's overall functioning. Furthermore, the financial burden of health-related costs can drain resources, limiting funding for other vital urban initiatives, such as creating a sustainable Eco-City or enhancing technological infrastructure in an AI-driven City.

Much like a weakened immune system exposes the body to greater risks, a struggling Healthy City makes the entire urban system more vulnerable. Without a strong foundation of health, the city's ability to innovate, adapt, and thrive is hindered—just as the body's ability to defend itself is impaired by a weakened immune system. The health of a city is the bedrock on which all other urban functions depend. If this pillar is compromised, the city's overall strength and resilience are diminished.

6.3.2 *Eco-City: The Immune System's Adaptation and Sustainability Mechanism*

The Eco-City focuses on environmental sustainability, promoting the optimal use of resources while minimizing ecological harm. This concept parallels the immune system's adaptive response, where the body evolves over time to better handle environmental threats, such as pathogens and stressors. Just as the immune system learns from encounters with harmful agents, the Eco-City adapts to the challenges of urban growth and environmental degradation.

To achieve sustainability, the Eco-City incorporates adaptive strategies such as renewable energy, waste recycling, water conservation, and sustainable building practices. These efforts aim to minimize the ecological footprint while striving for self-sufficiency, reducing external dependencies, and fostering an ecosystem that can regenerate itself. This mirrors the immune system's ability to adjust to evolving threats.

Just as the immune system protects the body from harmful pathogens, the Eco-City works to preserve the environment by managing resources responsibly and promoting sustainability. If the environment deteriorates—through pollution, poor waste management, or unsustainable resource use—the entire urban system is at risk.

For example, an AI-driven City, which depends on clean energy and stable resources for its technological infrastructure, would struggle if the environment becomes polluted or resources are depleted. Pollution and resource scarcity could undermine data centers, slowing technological progress and causing disruption. Similarly, a Resilient City requires a healthy environment capable of absorbing shocks such as floods or heatwaves. Without a robust ecosystem, the city's capacity to endure such challenges weakens, leaving it vulnerable to environmental stresses.

When the Eco-City pillar fails, it jeopardizes not only the city's environmental health but also its long-term viability. The consequences extend beyond ecological degradation, affecting both the well-being of residents and the city's ability to withstand external challenges. Just as a weakened immune system compromises the body's defense, an unhealthy urban environment hinders the city's ability to thrive and adapt. Therefore, the Eco-City is crucial not only for protecting the environment but also for ensuring the long-term strength and sustainability of all urban functions.

6.3.3 *AI-driven City: The Immune System's Regeneration and Feedback Mechanism*

An AI-driven City leverages cutting-edge technologies, like artificial intelligence, to enhance efficiency and improve the quality of life for its residents. This mirrors the immune system's regenerative capacity, which continuously learns, adapts, and heals. Just as the immune system uses feedback mechanisms to regulate its responses—such as turning off inflammation once a threat has been neutralized—an AI-driven city can use data analytics and predictive models to optimize urban management and improve service delivery.

For instance, AI systems can monitor infrastructure health, predict and manage traffic flow, or optimize energy use, enabling the city to regenerate resources in real time. AI's adaptive nature allows it to detect shifts in urban patterns and make instant adjustments to ensure the city thrives. Much like the immune system's memory helps prevent future infections, AI-driven cities learn from past disruptions to better prepare for and mitigate future challenges, fostering continuous improvement.

However, just as the immune system requires a healthy body to function optimally, AI systems depend on a well-connected urban environment to perform at their best. In a Healthy City, where residents are well-informed, digitally literate, and connected, AI systems can be more effective. For example, AI can track public health data or optimize resource distribution using information generated by a functioning community.

But AI's potential can only be fully realized in an environment that supports it. AI technologies often rely on energy-intensive data centers, and if these centers contribute to pollution or are unsustainable, they undermine the goals of the Eco-City and can erode the city's environmental health.

AI can also enhance the Resilient City by anticipating crises—such as extreme weather, natural disasters, or social challenges—before they occur. However, this depends on the availability of accurate data and a healthy environment for AI to operate. If the relationship between AI, health, and the environment is compromised, the city as a whole becomes less resilient. Just as a malfunction in part of the immune system weakens the body's defenses, a disconnection between AI, health, and the environment reduces the city's ability to function effectively. Thus, AI-driven systems should be carefully integrated into the larger urban ecosystem to ensure the city's long-term success and sustainability.

6.3.4 Resilient City: The Immune System's Integration and Recovery Mechanism

The Resilient City is designed to endure and recover from a variety of shocks, such as natural disasters, economic downturns, or social crises. This mirrors the immune system's integration and recovery functions, where responses from various immune cells are coordinated to mount a robust defense. After neutralizing a threat, the immune system restores balance by repairing damaged tissues and reinforcing defenses for future challenges.

Similarly, the Resilient City integrates urban systems—such as infrastructure, healthcare, communication, and social networks—into a flexible structure that can adapt to and recover from shocks. It anticipates risks, prepares in advance, and ensures that critical functions continue even after a crisis, much like the immune system restores balance after an attack. Resilience, however, extends beyond recovery; it involves learning from past events to strengthen future defenses, much like how the immune system adapts to prevent future infections.

Resilience is not a standalone quality; it relies on the seamless functioning of other urban systems. For example, a Healthy City is essential for resilience. Just as a strong immune system enables recovery from illness, a healthy population is better equipped to withstand and recover from disasters—whether natural or social. When residents are physically, mentally, and emotionally well, they are more capable of supporting recovery efforts and rebuilding the city.

Likewise, a Resilient City cannot thrive without a functioning Eco-City. Environmental stability is key—if the city faces pollution, resource depletion, or climate change, its resilience diminishes. For example, the city's ability to absorb and recover from extreme weather events, like floods, depends on the health of its ecosystem. A weakened environment limits the ability to bounce back, just as a compromised immune system hampers the body's ability to defend itself.

Furthermore, the ability to adapt to new challenges is crucial for urban resilience. AI-driven technologies serve as the city's nervous system, processing data and enabling real-time responses to emerging risks. Just as the body reacts to stimuli through the nervous system, AI systems can predict and mitigate disasters before they escalate, ensuring a swift and efficient response. Without AI, a city's capacity to adapt to unforeseen challenges—such as extreme weather or social unrest—would be severely limited.

When any part of this urban system fails—whether it's poor health, environmental instability, or lack of adaptive technologies—the resilience of the entire city is weakened. Just as damage to the immune system hampers the body's defenses, a breakdown in one urban function exposes the city to greater risks. The interconnectedness of the Healthy City, Eco-City, and AI-driven City highlights the importance of a holistic approach to urban planning, where recovery, prevention, and adaptation work together to ensure long-term resilience.

6.3.5 *Tourism City: The Immune System's Recognition and Response to External Influences*

The Tourism City thrives on the influx of visitors, shaping its economy, culture, and identity around the tourism industry. This mirrors the immune system's ability to recognize and respond to external agents, distinguishing between harmful invaders and benign elements. Just as the immune system identifies threats, the Tourism City should identify and integrate visitors, leveraging their presence for economic and cultural enrichment without overwhelming the system.

Tourism fosters cross-cultural exchange, knowledge transfer, and economic development, akin to how the immune system tolerates beneficial interactions while neutralizing harmful agents. A successful Tourism City, like the immune system recognizing beneficial microbes, should balance external influences—ensuring they enhance the urban environment without destabilizing it. This balance fosters both growth and stability.

Just as the immune system protects the body's health, the success of a Tourism City depends on the well-being of its people and environment. A Healthy City lays the foundation for tourism by providing a safe, vibrant atmosphere that attracts visitors. Healthy, socially connected communities create a welcoming environment that encourages tourists to stay and explore. Similar to how a strong immune system resists illness, a Healthy City ensures that residents and visitors alike enjoy a high quality of life, making it a desirable destination.

Additionally, an Eco-City plays a crucial role by ensuring tourism activities do not overburden the environment. Just as the immune system defends the body from harmful elements, an Eco-City carefully manages resources and minimizes pollution, preserving natural attractions and ensuring the urban environment remains pristine. Without environmental sustainability, tourism will decline, as tourists are unlikely to visit destinations that are polluted or degraded.

A Resilient City is also vital for maintaining tourism during crises. Whether due to economic downturns, natural disasters, or other disruptions, resilience ensures the city can recover swiftly and continue offering experiences that attract tourists. Like the immune system's recovery after illness, a Resilient City helps the tourism sector rebound after setbacks, keeping it a reliable contributor to the local economy.

In summary, the success of a Tourism City depends on the interconnection and health of all other pillars. If any pillar falters—whether it's public health, environmental sustainability, or resilience—the tourism industry suffers, destabilizing the economy. Much like a body weakened by a compromised immune system, a Tourism City is only as strong as the health, sustainability, and resilience of its entire urban ecosystem.

6.3.6 Yearning City: The Immune System's Long-Term Memory and Evolutionary Potential

The Yearning City is a living, evolving organism, driven by the collective ambition to nurture happiness, well-being, and a deep sense of belonging for all its residents. Like the immune system's long-term memory, the city draws on past experiences—successes and failures alike—to learn and adapt. This process ensures that its people thrive in an environment that evolves alongside their needs. Just as the immune system remembers past threats to better protect the body, the Yearning City channels its history into creating a more inclusive, resilient, and fulfilling future.

At its heart, the Yearning City fosters collective growth by strengthening bonds between individuals, communities, and institutions. Its long-term memory, like that of the immune system, provides a wellspring of wisdom that supports continuous improvement. This foundation enables the city to build a resilient culture of well-being, where the happiness of each person contributes to the flourishing of the whole.

The city's evolutionary potential mirrors the immune system's capacity to adapt to new challenges and changing environments. As the immune system adjusts to novel threats, the Yearning City transforms to meet its people's shifting needs, integrating sustainability, equity, and purpose into its growth. Each iteration of change deepens the city's alignment with its inhabitants' aspirations, ensuring that progress benefits everyone and no one is left behind.

In this way, the Yearning City becomes more than a physical space—it embodies the immune system's ability to promote long-term health, resilience, and harmony. Its focus extends beyond immediate challenges to nurturing the mental, emotional, and social well-being of all its residents. For instance, the Yearning City prioritizes holistic health systems that treat not just illness but also mental and emotional challenges, fostering a more fulfilling quality of life. It invests in technologies that enhance accessibility, convenience, and connection, and it adopts community-centered strategies to strengthen unity and resilience.

When progress in any one area falters, however, the balance of the Yearning City is disrupted. Neglecting environmental sustainability, for example, harms not only physical health but also people's sense of harmony with nature. Likewise, ignoring social inequities can erode the shared sense of fairness and belonging that binds communities together. Such setbacks ripple through the urban system, weakening its overall ability to thrive.

Yet, the Yearning City continuously strives for improvement, refusing to remain stagnant. It designs inclusive public spaces that foster interaction, supports community initiatives that nurture meaningful connections, and leverages technology to enhance quality of life for all. Every step forward contributes to a larger effort to create a more harmonious and fulfilling urban experience.

This relentless pursuit of betterment sparks a ripple effect, where progress in one area fuels growth in others. By prioritizing well-being, belonging, and harmony, the Yearning City ensures that all aspects of life work together seamlessly. It stands as a testament to the power of evolution and collective ambition, becoming a beacon of happiness, sustainability, and shared prosperity—a place where people and communities truly flourish.

6.3.7 Interconnected Pillars of the HEARTY City

The interconnectedness of the HEARTY City's pillars mirrors the intricate, adaptive network of the immune system. Each pillar is essential in maintaining the city's health and resilience, much like each component of the immune system protects and regenerates the body. These pillars represent different aspects of urban life, and, just like the immune system's responses, their interactions can lead to both co-benefits and dilemmas—where benefits in one area may create challenges in another, much like how immune responses can sometimes lead to overreaction or unintended consequences. Just as the immune system relies on the balanced interaction of its components to ensure survival, the HEARTY City thrives through the harmony between its six pillars, adapting to an ever-changing world. The following presents examples of the interconnected pillars, analogous to the human immune system.

6.3.7.1 Healthy City (Defense) and Eco-City (Adaptation): Innate Immunity (Macrophages) and Adaptive Immunity (T Cells)

The Healthy City (Defense) and Eco-City (Adaptation) can be compared to the innate and adaptive immune systems, respectively. Just as macrophages, the body's first responders, patrol and neutralize harmful pathogens before they cause damage, the Healthy City provides immediate health protections to ensure safety and well-being. The Eco-City, in turn, benefits from this protection, evolving like T cells in the adaptive immune system, which learn from past infections and adapt their strategies to prevent future harm. A strong defense system in the Healthy City enables the Eco-City to thrive and adapt sustainably, similar to how a robust innate immune response supports the adaptability of the adaptive immune system. However, there is a dilemma: overreliance on innate defense mechanisms—such as macrophages constantly cleaning up debris—could hinder the Eco-City's ability to adapt. This mirrors how innate immunity can suppress the adaptive immune system's ability to respond to novel threats. The risk is that too much focus on immediate defense could undermine the long-term ecological evolution of the city.

6.3.7.2 Healthy City (Defense) and AI-driven City (Regeneration): Innate Immunity (Neutrophils) and Immunological Memory (Memory B Cells)

The Healthy City (Defense) and AI-driven City (Regeneration) can be likened to innate immunity (neutrophils) and immunological memory (memory B cells). The Healthy City, like neutrophils in the immune system, provides rapid, essential responses to immediate health threats. Meanwhile, the AI-driven City mirrors memory B cells, which store data from past infections to create more effective responses in the future. Just as memory B cells enhance the immune system by recognizing and responding to pathogens more efficiently over time, AI can learn from urban data to predict threats and optimize recovery. However, the dilemma arises when an over-focus on the quick responses of neutrophils limits the ability of memory B cells (or AI-driven insights) to adapt and innovate. The risk is that AI could become too rigid, relying on past data rather than adapting to novel, unforeseen challenges in the city's evolution.

6.3.7.3 Healthy City (Defense) and Resilient City (Recovery): Innate Immunity (Dendritic Cells) and Tissue Repair (Fibroblasts)

The Healthy City (Defense) and Resilient City (Recovery) can be compared to innate immunity (dendritic cells) and tissue repair (fibroblasts). Dendritic cells act as messengers in the immune system, alerting the adaptive response after detecting a threat, similar to how the Healthy City identifies and mitigates health risks. The Resilient City functions like fibroblasts, which facilitate tissue repair and recovery after injury. Just as dendritic cells strengthen the immune response, a resilient city with strong foundational defenses can quickly repair and restore urban functions after crises. However, a focus on initial defense—such as dendritic cells engaging with a pathogen early on—may sometimes overshadow the deeper, longer-term tissue repair needed by the Resilient City, much like how inflammation can hinder tissue regeneration. Overemphasis on defense could impede the city's ability to fully recover and rebuild from systemic issues.

6.3.7.4 Healthy City (Defense) and Tourism City (Recognition): Innate Immunity (Natural Killer Cells) and Antigen Recognition (Dendritic Cells)

The Healthy City (Defense) and Tourism City (Recognition) can be compared to innate immunity (natural killer cells) and antigen recognition (dendritic cells). Natural killer (NK) cells provide swift, broad defense by performing surveillance to eliminate compromised cells and potential threats, much like how the Healthy City offers immediate protection. The Tourism City, on the other hand, functions like dendritic cells, which recognize foreign antigens—such as tourists and cultural inputs—and initiate a response that integrates these elements into the system, fostering cultural and economic growth. However, the dilemma arises when excessive tourism, like overactive NK cells, overstimulates the city's resources, potentially damaging infrastructure and local communities. An uncontrolled influx of tourists could overwhelm the city's health and safety systems, just as overactive NK cells can damage healthy tissues.

6.3.7.5 Healthy City (Defense) and Yearning City (Memory and Evolution): Innate Immunity (Macrophages) and Immunological Memory (Memory T Cells)

The Healthy City (Defense) and Yearning City (Memory and Evolution) can be likened to innate immunity (macrophages) and immunological memory (memory T cells). Macrophages play a crucial role in maintaining an immediate response to pathogens and removing cellular debris, much like how the Healthy City ensures basic health protections. The Yearning City mirrors memory T cells, which store information about pathogens, enabling the immune system to respond more effectively in the future. In this way, the Healthy City creates a foundation for the Yearning City to build upon, allowing it to learn from past experiences and evolve. However, the dilemma arises when the Healthy City's innate defenses, like macrophages, prevent the city from evolving culturally, socially, or environmentally. Focusing too much on maintaining baseline protection may hinder the development and growth of new ideas, much like how innate immunity can suppress the evolution of the adaptive immune system.

6.3.7.6 Eco-City (Adaptation) and AI-driven City (Regeneration): Homeostasis (Regulatory T Cells) and Surveillance (Cytokines)

The Eco-City (Adaptation) and AI-driven City (Regeneration) can be compared to homeostasis (regulatory T cells) and surveillance (cytokines). Regulatory T cells maintain immune balance and prevent overreaction, much like the Eco-City, which seeks ecological balance by adapting to external environmental changes. The AI-driven City functions similarly to cytokines, which act as signaling molecules to guide the immune system's response. Just as cytokines coordinate immune activity by sending signals to cells, AI helps the city regenerate by responding to urban needs through data and predictive models. However, the dilemma arises when an overemphasis on stability, as seen with regulatory T cells, may hinder rapid innovation. Similarly, dysregulated cytokines can cause excessive signaling or inflammation, just as the AI-driven City might become overly reliant on established patterns, slowing its adaptation to new and emerging issues.

6.3.7.7 Eco-City (Adaptation) and Resilient City (Recovery): Homeostasis (Mast Cells) and Tissue Repair (Fibroblasts)

The Eco-City (Adaptation) and Resilient City (Recovery) can be compared to homeostasis (mast cells) and tissue repair (fibroblasts). Mast cells help maintain balance by releasing signals that regulate the immune response, much like how the Eco-City adapts to environmental shifts to sustain ecological balance. The Resilient City mirrors fibroblasts, which play a crucial role in repairing and regenerating tissue after injury. A well-functioning Eco-City enables the Resilient City to recover more efficiently from crises, fostering a stable environment for both immediate recovery and long-term adaptation. However, the dilemma arises when an overemphasis on ecological balance, akin to mast cells promoting stability, may slow the Resilient City's ability to recover quickly after crises—similar to how fibroblasts can be slow to repair after persistent low-grade inflammation. Furthermore, ecological adaptation may conflict with the immediate urgency of recovery.

6.3.7.8 Eco-City (Adaptation) and Tourism City (Recognition): Homeostasis (Macrophages) and Antigen Recognition (Dendritic Cells)

The Eco-City (Adaptation) and Tourism City (Recognition) can be compared to homeostasis (macrophages) and antigen recognition (dendritic cells). Macrophages maintain balance in the immune system by removing harmful agents, much like the Eco-City, which adapts to ecological pressures to sustain its environment. The Tourism City functions like dendritic cells, which recognize and process external information—such as new tourists or cultural inputs—allowing the city to integrate and benefit from these influences. However, the dilemma arises when an overemphasis on maintaining ecological balance, like macrophages focusing on removing harmful agents, may cause the Eco-City to resist necessary changes brought by tourism. Similarly, over-tourism, like excessive antigen recognition, could strain resources and disrupt the city's ecological balance.

6.3.7.9 Eco-City (Adaptation) and Yearning City (Memory and Evolution): Homeostasis (Regulatory T Cells) and Immunological Memory (Memory B Cells)

The Eco-City (Adaptation) and Yearning City (Memory and Evolution) can be compared to homeostasis (regulatory T cells) and immunological memory (memory B cells). Just as regulatory T cells help the immune system maintain balance, the Eco-City adapts to its environment to ensure long-term sustainability. The Yearning City functions like memory B cells, which retain past knowledge to improve responses to future threats. A stable Eco-City creates a foundation that allows the Yearning City to evolve by learning from its history, making it more resilient to future changes. However, the dilemma arises when too much focus on ecological balance might suppress the cultural or social changes necessary for growth—similar to how regulatory T cells can inhibit adaptive immunity by preventing vital immune responses. If ecological priorities dominate, the Yearning City may struggle to evolve culturally and socially.

6.3.7.10 AI-driven City (Regeneration) and Resilient City (Recovery): Surveillance (Cytokines) and Tissue Repair (Fibroblasts)

The AI-driven City (Regeneration) and Resilient City (Recovery) can be compared to surveillance (cytokines) and tissue repair (fibroblasts). Cytokines play a crucial role in activating the immune system and directing cells to areas in need of recovery, much like how the AI-driven City uses data to guide regeneration and recovery strategies. The Resilient City functions like fibroblasts, facilitating the healing process by rebuilding infrastructure and systems after a crisis. However, the dilemma arises when cytokine signaling becomes overly active or misdirected, potentially causing tissue damage or delaying recovery. Similarly, AI-driven decision-making could become too focused on optimization, neglecting to address emerging or unforeseen needs in the recovery process.

6.3.7.11 AI-driven City (Regeneration) and Tourism City (Recognition): Surveillance (Cytokines) and Antigen Recognition (Dendritic Cells)

The AI-driven City (Regeneration) and Tourism City (Recognition) can be compared to surveillance (cytokines) and antigen recognition (dendritic cells). Cytokines enable immune cells to target and address infections, much like how AI-driven systems analyze urban data to detect and respond to external threats and opportunities. The Tourism City functions like dendritic cells, identifying and integrating new and beneficial inputs—such as tourists or cultural influences—into the city's growth. However, the dilemma arises when an overreliance on cytokines (AI) could stifle the natural cultural processes that allow the Tourism City to thrive. If AI becomes too focused on optimization, it may hinder the city's ability to embrace spontaneous cultural developments, much like how excessive immune signaling can lead to unwanted inflammation.

6.3.7.12 AI-driven City (Regeneration) and Yearning City (Memory and Evolution): Surveillance (Cytokines) and Memory Cells (Memory T Cells)

The AI-driven City (Regeneration) and Yearning City (Memory and Evolution) can be compared to surveillance (cytokines) and memory cells (memory T cells). Cytokines guide immune responses through data and signaling, much like how AI in the city processes data to direct regeneration efforts. The Yearning City functions like memory T cells, drawing on past experiences to guide future development, learning, and evolving as a result. However, the dilemma arises when the AI-driven City becomes too reliant on data-driven predictions, potentially neglecting the organic, spontaneous learning and cultural evolution of the Yearning City. This mirrors the issue in the immune system of focusing too much on past memory (like memory T cells), making it less flexible in adapting to new, unforeseen challenges.

6.3.7.13 Resilient City (Recovery) and Tourism City (Recognition): Tissue Repair (Fibroblasts) and Antigen Recognition (Dendritic Cells)

The Resilient City (Recovery) and Tourism City (Recognition) can be compared to tissue repair (fibroblasts) and antigen recognition (dendritic cells). Fibroblasts aid in tissue repair and regeneration after injury, much like the Resilient City rebuilds after disruptions. The Tourism City functions like dendritic cells, identifying new external threats and opportunities—such as tourism—and ensuring the city grows culturally and economically. However, the dilemma arises when tourism overwhelms the city, much like over-exposure to antigens, disrupting the Resilient City's recovery process and leading to overstretched resources and infrastructure damage.

6.3.7.14 Resilient City (Recovery) and Yearning City (Memory and Evolution): Tissue Repair (Fibroblasts) and Immunological Memory (Memory B Cells)

The Resilient City (Recovery) and Yearning City (Memory and Evolution) can be compared to tissue repair (fibroblasts) and immunological memory (memory B cells). The Resilient City, like fibroblasts, helps recover and regenerate after a crisis, while the Yearning City, like memory B cells, draws on its immunological memory to improve future responses to similar challenges. This ability to adapt and learn ensures long-term resilience and growth. However, the dilemma arises when focusing too much on short-term recovery, as it may undermine the city's long-term evolution—similar to how rapid tissue repair can result in scarring, limiting the body's adaptability over time. Likewise, an emphasis on recovery might overlook the need for cultural or systemic evolution.

6.3.7.15 Tourism City (Recognition) and Yearning City (Memory and Evolution): Antigen Recognition (Dendritic Cells) and Immunological Memory (Memory T Cells)

The Tourism City (Recognition) and Yearning City (Memory and Evolution) can be compared to antigen recognition (dendritic cells) and immunological memory (memory T cells). Dendritic cells play a pivotal role in recognizing external threats and integrating them into the body's immune response, much like how the Tourism City recognizes external cultural inputs and shapes them into the city's identity. The Yearning City functions like memory T cells, evolving over time and adapting its responses based on past experiences and inputs. However, the dilemma arises when tourism overwhelms the city, much like excessive antigen exposure can hinder the Yearning City from evolving organically. Excessive external influence can dilute the city's cultural identity, just as the immune system can become overwhelmed by too much foreign antigen exposure.

6.4 The HEARTY City: A Vision of Interdependence for Future Urban Growth

In its essence, the HEARTY City concept, when seen through the lens of the human immune system, presents a vision of urban life that rises above the conventional. It beckons us to shift our understanding of cities—not as mechanical, isolated entities bound by rigid systems, but as living, interconnected organisms, constantly evolving and adapting to the world around them. This analogy invites a radical shift in perspective, asking us to view urban systems as complex, dynamic, and inextricably linked, much like the human body. Within this intricate system, the health of each pillar—be it the Healthy City, the Eco-City, the AI-driven City, the Resilient City, the Tourism City, or the Yearning City—is inherently tied to the vitality of the whole. The flourishing of the urban body hinges not just on individual growth, but on a harmonious interplay between its various parts.

This interconnected vision of the HEARTY City challenges us to rethink the very fabric of urban development. It offers a profound ethical and philosophical challenge to our traditional notions of urbanism, urging us to move beyond the pursuit of wealth and technology for their own sake, and instead focus on the holistic well-being of the city as a living entity. It calls for a balance—a balance between the environment, the inhabitants, the infrastructure, and the resilience needed to withstand external pressures. Just as the immune system should remain vigilant, adaptive, and capable of responding to a host of ever-evolving threats, so too should the city. It should be agile, resilient, and sustainable—constantly recalibrating itself to meet the ever-shifting tides of both internal and external forces.

Yet, the path to realizing this vision is fraught with complexity. As the immune system is not without its vulnerabilities, so too is the concept of the HEARTY City.

The very interconnectedness that makes this vision so compelling also renders its practical implementation daunting. The symbiotic relationships between public health, environmental sustainability, technological infrastructure, and social resilience are fraught with tension. How do we strike a balance between a thriving economy and the need for environmental stewardship? How can technological advancements, especially in AI-driven cities, enhance social equality rather than widen existing gaps or degrade the environment?

Philosophically, the HEARTY City forces us to confront the true meaning of progress. It prompts us to reflect on what it means for a city to be truly "healthy." Health, in this context, transcends the mere absence of disease or dysfunction. It speaks of a flourishing—of balance and harmony—where each part of the city develops in concert with the whole. Health is a dynamic, ever-changing state, demanding constant care and attention. Much like the immune system, which should adapt in response to an evolving landscape of threats, so too should the HEARTY City continuously adapt to the shifting challenges of our world. In this way, the HEARTY City is not a static model to be built, but a process—a perpetual becoming—where the success of each pillar is inextricably bound to the health of the whole.

The challenges before us are indeed vast. Realizing the HEARTY City vision requires more than a set of technical solutions or policy reforms. It demands a fundamental shift in our collective consciousness—a shift toward understanding urban life as a shared, communal experience. It calls us to recognize that our well-being is not an isolated pursuit, but something that thrives in connection with the well-being of others. The construction of such cities requires us to look beyond the present moment and consider the long-term consequences of our decisions—for both ourselves and the generations to come. It is a call to balance technological innovation with ecological preservation, economic vitality with social equity, and individual freedom with collective responsibility.

In this light, the HEARTY City challenges us to transcend the traditional view of urban spaces as mere collections of isolated domains—health, technology, environment, economy—and instead see them as a living organism, where the flourishing of each part is dependent upon the flourishing of the whole. Much like the human body, which thrives only when all its systems work together in harmony, the HEARTY City can only flourish when its various elements—health, sustainability, technology, resilience, and growth—are aligned in a delicate, ever-evolving balance.

In conclusion, the HEARTY City offers not only a compelling vision for the future of urban development but also a deep philosophical challenge. How can we build cities that are simultaneously sustainable, adaptable, resilient, and thriving in the face of the profound environmental, technological, and social challenges of our time? The answer lies not in a set of predefined solutions, but in a deeper understanding of what it means for a city to be "whole." As we move forward, we should embrace the complexity, interdependence, and adaptability that the HEARTY City demands, crafting urban environments that are not only sustainable but also living, breathing systems capable of responding to the ever-changing needs of both their residents and the world around them.

References

Abbas, A. K., Lichtman, A. H., & Pillai, S. (2023). *Basic immunology: Functions and disorders of the immune system* (7th ed.). Elsevier.
Ahad, M. A., Paiva, S., Tripathi, G., & Feroz, N. (2020). Enabling technologies and sustainable smart cities. *Sustainable Cities and Society, 61*, 102301.
Dai, J., & Azhar, A. (2024). Collaborative governance in disaster management and sustainable development. *Public Administration and Development, 44*(4), 358–380.
Delves, P. J., Martin, J. E., Burton, D. R., & Roitt, I. M. (2017). *Janeway's immunobiology* (9th ed.). Garland Science.
HakemZadeh, F., & Rousseau, D. M. (2024). Evidence-based decision-making is a social endeavor. *Behavioral Science & Policy, 10*(1), 27–33. https://doi.org/10.1177/23794607241265206
Janeway, C. A., Travers, P., Walport, M., & Shlomchik, M. J. (2001). *Immunobiology: The immune system in health and disease* (5th ed.). Garland Science.
Murphy, K. M., Weaver, C., & Berg, L. J. (2022). *Janeway's Immunobiology* (10th ed.). W. W. Norton & Company.
UN. (2020). World social report 2020: Inequality in a rapidly changing world. https://www.un.org/development/desa/dspd/wp-content/uploads/sites/22/2020/02/World-Social-Report2020-FullReport.pdf
Vogel, R., Göbel, M., Grewe-Salfeld, M., Herbert, B., Matsuo, Y., & Weber, C. (2022). Cross-sector partnerships: Mapping the field and advancing an institutional approach. *International Journal of Management Reviews, 24*(3), 394–414.
Zhang, J. (2025). HEARTY City. In J. Zhang, L. Zhang, & Y. Jiang (Eds.), *COVID-19 & pandemics, lifestyles, human mobilities, and cities: A perspective of planetary health*. Springer.

Chapter 7
The DIRECT Approach: Navigating Policymaking Process Management

7.1 The Art of Shaping Cities and Society: A Process-Focused Journey

At the heart of every thriving city and its society, there is a pulse—a rhythm of decisions, actions, and visions that guide its direction. Policymaking about cities and society, at its core, is the process of shaping the future for people. It's a massive responsibility, driven by ambition, necessity, and accountability. In cities, where challenges are as diverse as the people living in them, policymaking is not just a technical task but a deeply moral duty. Every policy created and every decision made impacts the very fabric of life, influencing how people live, work, and interact with each other and their environment (OECD, 2017, 2019).

To imagine cities and society that are more than just places to survive—cities and society that are resilient, sustainable, and inclusive—requires balancing the present with the future. Policymakers should carefully consider the needs of today while planning for the aspirations of tomorrow, guiding cities and society to be adaptable to change and resilient in the face of modern challenges. This is a gradual process, unfolding in steps, each one bringing us closer to a vision of urban and social life that is not just livable, but thriving in all its complexity.

The author's journey into this emerging discipline began as a response to the unprecedented challenges of the COVID-19 pandemic. In early 2020, invited to co-chair the WCTRS COVID-19 Task Force, he was prompted to reflect deeply on how to address the crisis's complex, cascading effects. He initially developed the PASS framework to mitigate the pandemic's impact (Zhang, 2021). Yet, as the virus spread, the author realized the critical importance of managing the policymaking process itself. This insight led to the creation of the DIRECT approach, which he explores in this chapter as a conceptual embodiment of the human immune system's "gene." From these reflections, the author (Zhang, 2025) proposed the novel discipline of "*Urbanimmunology*."

J. Zhang, *Urbanimmunology*, Urban Sustainability,
https://doi.org/10.1007/978-981-95-4878-1_7

7.1.1 Setting the Stage: Agenda-Setting and Identifying Urgency

Every journey begins with recognizing a need, a challenge that requires attention. In policymaking about cities and society, this first step is agenda-setting—the process of identifying the critical issues a city or society should address. This goes beyond technical know-how; it demands a deep understanding of the social, economic, and political factors that shape these problems. Whether it's affordable housing, overburdened infrastructure, the push for environmental sustainability, or the fight for social equity, the agenda should focus on what is not only urgent but also essential for the well-being of people. A clear agenda guides the way forward, ensuring that no challenge, no matter how difficult, is left unaddressed. Without this clarity, cities and society risk neglecting vital issues, which can lead to division, despair, and decay (Gudowsky et al., 2017).

7.1.2 Crafting Solutions: Policy Formulation

Once the challenges are clear, the next step in policymaking begins: the art of policy formulation. It is here that the abstract becomes tangible, that the city's needs are translated into concrete, actionable policies. Yet, this is not the work of a single mind, but a collaborative effort—a chorus of voices from all corners of society. Government officials, civil society leaders, experts, and private stakeholders all contribute their knowledge and experiences to shape policies that reflect the city's unique character. Policies should be crafted with care, grounding them in the lived realities of the people, ensuring that they are not merely theoretical constructs but solutions that resonate deeply with those they aim to serve. It is through this thoughtful, inclusive process that cities and society can craft policies that are not just effective but enduring, built to withstand the tests of time (Sørensen et al., 2014; Lathia & Dadhaniya, 2017).

7.1.3 Gaining Momentum: Political Will and Public Support

At the core of effective policymaking is decision-making—a moment when vision meets reality. It's here that the competing interests of economic growth, social inclusion, and environmental sustainability should be carefully balanced. This stage requires strong leadership—leaders who can navigate complex trade-offs, build consensus among diverse groups, and secure the political will to make policies a reality. But it's not enough for leaders to simply make decisions; they should also gain public support, ensuring that the policies resonate with the people, who are the heart of the city or society. Without this support, even the best-designed policies can fail to reach their full potential (Wittwer et al., 2023; Zhang, 2018; Tønnesen et al., 2022).

7.1.4 *Bringing Vision to Life: Policy Implementation*

Once decisions are made, the vision should be brought to life through action. The implementation phase is where ideas and policies are turned into actual projects, services, and infrastructure that shape the urban and social environment. This is where the abstract becomes real, and the true value of a policy is tested. It's not enough to have a plan; the plan should be carried out with care and precision. Successful implementation requires clear structures, enough resources, skilled people, and smooth coordination across different levels of government. This is the stage where the results of policymaking are either realized or fade away if neglected. For success, this phase depends on strong institutional frameworks, proper resources, qualified personnel, and solid cooperation at all levels of government (UNDP, 2017).

7.1.5 *Reflection and Growth: Evaluation and Monitoring*

Once policies are in action, the final step in the policymaking process begins: evaluation and monitoring. This stage isn't just about checking if policies have achieved their goals; it's about ensuring that cities and society stay adaptable to the changing challenges around them. As new issues arise, policies need to evolve, learning from both successes and failures. Monitoring systems act as the eyes of the city or society, tracking progress, measuring outcomes, and highlighting areas that need improvement. Regular evaluations keep the momentum of change going, ensuring accountability and helping policies stay relevant and effective as the urban and social landscape evolves. It allows cities and society to learn from their experiences, avoid repeating past mistakes, and make the necessary changes to keep up with new circumstances (UN-Habitat, 2020; Angelidou & Politis, 2024).

7.1.6 *Engaging the Public: Participation and Feedback*

At every stage of the policymaking process, it's essential that the voices of the people are heard. Public participation is the heart of any policy that genuinely aims to serve the people. Creating policies from above isn't enough; they need to reflect the needs, wants, and hopes of the people, especially those who are most vulnerable or marginalized. Through consultations, participatory planning, and inclusive governance, policymakers can make sure their decisions align with the needs of the people they aim to serve. This helps reduce resistance, promote unity, and ensures no one is left out. In this way, policymaking becomes a shared effort—a collective mission to build a city or society that mirrors the dreams and realities of all its people (Zhang, 2018).

7.1.7 Navigating Change: Adaptability and Resilience

In a world where cities and society are always changing—shaped by new technologies, environmental shifts, and economic changes—policies should be built with flexibility at their core. As the world evolves, cities and society should evolve too. Policymakers need to be ready to adjust, revising and updating policies as new challenges arise. This adaptability is essential in facing issues like climate change, economic uncertainty, and global health crises. By embracing change with resilience, cities and society can withstand challenges and make sure their policies stay strong, relevant, and effective, no matter how the world shifts (Liu et al., 2023; Zhang & Hayashi, 2022).

7.1.8 Looking Ahead: A Vision for the Future

Policymaking is the foundation on which the cities and society of tomorrow will be built. By approaching this process with foresight, inclusivity, and adaptability, cities and society can address the challenges of modern life—like housing shortages, climate change, transportation issues, and social inequality—while creating spaces that are not just livable but resilient. In this constantly changing world, policymaking is a journey that requires care, precision, and a strong commitment to the future. It is only through this process that we can create cities and society that thrive, grow, and endure—places that meet the needs of both today's and tomorrow's generations.

7.2 Using Human Immune Intelligence to Derive the DIRECT Approach

7.2.1 DIRECT in Its Original Context

The DIRECT approach was first introduced by the author to address the challenges of policymaking process management arising from the COVID-19 pandemic and broader risk management concerns (Zhang, 2021, 2022), while he served as the co-chair of the WCTRS COVID-19 Task Force.[1] The six components of "DIRECT" are as follows: (1) D: Detect, (2) I: Inform/Intervene, (3) R: React, (4) E: Enlighten/Enforce/Evaluate, (5) C: Collaborate, and (6) T: Transfer.

7.2.1.1 D: Detect—Detecting Infected Individuals and Risky Behaviors

Early detection of infected individuals and risky behaviors, such as crowding or traveling to endemic areas, is critical for controlling the spread of COVID-19. Identifying infected individuals promptly helps isolate them and reduce further transmission, while recognizing risky behaviors allows authorities to take preventive actions like issuing travel advisories or recommending social distancing in high-risk areas. Monitoring these behaviors and patterns enables health systems to respond more efficiently, allocate resources effectively, and provide timely guidance to the public, ultimately helping to manage the spread of the virus.

7.2.1.2 I: Inform/Intervene—Informing Detections and Intervening, if Necessary

Informed decision-making is crucial for guiding individuals and stakeholders in adjusting risky or unsustainable behaviors. Effective communication between policymakers and the public is key to conveying the risks identified through detection and ensuring that necessary interventions are understood and acted upon. By providing clear, accurate information about detected risks, authorities can help people make informed choices that reduce harm. Additionally, when necessary, timely interventions can be implemented to prevent further transmission or negative outcomes, fostering sustainable behaviors that contribute to both immediate pandemic control and long-term health and safety.

7.2.1.3 R: React—Reacting to the Above Two Components

The effectiveness of interventions is deeply influenced by how individuals and communities react to the detection of risks and the information shared by authorities. The success of public health measures depends on whether people recognize the importance of these risks and act accordingly. When individuals follow the guidance provided—such as practicing preventive behaviors, adhering to regulations, and adjusting personal habits—they help to curb the spread of the virus. Moreover, widespread, timely, and proactive responses, such as adhering to safety measures and participating in vaccination efforts, strengthen the effectiveness of interventions.

---

[1] https://wctrs-society.com/about/wctrs-covid-19-task-force/ (Accessed on February 1, 2025).

7.2.1.4 E: Enlighten/Enforce/Evaluate—Enlightening Voluntary Behavioral Change, Enforcing Compliance, and Evaluating All DIRECT Components

Promoting lasting behavioral changes often starts with encouraging voluntary adjustments through education and awareness, empowering individuals to make informed decisions. However, if voluntary changes prove insufficient, enforcement through legal, regulatory, or economic measures may be necessary to ensure compliance and achieve desired outcomes. Sustainable behavior change is best supported by fostering enlightened choices that align with long-term health goals. Equally important is the continuous evaluation of all interventions, grounded in scientific evidence, to assess their effectiveness, efficiency, and cost-efficiency. This comprehensive approach ensures that policies not only drive behavioral change but also contribute to optimal and sustainable public health outcomes.

7.2.1.5 C: Collaborate—Collaborating Among All Stakeholders

Collaboration among all stakeholders—governments, healthcare providers, businesses, communities, and individuals—is vital for effectively managing the COVID-19 pandemic and ensuring long-term resilience. By sharing resources, knowledge, and strategies, this collective effort improves response measures such as testing, vaccination, and support for vulnerable populations, while also fostering public trust. Additionally, such cooperation strengthens future preparedness and supports sustainable development, addressing health disparities and promoting a more equitable, resilient, and sustainable post-pandemic world.

7.2.1.6 T: Transfer—To Transfer Both Successful and Unsuccessful Experiences

Transferring both successful and unsuccessful experiences from the pandemic is critical for shaping sustainable post-pandemic development. By analyzing what worked and what didn't, we can apply valuable lessons to improve current practices, inform policy, and drive scientific innovation. Drawing from historical lessons and previous pandemics enhances our ability to address emerging challenges more effectively, ensuring that we build on past knowledge to prevent future crises. This process of knowledge transfer not only strengthens public health responses but also supports broader goals, such as fostering resilience, reducing inequalities, and advancing long-term sustainable development.

7.2.2 The Analogy of Human Immune Intelligence

The six components of "DIRECT" and their sequence bear a striking resemblance to the mechanisms of the human immune system, a system that, through complex processes of detection, defense, and adaptation, protects the body from external threats and internal imbalances (Janeway et al., 2001; Delves et al., 2017; Murphy et al., 2022; Abbas et al., 2023). In this analogy, the urban and social ecosystem, like the body, requires a system of vigilant awareness, coordinated action, and continual refinement to maintain health, resilience, and sustainability.

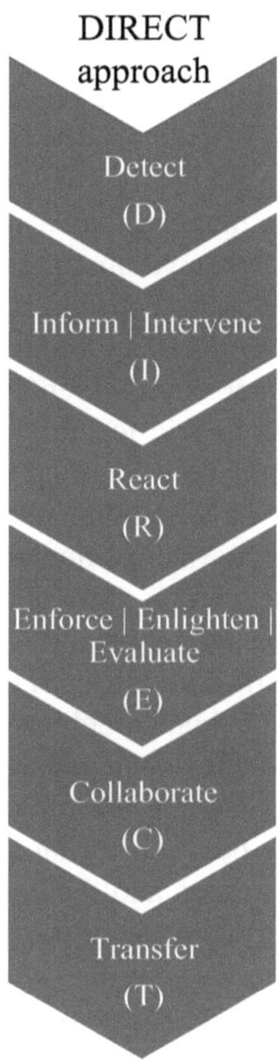

7.2.2.1 D: Detect—Recognizing Challenges (Recognition Phase of the Immune System)

In the natural order of things, the immune system's first line of defense against harm is the act of perception. It is through an exquisite array of specialized receptors—pattern recognition receptors (PRRs) on cells such as macrophages and dendritic cells—that the body discerns the presence of danger. These receptors, akin to vigilant sentinels, are attuned to pathogen-associated molecular patterns (PAMPs) and damage-associated molecular patterns (DAMPs), marking the beginning of a process of recognition. This phase is the immune system's discernment, where it distinguishes between what is benign and what is a threat. Furthermore, the involvement of the major histocompatibility complex (MHC) acts as a bridge, presenting foreign antigens to T cells, thus initiating a more sophisticated and adaptive immune response.

In the domain of policymaking about cities and society, a similar process unfolds. The *Detect* phase is akin to the immune system's ability to recognize what is out of balance, to identify risks, and to observe unsustainable practices within the urban and social environment. Much like the body detects pathogens, cities and society should detect inefficiencies—whether it is an unresponsive transportation system, the creeping onset of pollution, or environmental degradation that threatens the health of the whole. In this age of rapid technological advancement, smart urban and social systems—equipped with sensors, data analytics, and surveillance technologies—serve as the modern-day immune receptors, designed to detect and understand the urban and social ecosystem's fluctuating dynamics. These technologies—always vigilant, always scanning—track patterns, gather data, and provide the crucial information that enables timely and informed interventions.

Yet, just as the immune system should discern threats with precision, urban and social detection systems should navigate a delicate balance between vigilance and respect for privacy. The same ethical and legal considerations that govern the immune system's integrity should guide the urban and social system's ability to detect. In this intricate interplay between technology and human rights, the goal remains the same: to detect the unseen and the emerging, while ensuring that privacy is safeguarded, and that the pursuit of public good never comes at the expense of personal dignity. As in the body's immune system, it is not enough simply to detect; it is the manner in which this detection is executed that will determine the health of the system as a whole.

7.2.2.2 I: Inform/Intervene—Activating Solutions (Activation Phase of the Immune System)

In the symphony of biological defense, once a pathogen is recognized, the immune system springs into action, a harmonious orchestration of response. The process is initiated when antigen-presenting cells (APCs)—such as macrophages and dendritic cells—encounter and process the pathogen, absorbing its essence and presenting its antigens to the ever-vigilant helper T cells (CD4+). This act of presentation is more than mere communication; it is the spark that ignites the flame of the immune response. With this revelation, the innate immune system, with its steadfast macrophages and natural killer cells, and the adaptive immune system, with its more refined helper T cells and cytotoxic T cells, are activated in concert. In this moment of recognition, the immune system "informs" itself of the nature of the threat and summons an appropriate, tailored defense.

This precise and measured activation of response mirrors a parallel in the realm of urban and social policies. The *Inform/Intervene* phase is a moment of collective awakening—a time to disseminate knowledge, to inform the public and all stakeholders of the risks that loom on the horizon. Here, the role of communication is paramount. Just as the immune system relies on the timely presentation of information to mobilize its defenses, urban policymakers too should relay the crucial information that shapes the course of intervention. The message is not merely a transmission of facts, but a call to action—a guide to modifying behavior and making the necessary adjustments in the face of mounting challenges.

The effectiveness of this phase hinges upon clarity. Much like the immune system's fine-tuned recognition and activation, the communication within urban systems should be clear, compelling, and actionable. When stakeholders—be they policymakers, communities, or individuals—receive well-articulated information about potential risks, they are more likely to engage and participate in sustainable practices. In this way, the success of urban and social policies rests upon an informed populace, one whose actions are rooted in understanding, just as the immune system's response is grounded in the profound awareness of the threat it faces. Here, the coordination of knowledge is not simply an academic exercise, but the very force that propels society toward collective action and meaningful change.

7.2.2.3 R: React—Responding to Issues (Response Phase of the Immune System)

Upon the recognition of a pathogen, the immune system, that intricate guardian of life, sets into motion a carefully orchestrated defense. This is not a singular, isolated event but a collaborative unfolding, beginning with the antigen-presenting cells (APCs)—macrophages and dendritic cells—that first encounter the invader. These cells, by processing the pathogen and presenting its antigens to the ever-vigilant helper T cells (CD4+), set into motion a cascade of communication, a dialogue of defense. In this dialogue, the immune system is informed of the nature of the threat, activating both its innate and adaptive arms. The innate immune response—swift, like the decisive strike of a warrior—comprises macrophages and natural killer (NK) cells, responding with immediate force. Meanwhile, the adaptive immune response, more refined and specific, brings forth the help of helper T cells and cytotoxic T cells, tailoring its defense to the precise nature of the invader. Signaling molecules—cytokines—act as the messengers, ensuring that immune cells are harmonized in their purpose, thereby amplifying and enhancing the response. Through this concerted effort, the immune system intervenes not haphazardly, but with precision and purpose, securing the body against the pathogen's encroachment.

In the realm of urban and social policies, the *React* phase mirrors this immune response, as it is the moment where intervention is put to the test in the crucible of real-world reactions. Here, it is not merely the policies themselves that matter but how individuals, communities, and organizations respond to these interventions. Just as immune cells like macrophages and NK cells spring into action, so too should the urban and social populace react to the policies designed for their welfare. The effectiveness of these interventions hinges upon the complex interplay of social, psychological, and contextual factors: awareness, attitudes, and trust. It is these intangible, yet profoundly influential, elements that determine whether policies take root or wither.

Much like the immune system's dynamic adaptation to ever-evolving threats, urban and social policies should possess a similar flexibility—an ability to shift, adjust, and grow in response to the realities that unfold in the urban environment. A policy that cannot evolve with the changing times is like an immune system that refuses to learn from past battles. The *React* phase, then, becomes a crucial moment for reflection and recalibration. It is the feedback loop that informs the next steps, ensuring that the interventions remain not only relevant but also resilient in the face of new and unforeseen challenges. Just as the immune system refines its defense strategies over time, urban and social policies too should refine their approach, learning from real-time feedback and evolving to meet the dynamic, ever-shifting needs of urban life. Only in this adaptive response can cities and society hope to remain sustainable, flexible, and capable of confronting the complexities of the future.

7.2.2.4 E: Enlighten/Enforce/Evaluate—Learning and Adapting (Memory, Activation, and Adaptive Functions of the Immune System)

The immune system is not merely a reactive force, but a keeper of memory—a guardian that learns from its encounters, adapting and growing stronger with each battle. Memory B cells and memory T cells hold the traces of past infections, enabling the body to mount a swift and precise defense upon the return of an old adversary. In this manner, the immune system's strength lies not only in its present response but also in its accumulated wisdom, learned through experience. Similarly, urban and social policies, to be truly effective, should be informed by the lessons of history, drawing upon the knowledge of past successes and failures to guide present action. This *Enlighten* function in policymaking beckons a wise reflection upon the past, fostering awareness. Only through understanding where we have faltered and where we have flourished can we shape a more informed and sustainable future. The wisdom of experience becomes the catalyst for wiser choices, as stakeholders are encouraged to adopt practices that reflect not only current needs but also the accumulated insights of previous generations. Just as the immune system strengthens with each encounter, urban and social policies can enhance its effectiveness, building on the lessons of the past to craft a more responsive and targeted approach to present challenges.

Yet, there comes a time when enlightenment alone will not suffice—when knowledge should be transformed into action, when intention should be matched with tangible change. This is where the *Enforce* function mirrors the immune system's activation. When faced with a threat that cannot be swayed by mere persuasion, the immune system calls forth its forces: cytotoxic T cells, antibodies, and other agents of defense. In the realm of urban and social policies, when voluntary compliance proves insufficient, the enforcement of laws and economic incentives becomes essential. These mechanisms serve as the body's way of mobilizing its defenses—compelling the necessary changes, ensuring that threats are mitigated and desired behaviors are fostered. Just as the immune system's cytokines amplify the response, urban and social policy interventions intensify when the stakes are raised. When confronted with escalating challenges—such as environmental degradation, social inequities, or public health crises—the force of policy should be amplified, ensuring that interventions become more powerful, urgent, and deeply embedded in the fabric of urban and social life.

But the story does not end with action; it should be tempered by reflection and adaptation. This is where the *Evaluate* function completes the cycle, echoing the adaptive nature of the immune system. Just as the immune response is refined with each successive exposure, urban and social policies should undergo continuous evaluation and adjustment. The immune system, in its wisdom, does not simply react but constantly reassesses and refines its strategies, learning from each encounter. Similarly, urban and social policies should be subjected to rigorous evaluation, using data and scientific evidence to measure their effectiveness. The process of evaluation is not passive; it is a dynamic act of adaptation. As urban and social challenges evolve, so too should the policies that address them, adjusting and refining strategies to meet emerging needs and shifting circumstances. In this way, both the immune system and urban and social policies share a common essence: they are not static forces, but dynamic processes of ongoing refinement and evolution. They adapt, they grow, they learn, ensuring that the defenses of the city or society remain robust, relevant, and capable of meeting the challenges of tomorrow.

7.2.2.5 C: Collaborate—Collective Action (Cooperation of Immune Cells and Organs)

The immune system, in its profound wisdom, teaches us that no single entity can stand alone in the face of adversity. The defense of the body is a collective endeavor, a harmonious orchestration of diverse cells and organs, each playing its part to protect the whole. Central to this collaboration are the helper T cells (CD4+), whose guiding hand steers the entire immune response, while B cells produce the antibodies necessary to neutralize the invading threat. Macrophages and dendritic cells serve as the sentinel, presenting the antigen to ignite the immune response, while cytokines, the signaling molecules, facilitate the communication that ensures the seamless flow of information. The lymphoid organs, such as the lymph nodes and spleen, act as the nexus where this collaboration unfolds, refining the body's response to the pathogen. Together, these various components form a unified, adaptive defense, each aware of its role, working in concert to protect the integrity of the organism. The major histocompatibility complex (MHC), too, plays a crucial role, ensuring that the immune system recognizes what is self and what is foreign, thus maintaining harmony and avoiding self-destruction.

Similarly, in the realm of urban and social policymaking, the act of collaboration is not merely an ideal, but a necessity. The creation of policies that are not only effective but sustainable demands the cooperative effort of a diverse array of stakeholders: local governments, private sectors, community leaders, and the citizens themselves. Each brings its own perspective, its own expertise, and its own sense of responsibility to the table. Just as the immune system thrives on the coordination of its various cells and organs, so too should urban and social policies benefit from a seamless flow of communication and trust between its participants. The integration of diverse viewpoints, the sharing of data, and the coordinated action of all parties is the bedrock upon which resilient and adaptive urban and social policies are built.

Much like the immune system's network of cytokines ensures a unified and coordinated defense, urban and social collaboration fosters a network of shared responsibility and mutual understanding. By pooling resources, knowledge, and experiences, cities and society can address the multifaceted challenges they face with a comprehensive approach that encompasses environmental, social, and economic dimensions. Collaboration ensures that no aspect of the urban ecosystem is left out of consideration, allowing for policies that are not only responsive but also inclusive and sustainable.

The strength of this collaborative approach lies not only in its capacity to respond to immediate challenges but also in its ability to build resilience over time. As the immune system adapts and evolves in response to ever-changing threats, so too should urban and social policies evolve, learning from past experiences and anticipating future needs. In fostering cooperation, cities and society create an environment in which resilience is not just a reaction to crisis, but a proactive, ongoing process—one in which the whole is greater than the sum of its parts. Through collaboration, cities and society become not only more effective in their policy responses but more united in their purpose, ensuring a robust, sustainable, and thriving future for all.

7.2.2.6 T: Transfer—Building Long-Term Adaptation (Acquired Immunity and Adaptation)

The immune system imparts a profound lesson in the nature of adaptation and memory. Through the process of transfer, the body learns, grows, and evolves. When confronted with a pathogen for the first time, the immune system may initially struggle, but in this struggle, it gains invaluable knowledge. The memory cells—B cells and T cells—retain the essence of this encounter, allowing the body to respond with greater speed and precision should the pathogen return. This adaptive immunity becomes a refined tool over time, learning from prior exposures, and increasingly capable of confronting the ever-evolving threats that life presents. The production of antibodies by B cells, tailored to neutralize the specific pathogens encountered, is a testament to the immune system's ability to adapt, ensuring that subsequent responses are not only faster but more robust. The major histocompatibility complex (MHC), too, plays its role in the discernment between self and other, maintaining the delicate balance of the immune system's defenses.

Similarly, in the realm of urban and social policymaking, there exists an analogous process of learning and adaptation, one grounded in the insights gleaned from past experiences. Just as the immune system retains the memory of pathogens to respond more efficiently, cities and society, through the *Transfer* phase, should retain the memory of their past policies—both their successes and their failures. These lessons are not merely relics of the past but active, living knowledge that informs the decisions of the present. By reflecting on previous interventions, both positive and negative, urban and social policymakers are better equipped to refine their strategies, ensuring that their responses are ever more attuned to the needs of a constantly changing urban and social landscape.

This process of learning from experience, of allowing past lessons to shape future actions, is essential for the resilience of urban and social systems. Much like the immune system evolves to counter ever-more sophisticated pathogens, cities and society should evolve to address new and emerging challenges, whether they be environmental, social, or economic. The capacity for cities and society to respond swiftly and effectively to recurring issues—such as housing shortages, transportation inefficiencies, or climate-related crises—mirrors the immune system's ability to swiftly produce antibodies upon re-exposure to a familiar threat. In this way, cities and society, like the immune system, become more proficient over time, continually refining their responses to better address both persistent problems and new challenges.

Thus, the *Transfer* phase in urban and social policymaking is not simply a process of reactive adaptation; it is a proactive commitment to growth and resilience. By drawing upon the wellspring of past experiences, cities and society are not only empowered to navigate the complexities of the present but are also better prepared to face the uncertainties of the future. Just as the immune system's adaptation ensures the continued health of the body, the adaptive capacity of urban and social policies ensures the lasting sustainability and vitality of cities and society.

7.3 Applications of the DIRECT Approach

The DIRECT approach was initially proposed to address the impacts of the COVID-19 pandemic, with its potential for broader application in risk management also being explored. While its use in the context of COVID-19 has been previously discussed, this section illustrates two additional applications. One is its application to decarbonizing the transport sector, summarized based on Zhang (2021, 2022). The other is a completely new application focused on public health promotion. These examples highlight the versatility of the DIRECT approach in tackling various challenges beyond its initial pandemic-related context.

7.3.1 Decarbonizing the Transport Sector: The "6-domain & 6-step" Approach

The decarbonization of the transport sector requires extensive efforts due to its complex interconnections with various socio-economic sectors, as highlighted by the Transport-in-All-Policies approach (Zhang et al., 2025). The process involves numerous stakeholders, making it even more challenging.

First, transport demand is driven by people's participation in activities and socio-economic production, which determine the volume of transport needed. This, in turn, forms the basis of carbon emissions along with other pollutants.

Second, emissions are directly linked to energy consumption, and the sources and efficiency of energy are critical in determining the amount and type of emissions. Therefore, decarbonizing the transport sector requires a transformation of both the energy supply and consumption structure. Additionally, given the significant amount of unused space within transport systems, it is vital to explore how to leverage these spaces to generate energy for the sector. In other words, integrating transport and energy systems is essential. Specifically, utilizing road slopes, open spaces near interchanges, parking areas on expressways, and rooftops of transport terminals to generate solar and wind energy is key. Surplus energy can be used to produce hydrogen. Due to the instability of renewable energy sources, they should be integrated into electricity grids, necessitating energy storage solutions. Therefore, a "source-grid-load-storage" system should be established, tailored to the transport context. Regarding the "load," shifting to multimodal transport is essential, especially since the road sector accounts for the majority of emissions. Electrifying road systems is therefore crucial, but the source of electricity remains critical. Currently, electricity for the transport sector is largely generated from coal, raising concerns about the environmental impact of electric road systems.

Third, decarbonization requires all stakeholders to transform their lifestyles and business practices, encouraging voluntary behavioral changes toward low-carbon development. These changes are influenced by socio-economic factors that motivate individuals and organizations to adopt more sustainable practices.

Fourth, the role of population mobility in transport decarbonization should not be overlooked, as population movements—both in terms of size and movement between cities or regions—are fundamental in determining transport emissions.

To address these challenges within a unified framework, the "6-domain & 6-step" approach was proposed, combining the DIRECT approach with the mechanisms of transport decarbonization. For more details, refer to Zhang (2021, 2022).

7.3.2 Public Health Promotion: Process Management Driven by Health Promotion Identity

The idea behind the "6-domain & 6-step" approach outlined above can be applied to various urban and social issues. Here, public health promotion is targeted. In this context, a new framework—called the health promotion identity—is proposed to unify all promotion domains. First, AI technologies are leveraged to inform the health promotion process, capitalizing on their growing popularity and potential. Second, the health identity is explained through logical reasoning. Finally, a specific promotion domain is explored to demonstrate how the health promotion process can be more effectively managed within the DIRECT framework.

7.3.2.1 Leveraging AI Technologies to Inform the Process Management of Health Promotion

Below, we illustrate how AI technologies can be leveraged to enhance public health through the six steps outlined in the DIRECT approach. This comprehensive framework not only emphasizes the integration of advanced technologies but also highlights the importance of systematic and strategic actions to foster healthier urban and social environments, ultimately improving the quality of life for all residents.

[Step 1] Detect (D) — Promoting the Development of AI for Comprehensive National Health Assessments This policy proposes the establishment of a self-assessment initiative for national health utilizing artificial intelligence (AI). To achieve this, the development of large-scale language models specifically tailored for health—referred to as Health Large Language Models (LLMs)—should be prioritized. Additionally, data regarding local built environments should be integrated and analyzed to create health hazard maps. This comprehensive approach aims not only to identify health issues but also to foster a better understanding of the environmental factors that contribute to them, thereby enhancing community health outcomes.

[Step 2] Inform (I) — Promoting the Development of AI to Enhance Behavioral Change and Policy Formulation Policies should be enacted to ensure that Health LLMs are accessible to the entire society. Through these models, individuals and organizations should be informed about health issues, including potential underlying problems, along with relevant information about the surrounding built environment that can facilitate health improvements. Furthermore, the government should provide health hazard maps to the public. By disseminating this vital information, the initiative aims to encourage behavioral changes among individuals and organizations, while also motivating government agencies to be more proactive in their policymaking efforts.

[Step 3] React (R) — Promoting AI Development to Monitor Health Promotion Behavior Processes There is no guarantee that behavioral change will occur spontaneously. Therefore, AI should be employed to monitor the processes of behavioral change. This monitoring should involve analyzing the relationship between changes in the built environment and behaviors, as well as their effects on health outcomes. By identifying which improvements lead to desirable behavioral changes, the policy should facilitate a more effective response to health challenges.

[Step 4] Enlighten/Enforce (E) — Promoting AI Development to Voluntary Behavioral Changes For individuals and organizations struggling to implement improvements, this policy should employ a dual approach of enlightenment (carrots) and enforcement (sticks). It should also propose specific improvements to the built environment and examine which modifications can effectively encourage behavioral changes. This multifaceted strategy aims to support those in need while fostering an environment conducive to healthier choices.

[Step 5] Collaborate (C) — Promoting AI Development to Support Stakeholder Collaboration AI should be utilized to facilitate consensus among various stakeholders—individuals, organizations, and government entities—aimed at health promotion actions such as improving the built environment and encouraging behavioral changes. This policy emphasizes the importance of sustained collaboration and the establishment of mechanisms that enable effective cooperation among all parties involved in health initiatives.

[Step 6] Transfer (T) — Promoting AI Development to Support the Transfer of Policy Expertise This policy aims to accumulate best practices and knowledge related to improving built environments and promoting behavioral changes, ensuring that they address local challenges. By reflecting these unique issues, policies can be implemented on a national scale. Simultaneously, through international cooperation, this knowledge should be shared globally, enhancing worldwide efforts to improve health outcomes and sustainable development.

7.3.2.2 Health Promotion Identity

To effectively utilize the six steps outlined for health promotion, it is essential to ensure their application is strategically focused on key problem areas. This approach mirrors Zhang's (2022) method of addressing carbon reduction in the transport sector by breaking down carbon emissions into six specific components: (1) the carbon intensity associated with energy consumption in transport, (2) the overall energy consumption of transport, (3) the transport pressures resulting from life and economic activities, (4) the activities driven by the needs of daily life and business, (5) the per capita needs in life and business, and (6) population dynamics. As analogous to the above carbon treatment, the author argues that health status can be expressed in the form of the following identity.

$$H = \frac{H}{UHB} \times \frac{UHB}{FHB} \times \frac{FHB}{SHFHB} \times \frac{SHFHB}{CC} \times CC$$

In this context, 'H' represents health status, 'UHB' denotes unhealthy behaviors, 'FHB' refers to factors that influence healthy behaviors, 'SRFHB' indicates the stakeholders responsible for addressing these influential factors, and 'CC' represents the common contexts that influence the decisions of these stakeholders.

Unhealthy behaviors include poor diet, physical inactivity, smoking, excessive alcohol consumption, substance abuse, inadequate sleep, neglecting mental health, high stress levels, unsafe sexual practices, and skipping medical check-ups, etc.

Factors that impede healthy behaviors can be classified into four categories: medical treatments, gene, behaviors, and environment. Regarding health, taking the United States as an example, while 95% of the health budget is allocated to healthcare, the contribution of healthcare to health and well-being is only 10–20%. In contrast, non-medical determinants account for 60–70% of health outcomes, with behavioral factors contributing 40–50% and social and physical environments contributing 20% (Hayes & Delk, 2018).

Stakeholders responsible for addressing the factors that impede health include individuals and their families, communities, workplaces, schools, businesses, and government bodies. Each stakeholder may take full responsibility for specific impeding factors, while multiple stakeholders may share collective responsibilities.

In terms of common contexts, several significant factors can profoundly influence stakeholders' decision-making processes. Climate change is a critical concern that affects public health and resource availability, prompting stakeholders to consider sustainable practices in their operations. Economic issues also play a vital role, as financial constraints and economic disparities can impact the allocation of resources for health initiatives. Additionally, the aging population presents unique challenges, necessitating adjustments in healthcare services, community support, and workforce planning. Together, these interconnected contexts compel stakeholders to collaborate and devise comprehensive strategies that address both immediate health concerns and broader societal impacts.

7.3.2.3 Health Promotion Identity-Driven Process Management

Aligned with the previously discussed identity of public health promotion, the following D-I-R-E-C-T measures are proposed, specifically focusing on the built environment—a key aspect of the social and physical environment—as a significant factor influencing healthy behaviors (FHB).

Detect — Use AI to Create Health Hazard Maps and Identify Areas with High Health Risks By leveraging artificial intelligence, health hazard maps can be generated to pinpoint areas that pose significant health risks to the community. These maps can visualize various factors, such as pollution levels, access to healthcare facilities, and the prevalence of unhealthy behaviors, thereby providing a comprehensive overview of public health challenges. Identifying these high-risk zones enables targeted interventions that can lead to improved health outcomes.

Inform — Communicate Information to Communities and Governments, Advocating for Necessary Improvements It is essential to disseminate the findings from the health hazard maps to both community members and government officials. This communication can take the form of community meetings, informational campaigns, and public reports that not only raise awareness but also encourage active participation in health initiatives. Advocating for necessary improvements fosters a collective responsibility to address identified health risks and implement effective solutions.

React — Communities and Governments Take Action to Implement Improvement Measures Once the relevant health risks have been identified and communicated, it is imperative for both communities and government entities to take concrete actions to address these issues. This may include upgrading infrastructure, enhancing public services, or enacting policies that promote healthier living environments. Collaborative efforts between these stakeholders can lead to more effective and sustainable health interventions.

Enlighten-Enforce — Implement Awareness and Reinforcement Actions to Encourage Communities and Governments to Adopt Improvement Measures To ensure that communities and governments are motivated to adopt improvement measures, targeted awareness campaigns and reinforcement strategies should be employed. These initiatives can educate stakeholders on the benefits of making changes to the built environment and provide incentives for participation. By emphasizing the positive impact of these actions on community health, stakeholders are more likely to engage in and commit to necessary improvements.

Collaborate — Utilize AI to Create a Framework That Enables Stakeholders to Collaborate on Improving the Built Environment Harnessing the power of AI can facilitate the development of systems that support collaboration among various stakeholders—such as community members, local businesses, and government agencies—focused on enhancing the built environment. These collaborative frameworks can streamline communication, share resources, and promote joint initiatives aimed at creating healthier urban spaces, ultimately leading to more effective health promotion efforts.

Transfer — Preserve This Knowledge and Transfer it to Other Regions to Expand Health Promotion Actions It is crucial to document and share the insights gained from these initiatives, ensuring that successful strategies for health promotion can be replicated in other regions. By establishing networks for knowledge transfer, communities can learn from each other's experiences, thereby broadening the impact of health promotion actions. This collective effort not only improves public health but also fosters a culture of collaboration and innovation in addressing health challenges.

The illustration above highlights factors that affect health behavior. Other health promotion domains can be explored in a similar manner, using the same framework to address their unique challenges. By incorporating all domains of the health promotion identity into the D-I-R-E-C-T measures, we can develop a 5x6 policymaking matrix, similar to the one proposed by Zhang (2022). This approach facilitates the creation of a comprehensive strategy that effectively addresses the complex and multifaceted challenges of public health. The matrix ensures that interventions are not only effective but also context-specific, promoting a more integrated and holistic understanding of health across various sectors. It also underscores the importance of seamless, continuous monitoring of public health by connecting with all relevant sectors, aligning with the Health-in-All-Policies framework (WHO, 2022).

7.4 Mastering the Urban Policymaking Process: A Forward-Looking Approach

In the heart of every great city and its society lies a living, breathing organism—a complex web of interconnected systems, each vulnerable to the external forces that seek to disrupt its balance. It is here, at the intersection of human ingenuity and natural processes, that the DIRECT approach finds its place, offering a profound and far-reaching framework for the evolution of urban and social policymaking. Inspired by the sophisticated, adaptive mechanisms of the human immune system, this approach challenges cities and society to move beyond the mere response to crises. It urges them to shift from the realm of reaction to that of proactive foresight—where early detection, informed intervention, adaptive responses, and continuous collaboration become the cornerstone of urban and social governance.

The promise of the DIRECT approach lies in its invitation to rethink the very nature of urban and social policymaking. It calls not only for the swift action demanded by the crises and society of today but for a deeper, more reflective process of decision-making. This is a call for cities and society to look back upon their past experiences, to learn from their mistakes, and, from that well of wisdom, chart a path forward. It is through this continuous cycle of reflection and adaptation that urban and social environments can build the resilience and vitality needed to navigate the complexities of an ever-changing world.

In the context of health crises and public well-being, the DIRECT framework serves as a bulwark against the unpredictable. It ensures that urban and social systems remain not static, but fluid and ready—capable of facing not only the immediate risks of pandemics but also the long-term challenges that lie on the horizon. Health, in this vision, is not an isolated pursuit but one that is deeply entwined with the broader ecological and social fabric of the city or society. To address health is to address all aspects of urban and social life, for the health of one is bound to the health of all. It is here that the DIRECT approach finds its greatest strength: in its call for collective action, for collaboration among diverse stakeholders, and for the shared responsibility that binds communities together in times of crisis. It is only through these cooperative efforts that cities and society can craft policies that are not only effective in the moment but enduring in their impact.

Yet the influence of the DIRECT approach reaches far beyond the confines of health crises. Its principles hold within them a wider promise, one that can reshape how cities and society respond to the full spectrum of urban and social challenges—be it the slow burn of climate change, the fractured networks of transportation inequities, or the deepening chasm of socio-economic inequality. The approach encourages urban and social policymakers to tap into an adaptive, evidence-based framework that is not simply reactive but anticipatory—ready to evolve alongside the shifting demands of the world around them.

In this way, the DIRECT approach does not just prepare cities and society for the crises of today. It strengthens the very fabric of urban governance itself. It enables cities and society to respond with dexterity, to flex with the winds of change, and to adapt to the unknown futures that lie ahead. By integrating these principles into the heart of policymaking, cities and society can ensure they are not just surviving but thriving—building a future that is resilient, sustainable, and ultimately more humane. For in this ever-changing world, where uncertainty looms large, it is only through such adaptive, forward-thinking frameworks that cities and society can safeguard the well-being of their inhabitants and continue to evolve in the face of all that lies ahead.

References

Abbas, A. K., Lichtman, A. H., & Pillai, S. (2023). *Basic immunology: Functions and disorders of the immune system* (7th ed.). Elsevier.

Angelidou, M., & Politis, C. (2024). An integrated framework to monitor progress of sustainable urban development strategies, addressing societal, ecological and innovation dilemmas. *Societal Impacts, 4*, 100075.

Delves, P. J., Martin, J. E., Burton, D. R., & Roitt, I. M. (2017). *Janeway's immunobiology* (9th ed.). Garland Science.

Gudowsky, N., Sotoudeh, M., Capari, L., & Wilfing, H. (2017). Transdisciplinary forward-looking agenda setting for age-friendly, human centered cities. *Futures, 90*, 16–30.
Hayes, T.O., Delk, R. (2018). Understanding the Social Determinants of Health. https://www.americanactionforum.org/research/understanding-the-social-determinants-of-health/ (Accessed on September 1, 2025)
Janeway, C. A., Travers, P., Walport, M., & Shlomchik, M. J. (2001). *Immunobiology: The immune system in health and disease* (5th ed.). Garland Science.
Lathia, R. V., & Dadhaniya, S. (2017). Policy formation for renewable energy sources. *Journal of Cleaner Production, 144*, 334–336.
Liu, Z., Fang, C., Liao, X., Fan, R., Sun, B., & Mu, X. (2023). Adaptation and adaptability: Deciphering urban resilience from the evolutionary perspective. *Environmental Impact Assessment Review, 103*, 107266.
Murphy, K. M., Weaver, C., & Berg, L. J. (2022). *Janeway's immunobiology* (10th ed.). W. W. Norton & Company.
OECD. (2017). *National urban policy in OECD countries*. OECD Publishing. https://doi.org/10.1787/9789264271906-en
OECD. (2019). 20 years of urban policy at OECD. https://www.oecd.org/content/dam/oecd/en/topics/policy-issues/urban-development/urban-policy-oecd.pdf
Sørensen, C. H., Isaksson, K., Macmillen, J., Åkerman, J., & Kressler, F. (2014). Strategies to manage barriers in policy formation and implementation of road pricing packages. *Transportation Research Part A, 60*, 40–52.
Tønnesen, A., Hanssen, G. S., Hansen, K. B., & Valencia, S. C. (2022). Integrative climate leadership in multi-level policy packages for urban mobility—A study of governance systems in two Nordic urban regions. *Transport Policy, 128*, 309–317.
UNDP. (2017). Institutional and coordination mechanisms: Guidance note on facilitating, integration and coherence for SDG implementation. Accessed January 17, 2025, from https://sustainabledevelopment.un.org/content/documents/2478Institutional_Coordination_Mechanisms_GuidanceNote.pdf
UN-Habitat. (2020). Monitoring and evaluating National Urban Policy: A guide. Accessed January 17, 2025, from https://unhabitat.org/sites/default/files/2020/05/monitoring-and-evaluating-national-urban-policy_a-guide_web.pdf
WHO. (2022). Implementing health in all policies: A pilot toolkit. License: CC BY-NC-SA 3.0 IGO. https://iris.who.int/bitstream/handle/10665/366435/9789240057128-eng.pdf?sequence=1&isAllowed=y
Wittwer, S., Hofer, K., & Kaufmann, D. (2023). An urban take on sustainable development policies and corresponding positioning strategies. *NPJ Urban Sustainability, 3*, 1. https://doi.org/10.1038/s42949-022-00080-y
Zhang, J. (2018). Social capacity building for environmental management related to transport sector: A broader perspective. In J. Zhang & C. M. Feng (Eds.), *Routledge handbook of transport in Asia* (pp. 446–494). Routledge, Taylor & Francis.
Zhang, J. (2021). Integrated carbon reduction policies on transportation in the era of post-COVID-19. *Urban Transport of China, 19*(5), 43–52. (in Chinese). https://doi.org/10.13813/j.cn11-5141/u.2021.0034
Zhang, J. (2022). Governance for post-COVID-19 carbon reduction: A case study of the transport sector. In J. Zhang & Y. Hayashi (Eds.), *Transportation amid pandemics: Lessons learned from COVID-19, Chapter 34* (pp. 429–440). Elsevier.
Zhang, J. (2025). Urbanimmunology. In J. Zhang, L. Zhang, & Y. Jiang (Eds.), *COVID-19 & pandemics, lifestyles, human mobilities, and cities: A perspective of planetary health*. Springer.
Zhang, J., & Hayashi, Y. (2022). *Transportation amid pandemics: Lessons learned from COVID-19*. Elsevier.
Zhang, J., Jiang, Y., & Zhang, Y. (2025). Transport-in-all-policies approach. In J. Zhang, Y. Jiang, & Y. Hayashi (Eds.), *Research handbook on transport and COVID-19*. Edward Elgar Publishing.

Chapter 8
Urbanimmunology: Catalyst for the Evolution of Cities and Society

8.1 Immunizing Futures: Moving Beyond Sustainability

On the cusp of an ever-shifting world, where change unfolds at a rapid and often unsettling pace, a new concept emerges—one that challenges us to rethink the very essence of urban and social systems: *Urbanimmunology*. This discipline goes beyond traditional urban and social planning; it is a transformative vision that calls us to see cities and society not as mere collections of buildings and roads, but as dynamic, living ecosystems capable of self-preservation, regeneration, and growth. In an era marked by accelerating urbanization, the profound impacts of climate change, and the unsettling tremors of geopolitical conflict, *Urbanimmunology* provides a fresh lens through which to confront the challenges of our time. The growing divisions in the global fabric, driven by economic and political divides, only emphasize the urgency of creating self-sufficient, adaptable urban and social systems. In a world defined by dynamic uncertainty and complexity, *Urbanimmunology* offers a path forward—drawing from human wisdom to navigate a diverse range of challenges.

At the core of this vision is the idea that cities and society, much like the human body, are complex, dynamic, and interconnected—interwoven webs of systems that learn, adapt, and evolve. Central to this vision is an integrated approach that draws from the adaptive, responsive, and regenerative principles of the immune system—incorporating resilience, sustainability, adaptability, and innovation. This enables cities and society to not only address a wide range of challenges but thrive—adaptively, autonomously, and intelligently. Inspired by the immune system's ability to sense and respond to threats, *Urbanimmunology* envisions cities and society that can quickly detect and counter environmental, social, economic, and political risks, while maintaining harmony with the vast, interconnected elements of urban and social systems.

J. Zhang, *Urbanimmunology*, Urban Sustainability,
https://doi.org/10.1007/978-981-95-4878-1_8

Urbanimmunology introduces a paradigm that reimagines cities and society as complex, adaptive ecosystems—much like the human body. In this view, cities and society are not isolated components, but interconnected systems, each part reliant upon the other for the health and prosperity of the whole. The idea is clear: cities and society should be equipped with adaptive systems capable of addressing monumental challenges—climate change, geopolitical risks, health crises, energy demands, and infrastructure needs. These urban systems should not merely endure; they should continuously learn, grow, and heal, adapting to new stresses while safeguarding their core functions. The goal is for cities and society to respond to external shocks—whether natural disasters, pandemics, or political upheavals—in a seamless, autonomous way, akin to the immune system's ability to recognize and respond to threats.

Scientific insights from *Urbanimmunology* underscore the critical importance of interconnectedness within urban and social systems. Key areas should work in harmony to collectively drive urban and social development toward multifaceted goals. The HEARTY City framework embodies this spirit, advocating for an all-encompassing approach to urban and social development that aligns with the principles of *Urbanimmunology*. It asserts that a city or society should be healthy, eco-friendly, smartly adaptive and responsive, resilient, comfortable, and foster a sense of attachment and belonging.

The role of disruptive technologies in this framework is pivotal. *Urbanimmunology* embraces the potential of these innovations to radically transform urban and social environments. It promotes the development of technologies that are not only evolutionary but sustainable, safe, and highly efficient—offering long-term value and the power to reshape cities and society into more unified, resilient systems. In this vision, technology does not disrupt the natural harmony of urban and social life; instead, it enhances it—making cities and society smarter, more adaptable, and more attuned to the needs of both residents and the environment.

A key lesson from *Urbanimmunology* is the necessity of rethinking urban and social governance. It advocates for governance structures that are not rigid and static, but flexible and responsive to the unique needs of different regions, social groups, and sectors. This approach ensures that policies are equitable, efficient, and adaptable to the shifting demands of an interconnected world. It calls for a more dynamic, holistic approach to urban and social development, one that recognizes the complexity of our challenges and the interconnectedness of the systems that make up our cities and society.

In short, this work urges us to envision cities and society not as static entities or isolated domains, but as dynamic, interwoven organisms—capable of resilience, regeneration, and growth. *Urbanimmunology* is a call to build cities and society that can not only endure the crises of today but thrive in the face of those yet to come. It challenges us to look beyond traditional urban and social development boundaries and see cities and society as living organisms—constantly adapting, evolving, and regenerating. Through this vision, we may discover a path forward—one that leads to a future where cities and society are not merely places to live, but thriving ecosystems capable of nurturing life amidst all its uncertainties and complexities.

8.2 Implications to Urban and Social Policymaking for Tomorrow

Urbanimmunology provides a groundbreaking framework that challenges conventional thinking, offering a path to policies that can adapt to and thrive in an unpredictable world. It calls for a fundamental shift in how we design, manage, and govern cities and society, ensuring that they are not only prepared to meet the needs of their populations, but also equipped to safeguard the future of the planet. This is a call to action for a radically new urban and social paradigm—one that is built for sustainability, resilience, and long-term survival in an ever-changing global landscape.

8.2.1 Cities and Society as Living, Evolving Organisms

The idea that cities and society function similarly to living organisms opens up a new realm for policymaking. Rather than focusing solely on managing cities and society as collections of isolated systems, we should treat cities and society as *integrated organisms* capable of learning, evolving, and self-regulating. This means embedding self-monitoring and self-adjusting mechanisms in urban and social systems, from energy grids that optimize themselves in real time to urban and social health networks that dynamically respond to changing environmental conditions. Cities and society should not be viewed as static entities, but as dynamic, living systems that continuously adapt and evolve in response to both predictable and unpredictable internal and external shocks.

8.2.2 Immune-Inspired, Hybrid Governance Models: Balancing Decentralization with Centralized Coordination

Traditional top-down governance often fails to respond swiftly or effectively to emerging challenges. Drawing inspiration from immune systems, where responses come from distributed components that work in concert, cities could adopt decentralized governance models (Nath & Madhoo, 2024). This involves empowering local actors—neighborhood-level organizations, community leaders, and tech-driven platforms—to take autonomous action when local needs arise. However, this decentralization should not come at the expense of centralized mechanisms that provide overall coordination, strategic direction, and resource allocation. By implementing a hybrid governance approach that balances decentralization with strong centralized frameworks, cities and society can become more resilient and adaptive to crises. This enables cities and society to "respond" at multiple levels simultaneously, leveraging both local autonomy and centralized oversight to address complex challenges effectively.

8.2.3 Feedback Loops Between Urban Systems and Global Development Balance

A transformative approach is required to align urban and social development with the interconnected socio-economic and environmental systems that sustain life on Earth, a concept we can frame as "global development balance." Instead of viewing urban and social health and planetary well-being as separate concerns, *Urbanimmunology* envisions a feedback loop where cities and society actively contribute to and derive benefits from this balance. This perspective emphasizes designing urban and social systems—such as energy, waste, and food networks—to bolster ecosystems while generating sustained positive outcomes. Policymaking should prioritize adaptive feedback mechanisms, enabling cities and society to respond to environmental changes and continuously refine their sustainability practices. Under this model, cities become both guardians and beneficiaries of global development balance, fostering a reciprocal relationship between urban and social systems and the planet's ecological integrity.

8.2.4 Immune-Inspired "Auto-regenerative" Urban Infrastructure

Rather than relying on costly and potentially unsustainable fixes when urban and social systems fail—such as power outages, transport disruptions, or health crises—cities and society should be designed with auto-regenerative capabilities. Drawing inspiration from immune system processes, where damaged tissues or cells are repaired or replaced automatically, this approach applies to not only urban and social infrastructure but also services, functions, and mechanisms that support cities and society. For example, self-healing energy systems (such as power grids that recover from failures) (Kumar & Bansal, 2019), transportation networks with adaptive mechanisms to manage congestion or accidents, and health systems that quickly mobilize resources in times of crisis. By integrating regenerative capabilities across these systems, cities and society become more resilient, reducing the risk of catastrophic failures (Reis et al., 2014) and ensuring that essential functions continue seamlessly even under stress.

8.2.5 Visionary Technologies as Immune Responses to Urban Threats

Visionary technologies should go beyond being mere efficiency tools and evolve into key components of urban and social "immune systems" capable of actively guiding and shaping the future of cities and society. These technologies, including AI, big data, machine learning, and IoT, should not only predict and assess risks but also autonomously mitigate and neutralize emerging threats. By simulating and modeling responses to potential stressors—such as climate change, economic disruptions, or health crises—these technologies can enable cities and society to proactively deploy solutions in real time. Rather than functioning solely as early warning systems, they should be embedded within a dynamic, self-regulating urban framework that continuously monitors, responds to, and evolves with emerging challenges. Urban and social policymakers should view these technologies as forward-thinking catalysts (Gregersen & Bianzino, 2023; Robertsone & Lapiņa, 2023), capable of steering cities and society toward a resilient and sustainable future, dynamically reconfiguring urban and social systems to meet both current and unforeseen demands.

8.2.6 Robustilience: Fostering Interconnected and Evolutionary Systems

Drawing on the principles of complex systems and immune responses, it is argued here that the concept of "robustilience" should replace the traditional notion of resilience. "Robustilience" is defined as a quality of being both strong and adaptable, capable of withstanding and recovering from disruptions while maintaining functionality and integrity. Rather than relying on isolated components, cities and society should focus on nurturing interconnected systems—social, economic, environmental, technological, and cultural—that not only maintain their strength but evolve continuously in response to changing conditions. Urban and social policies should empower these systems by providing the resources and frameworks necessary to support their growth and adaptability, ensuring they can effectively address both current stresses and future uncertainties. This approach involves fortifying the mechanisms, services, and functions that sustain urban and social life, all of which contribute to the "robustilience" of cities and society. By embedding adaptive and regenerative principles across all sectors of urban and social life (Zhang,

2017), cities and society can not only navigate the complexities of today but shape their capacity to thrive amidst the unpredictable challenges of tomorrow.

8.2.7 Urban and Social Systems as "Proactive Immune Systems"

Urban and social systems should evolve from traditional reactive approaches to challenges and embrace a proactive, immune-inspired model. Rather than waiting for crises such as traffic congestion, health emergencies, or energy shortages to unfold, cities and society should be designed to anticipate and prevent these issues. Drawing from the principles of immune responses, urban and social systems can function as dynamic, self-monitoring entities, continuously assessing potential threats and neutralizing them before they escalate. This shift goes beyond just technological solutions like sensors and predictive algorithms; it involves fundamentally rethinking how cities and society interact with their environments—creating urban ecosystems that can evolve in real time, learn from patterns, and respond autonomously. By transitioning from crisis management to anticipation and prevention, cities and society can safeguard infrastructure, support populations, and promote long-term, sustainable growth in an unpredictable world. This transformation calls for cities and society to become more than just spaces for habitation—they should develop into adaptive, anticipatory, and self-regulating systems that continuously evolve to protect both their people and the environment.

8.2.8 Harnessing Collective Wisdom for the Evolution of Urban Immunity

Urbanimmunology calls for a profound shift in how cities and society learn from past challenges and external forces, focusing on a dynamic, evolutionary process. Rather than merely exchanging information, cities should develop a system of wisdom integration (Intezari, 2015)—a continuous, proactive incorporation of deep insights and experiences that not only build on past lessons but also anticipate and counter future challenges. This approach transforms cities and society into self-improving entities, constantly evolving and enhancing their ability to manage and mitigate emerging risks. By embedding wisdom-driven evolution into urban and social systems, cities and society create an interconnected framework where learning is ingrained at every level, empowering them to strengthen their capacity to confront disruptions and thrive amidst uncertainty. This process establishes cities and society as ever-evolving organisms, driven by an immune-like feedback loop that amplifies their ability to confront the complexities of a rapidly changing world.

8.2.9 *Empowering Invisible, Quiet-yet-Impactful Actions: Harnessing Local Voluntary Changes and Engagement*

In immune systems, many responses occur at a microscopic level, invisible to the naked eye. Similarly, urban and social systems can thrive by fostering small-scale, decentralized actions that, while subtle, have the potential to ignite powerful, widespread transformations when magnified across cities and societies. These "quiet yet impactful" initiatives—such as community-driven gardens, local renewable energy projects, or neighborhood-led efforts—may appear insignificant on their own, but when nurtured and scaled, they become crucial components of the overall vitality and performance of cities and society. Rather than competing with larger systems, these grassroots actions (Troullaki & Rozakis, 2024) can enhance and fortify broader urban and social strategies. Policymakers should actively support these bottom-up innovations, creating fertile ground where local actors—residents, community leaders, and small-scale innovators—are empowered to lead change. By cultivating a network of decentralized, voluntary actions, cities and society can harness the collective strength of these "micro-level immune cells," forging a more adaptive, interconnected urban ecosystem that functions harmoniously as a unified whole.

8.3 An Uncharted Journey of Exploration

Like any pioneering field, *Urbanimmunology* encounters its own set of challenges, both practical and scientific, that require thoughtful reflection and strategic navigation. Rather than viewing these challenges as obstacles, we should embrace them as crucial points of tension and inquiry—opportunities to deepen our understanding and refine this emerging discipline. It is through grappling with these complexities, as part of an ongoing journey of scientific exploration, that we uncover pathways for growth and advancement.

8.3.1 *Practical Navigation: Guiding the Implementation of Urbanimmunology*

From a practical standpoint, *Urbanimmunology* offers profound potential for transformative change in urban and social planning, policy, and technology. However, its translation from abstract theory to actionable practice is not without its hurdles. One of the first and most evident challenges lies in the application of the immune system's complexity to the urban and social environment. While the analogy between immune responses and urban and social systems is compelling, cities and society are vastly more intricate and interconnected (Batty, 2009) than the relatively self-contained biological organisms. The social, economic, political, and environmental spheres of a city or society (Healey, 2006) cannot be reduced to the simplicity of biological models; they are shaped by deeply rooted histories, cultural norms, and power structures. While the analogy is useful, it risks oversimplifying the unique complexities of urban and social systems and may fall short in addressing the full spectrum of challenges cities and society encounter in their pursuit of resilience. Previous examinations of existing urban theories have explored ways to mitigate the potential shortcomings of this analogy, but these efforts have only begun to scratch the surface. Further in-depth investigation is needed.

Moreover, while *Urbanimmunology* offers a conceptual framework, its practical implementation requires substantial empirical validation. The ideas presented here are nascent and untested in practice, and their full realization will demand years of rigorous fieldwork, data collection, and iterative testing. This is not a quick fix, but rather a long-term journey of reflection, adaptation, and reevaluation. We find ourselves in the early stages of this intellectual journey, and thus the true potential of *Urbanimmunology* can only be understood through lived experience—through the testing, adapting, and evolving of policies in the real-world contexts of diverse urban and social environments.

Another limitation in practical terms is the issue of *scale*. How do we translate the principles of immune-like responses to cities and society that differ drastically in size, resources, and cultural context? A central question here is this: how do we ensure that these principles are not only scalable but also sensitive to the unique needs of different communities? The policies and technologies that work well in large, resource-rich cities of the Global North may be entirely inappropriate for smaller, resource-constrained cities in the Global South. This gap in development capacity (Kuhl, 2009) should be addressed if *Urbanimmunology* is to offer equitable benefits across the world. Cities and society of the Global South are often caught in a cycle of underdevelopment and infrastructural fragility, leaving them particularly vulnerable to the threats of climate change, pandemics, and social instability. *Urbanimmunology* should be mindful of these gaps, ensuring that its applications do not inadvertently reinforce global inequalities but, instead, offer pathways for inclusive development (Gupta & Vegelin, 2016), fostering technological leapfrogging and creating resilient systems that can respond to local contexts.

8.3.2 Scientific Navigation: Theoretical Pathways

From a scientific perspective, *Urbanimmunology* also faces limitations, particularly in its interdisciplinary nature. The complexity of the urban immune metaphor depends on drawing connections between fields as varied as urban and social planning, biology, health sciences, engineering, and social theory. While interdisciplinary approaches have the potential to unlock new insights, they also introduce a number of challenges—most notably, the difficulties inherent in bridging the gaps between these disparate disciplines (Nyahunda & Tirivangasi, 2021;

Peek & Guikema, 2021). Each discipline comes with its own language, methodologies, and paradigms, which can make collaboration both difficult and slow. Furthermore, the application of immune system principles to urban and social environments is still largely theoretical, and the empirical evidence needed to validate these models is sparse. *Urbanimmunology* is still in its infancy as a scientific discipline, and its claims require rigorous testing, refinement, and validation across a wide array of urban and social contexts.

There may also be a risk that, in trying to draw parallels between immune systems and urban and social life, we may overlook the unique aspects of human society that do not fit neatly into the biological metaphor. Human societies are driven by values, ethics, and social dynamics (Luhmann, 1996) that cannot be captured solely through biological or ecological models. The profound philosophical questions surrounding the nature of the city or society—its role in nurturing justice, democracy, and well-being (Harvey, 1973)—cannot be simply reduced to concepts like resilience, adaptability, or immune-like responses. As such, while *Urbanimmunology* offers a scientific framework for thinking about cities and society as living, responsive systems, it should be continually enriched with human-centered insights that account for the complexity of the social, political, and cultural dimensions of urban and social systems.

Moreover, the integration of diverse scientific fields into *Urbanimmunology* raises questions about the limits of knowledge itself. How do we reconcile the empirical rigor of the natural sciences with the more interpretive, qualitative aspects of social theory? How do we balance the precision of data-driven approaches with the messiness of lived experience? These questions highlight the need for a more holistic, integrative approach to scientific inquiry—one that not only combines disciplines but also reflects a deeper philosophical understanding of the limits and possibilities of knowledge itself. *Urbanimmunology* should not only strive to build bridges across different scientific domains but also confront the fundamental limits of our understanding, acknowledging that the urban and social environment is too complex to be captured by any single disciplinary lens.

8.3.3 Toward New Forms of Knowledge Integration: Going Beyond Interdisciplinary and Transdisciplinary Approaches

The limitations of *Urbanimmunology* may not be merely scientific or practical but fundamentally epistemological. As Kuhn et al. (2000) have argued, the very frameworks through which we construct knowledge—our paradigms—shape both what we can know and how we know it. To move beyond the constraints of traditional interdisciplinary or transdisciplinary approaches, we should embrace modes of thought that disrupt the rigid boundaries of disciplines, institutions, and methodologies. *Urbanimmunology* should be viewed as an open invitation to rethink the nature of knowledge itself. It calls for the integration of diverse epistemic traditions, alternative worldviews, and non-linear thinking that transcend the limits of conventional paradigms. By doing so, we adopt an intellectual stance that is open, dynamic, and attuned to the complexities of urban and social life, seeking solutions that are context-specific while remaining receptive to insights from a broad range of sources.

8.4 Steering Future Developments Beyond the SDGs

The United Nations' Sustainable Development Goals (SDGs) have illuminated a vital path, guiding humanity toward justice, equity, and sustainability. *Urbanimmunology* aligns with this noble pursuit while pushing further, urging cities and society to evolve beyond mere survival, advocating for dynamic, regenerative systems capable of adapting to challenges and healing over time. It envisions cities and society not as rigid, static constructs, but as living, breathing entities—constantly shifting, regenerating, and healing from disruptions.

As we stand on the threshold of the post-SDG era, beyond the horizon of 2030, it becomes clear that the world requires a profound transformation of thought—one that moves beyond the imperative of survival to the bold aspiration of thriving on a global scale. While the SDGs have set essential milestones, the rising complexity of global issues demands a framework of deeper, more integrated understanding. With a burgeoning global population and the finite nature of our resources, humanity is called to reimagine its approach to development in a way that is as holistic as it is harmonious.

What lies beyond the SDGs? Is it time to turn our focus toward harmonized development goals (HDGs)—ones that balance human progress with the planet's ecological capacity, reconcile geopolitical powers with local prosperity and independence, and strive for a thriving, equitable global society? (Zhang, 2024). In this vision, harmonization transcends the concept of sustainability, evolving into a transformative process that nurtures diversity, mitigates conflict, and ensures a dynamic equilibrium of all things. Or, have we already glimpsed the seeds of new frameworks capable of fundamentally reshaping the world for its enduring prosperity?

To truly address the future of urban and social systems and the Earth itself, we should transcend the traditional paradigm of growth and reimagine progress within the context of a living, interconnected globe. The relentless pursuit of constant expansion should give way to a vision where cities and society recognize their place within a larger, boundless ecosystem—one that honors the limitations of finite resources. Growth should no longer be synonymous with endless accumulation, as Wijburg (2024) so poignantly warns, but with the ability to adapt, regenerate, and renew in ways that cultivate both human well-being and the vitality of the planet.

This new paradigm demands a radical reframing of urban and social evolution—not as a race against the environment, but as a symbiotic interplay, in which urban and social systems and ecosystems evolve in harmony. Growth, in this vision, becomes synonymous with resilience and regeneration—ensuring that systems can respond to disruptions in ways that enhance quality of life without exceeding the planet's thresholds (O'Neill et al., 2018). It is a shift from a static, growth-driven model to one that celebrates dynamic adaptation, where cities and society emerge as living, interconnected organisms—capable of responding to both human needs and the imperatives of the Earth.

In this envisioned future, human society's journey is not about scaling up relentlessly, but about scaling in ways that bolster resilience, creativity, and interdependence between humanity and the natural world. This vision moves beyond sustainability, embracing the notion of regeneration—creating environments that continually renew themselves, fostering collective well-being and harmonious development. It is an invitation to redefine success, not as unchecked growth (López-Gay et al., 2020), but as the mutual flourishing of people and planet. This shift in perspective sets the stage for a global trajectory rooted in balance, evolution, and the shared responsibility of all living systems, in pursuit of a thriving, regenerative future (Camrass, 2023).

8.5 The Philosophical Awakening: Reimagining Cities and Society Through *Urbanimmunology*

As the seeds of *Urbanimmunology* take root, we stand on the precipice of more than a theoretical shift. We stand at the threshold of a profound transformation in how we perceive cities and society, our place within them, and our role in the delicate web of life that binds us all. In a world increasingly shaped by environmental turmoil, technological upheaval, and social fragmentation, the very concept of urban and social life demands reimagination. The streets we walk upon, the buildings that surround us, and the spaces where we live, work, and dream should no longer be seen as mere backdrops to our existence. Instead, they beckon us toward a new way of thinking—one that rejects static, mechanistic models of planning and governance in favor of a fluid, dynamic approach that mirrors the complexity of living organisms.

Cities and society, once regarded as concrete emblems of human mastery over nature, now emerge as something far more alive—organic, responsive, and ever-evolving in tune with both human and ecological imperatives. The notion that cities and society could function as living systems—capable of self-regulation, adaptation, and transformation—marks a revolutionary departure from traditional urban design. Cities and society are no longer just places of habitation; they are systems that sustain, reflect, and, at times, challenge us to evolve alongside them.

In this new vision, urban and social development becomes an ongoing, relational process—maintaining equilibrium between the human and the non-human, the built and the natural. This revelation forces us to confront an ancient philosophical dilemma: the tension between permanence and change. Historically, cities and society have symbolized stability—monuments to human achievement and mastery over nature. Yet, *Urbanimmunology* shows that the essence of a city or society lies not in its permanence, but in its capacity for continuous evolution. Like the immune system, cities and society should remain adaptable, constantly learning and evolving in response to their environment. In the face of challenges, the city and society are not passive recipients of external forces; they are dynamic organisms that should defend, recover, and regenerate under stress.

This transformation demands a radical shift in philosophy—one that reimagines our relationship with the city and society. What if, instead of viewing the city and society as a mere backdrop to our lives, we recognized them as active participants in shaping our existence? Like the immune system, which anticipates and adapts to threats before they arise, cities and society could respond dynamically to our needs and challenges. This prompts profound questions about agency within urban and social spaces: are cities and society simply passive recipients of human decisions, or do they possess agency—an inherent capacity to shape how we live, work, and relate to one another?

Urbanimmunology proposes a radical inversion: the city and society become transformative agents, fostering the flourishing of both human and ecological systems. This vision goes beyond the practicalities of urban and social systems, challenging us to rethink existence itself. If cities and society are no longer mere places where we live, could they become living laboratories—spaces for human and ecological transformation, incubating new ways of living, interacting, and coexisting? The implications are profound, resonating deeply with the core of our collective human experience.

In this vision, the relationship between people and cities is one of symbiosis, not dominance. Just as the immune system is vital and vibrant, cities and society can become active participants in the larger harmony of life. They balance the stability of infrastructure with the fluidity of ecological and social systems, fostering resilience and adaptability. The city and society evolve into a living organism—responding to change, anticipating challenges, and regenerating in the face of adversity.

This vision celebrates resilience as a dynamic, proactive force. *Urbanimmunology* suggests that cities and society are not merely designed to withstand storms but to foster conditions for growth, innovation, and flourishing. In this reimagined world, robustilience is woven into the very fabric of the city and society. These spaces become places where new possibilities emerge, and where interconnectedness is not just acknowledged but embraced. By nurturing cities and society that are responsive, adaptive, and regenerative, we create the conditions for both human and ecological health to thrive.

Urbanimmunology is not just a new way of thinking about cities and society; it is a call to reframe our collective existence, rethink how we relate to the spaces we inhabit, and reconsider the society we wish to build. It challenges us to engage in a profound philosophical exploration of what it means to live in an urbanized world and asks us to reflect on not just the future of our cities and society, but the future of humanity and the planet itself.

The promise of *Urbanimmunology* is not merely in offering solutions to urban and social challenges, but in opening up a new way of being. It urges us to recognize that cities and society are not inert systems—they are vital organs within the living body of Earth. As we design and build cities today, we are shaping not only the future of human society but the future of the planet. The work of *Urbanimmunology* is not just about policy, design, or technological innovation; it is a moral and existential responsibility. It calls us to act with wisdom and foresight, building cities and society that are not just resilient to change, but active forces in shaping the world around them.

In the world envisioned by *Urbanimmunology*, cities are places of flourishing—not just for humans, but for the ecosystems, cultures, and societies they support. It challenges us to rethink not just the practicalities of urban design, but the kind of world we want to create. It calls us to build cities and society that evolve in harmony with the planet—growing, adapting, and flourishing in ways that honor both human creativity and the ecological systems on which all life depends.

In conclusion, *Urbanimmunology* challenges us to see cities and society not as static entities, but as living, evolving ecosystems—capable of adaptation, self-regeneration, and symbiosis with the environment. The future of cities and society does not lie in preserving the status quo, but in cultivating dynamic, regenerative systems that can rise to meet the demands of an ever-changing world. It calls us to reimagine cities and society as vibrant spaces where life thrives—where the health of urban and social systems is inextricably linked to the health of the planet, and the flourishing of both is woven into a complex, beautiful web of interdependence.

References

Batty, M. (2009). Cities as complex systems: Scaling, interaction, networks, dynamics and urban morphologies. In R. Meyers (Ed.), *Encyclopedia of complexity and systems science*. Springer. https://doi.org/10.1007/978-0-387-30440-3_69

Camrass, K. (2023). Regenerative futures: Eight principles for thinking and practice. *Journal of Futures Studies, 28*(1), 89–99.

Gregersen, H., & Bianzino, N. M. (2023). AI can help you ask better questions — And solve bigger problems. Technology and analytics. *Harvard Business Review*. https://hbr.org/2023/05/ai-can-help-you-ask-better-questions-and-solve-bigger-problems?ab=hero-subleft-2

Gupta, J., & Vegelin, C. (2016). Sustainable development goals and inclusive development. *International Environmental Agreements: Politics, Law and Economics, 16*, 433–448.

Harvey, D. (1973). *Social justice and the city*. University of Georgia Press.

Healey, P. (2006). *Urban complexity and spatial strategies: Towards a relational planning for our times*. Routledge.

Intezari, A. (2015). Integrating wisdom and sustainability: Dealing with instability. *Business Strategy and the Environment, 24*(7), 617–627.

Kuhl, S. (2009). Capacity development as the model for development aid organizations. *Development and Change, 40*(3), 551–577.

Kuhn, D., Cheney, R., & Weinstock, M. (2000). The development of epistemological understanding. *Cognitive Development, 15*(3), 309–328.

Kumar, S. A. R., & Bansal, R. C. (2019). Multiagent-based autonomous energy management system with self-healing capabilities for a microgrid. *IEEE Transactions on Industrial Informatics, 15*(12), 6280–6290.

López-Gay, A., Cocola-Gant, A., & Russo, A. P. (2020). Urban tourism and population change: Gentrification in the age of mobilities. *Population, Space and Place, 27*(1), e2380. https://doi.org/10.1002/psp.2380

Luhmann, N. (1996). *Social systems*. Stanford University Press.

Nath, S., & Madhoo, Y. N. (2024). Centralized climate clubs and decentralized governance: A model of global CO2 mitigation. *The American Journal of Economics and Sociology, 84*(1), 59–68. https://doi.org/10.1111/ajes.12586

Nyahunda, L., & Tirivangasi, H. M. (2021). Interdisciplinary approach to climate change: Intersecting environmental social work and sociology in climate change interventions from an Afrocentric perspective. In W. Leal Filho, J. Luetz, & D. Ayal (Eds.), *Handbook of climate change management*. Springer. https://doi.org/10.1007/978-3-030-22759-3_282-1

O'Neill, D. W., Fanning, A. L., Lamb, W. F., & Steinberger, J. K. (2018). A good life for all within planetary boundaries. *Nature Sustainability, 1*, 88–95.

Peek, L., & Guikema, S. (2021). Interdisciplinary theory, methods, and approaches for hazards and disaster research: An introduction to the special issue. *Risk Management, 41*(7), 1047–1058.

Reis, S., Hu, Y., Babino, A., et al. (2014). Avoiding catastrophic failure in correlated networks of networks. *Nature Physics, 10*, 762–767.

Robertsone, G., & Lapiņa, I. (2023). Digital transformation as a catalyst for sustainability and open innovation. *Journal of Open Innovation: Technology, Market, and Complexity, 9*(1), 100017.

Troullaki, K., & Rozakis, S. (2024). Grassroots innovation: A review and a meta-theoretical sustainability assessment framework. *Environmental Innovation and Societal Transitions, 50*, 100822.

Wijburg, G. (2024). On undevelopment and de-development: A geographical critique on perpetual growth and resource-based accumulation. *Royal Geographical Society, 57*(2), e12988. https://doi.org/10.1111/area.12988

Zhang, J. (2017). *Life-oriented Behavioral research for urban policy*. Springer.

Zhang, J. (2024). Interdisciplinary research on future cities and transportation systems. In *Selected speeches of 2023 world cities day events*. Shanghai Scientific and Technical Publishers.

Afterword

Though this book exceeds 120,000 words, the author is convinced that the proposed framework of *Urbanimmunology* is far from complete. This conversation—one that touches upon the very fabric of our cities and society—has only just begun.

In our contemporary world, urban and social issues are manifold and complex, with many problems repeating and persisting through time. The creation and resolution of these challenges often come at the cost of precious natural resources. Despite the progress of technology and social development, we are confronted with the profound question of how to avoid the waste of these irreplaceable resources, how to avoid wasteful duplication of efforts, and how to achieve genuine sustainability and long-term prosperity for humanity.

Consider, for instance, electric vehicles—hailed as revolutionary in their potential to reduce carbon emissions in transportation. The promise is vast, but it is not without its complications. Electric vehicles depend heavily on scarce resources such as cobalt, nickel, and lithium, which are concentrated in only a few regions and exposed to the volatility of global supply chains. This stark reality points to an inconvenient truth: while technological progress promises great benefits, it also introduces new vulnerabilities. Moreover, though the recycling of existing battery materials and the development of alternative materials are crucial to breaking our dependence on these scarce resources, progress in both areas remains unsatisfactory. Thus, we are reminded that innovation alone does not guarantee a solution—it may well amplify our reliance on the very resources we seek to conserve.

In this context, balancing technological innovation with the responsible use of resources becomes a pressing matter that requires our immediate attention. In response to this, the author proposes the concept of "integration of transport, energy, and materials"—an ambitious vision of harmonizing sustainable transportation, renewable energy, and recyclable materials. This framework aims not only to optimize resource usage but also to protect the environment. Yet, this is just one possibility among many. Across fields, the potential for similar innovations is vast. How can we, for example, merge green architecture, smart building systems, and

J. Zhang, *Urbanimmunology*, Urban Sustainability,
https://doi.org/10.1007/978-981-95-4878-1

renewable energy to enhance energy efficiency in the built environment? How can industry transition from a model of linear consumption to a circular economy, where resources are reused and recycled in perpetuity? These are the avenues that await further exploration. Every sector holds untapped potential, and through cross-disciplinary collaboration, we can uncover new perspectives that fundamentally alter how we address urban and social issues.

This book, in conjunction with the author's redefined concept of "planetary health," opens the door to an initial exploration of the deep interconnection between *Urbanimmunology* and resource conservation. Yet, this dialogue is only in its infancy, and much more remains to be unpacked.

The book also underscores an essential truth: solving urban and social problems cannot be reduced to a matter of technological innovation alone. Despite the rapid advancements in technology that offer more efficient solutions, we must ask ourselves how to prevent the waste of resources in this fast-paced development. How do we fully leverage existing built environments and technologies to tackle urban and social challenges? The overzealous approach of "tear down and rebuild" often squanders valuable resources without addressing the root causes. Human progress, at its core, is an ongoing process of adapting to and transforming nature. Along the way, we have accumulated vast amounts of built resources, both positive and negative. This accumulation reminds us of the urgent need to not only pursue technological progress but to make wise use of what we already have—avoiding unnecessary waste and redundant reconstruction.

Urbanimmunology, then, extends far beyond technological innovation. It requires us to grasp the deep, intertwined relationships between humans and nature, technologies, and resources, through multiple lenses. In our pursuit of progress, we must critically reflect on the long-term implications of each decision and innovation. While technologies may pave the way to solutions, the real challenge lies in ensuring these solutions are sustainable, equitable, and just. Only through this deeper inquiry can we avoid perpetuating new dependencies and vulnerabilities.

Urbanimmunology offers not just a framework for addressing the urban and social issues of today but a roadmap for applying this concept across many perspectives. When confronted with complex urban and social challenges, we cannot afford to rely on isolated, single solutions. Instead, we must embrace transdisciplinary collaboration—crossing the boundaries between sociology, environmental science, ethics, economics, and beyond. Each discipline holds the potential to offer new insights, fresh ideas, and transformative solutions.

The value of *Urbanimmunology* lies not just in conserving resources, but in compelling us to reconsider the intricate, often hidden relationships between humanity, nature, technologies, and society. Through profound reflection, we can better utilize our existing resources, avoid unnecessary duplication, and better prepare ourselves to face the future. The application of *Urbanimmunology* reaches far beyond the technical or resource-focused domain; it demands the integration of cultural, ethical, and social considerations. To resolve the complex urban and social challenges of our time, we must move beyond narrow technological fixes and adopt more comprehensive, inclusive strategies. Only by approaching these issues from diverse perspectives can we address the deep-seated contradictions within cities and society, ultimately creating a more resilient, sustainable urban ecosystem.

Through innovative thinking and breaking the barriers of disciplines, we can not only solve today's urban and social problems, but also lay a strong foundation for future progress—building a more harmonious, livable, and just urban and social environment for all.

September 28, 2025

A day celebrating the wisdom of Confucius

MIX
Papier aus verantwortungsvollen Quellen
Paper from responsible sources
FSC® C105338

If you have any concerns about our products,
you can contact us on
ProductSafety@springernature.com

In case Publisher is established outside the EU,
the EU authorized representative is:
Springer Nature Customer Service Center GmbH
Europaplatz 3, 69115 Heidelberg, Germany

Printed by Libri Plureos GmbH
in Hamburg, Germany